No.	Chapter Ref.	Pole-zero	$F(s)$	Response	$f(t)$
8	Chap. 15 14-2		$\dfrac{s}{(s-j\omega_1)(s+j\omega_1)} = \dfrac{s}{s^2+\omega_1^2}$		$\cos \omega_1 t$
9	14b 14-13		$\dfrac{1}{(s+\alpha)(s^2+\omega_1^2)}$		$\dfrac{\sin(\omega_1 t - \theta) + e^{-\alpha t}\sin\theta}{\omega_1\sqrt{\alpha^2+\omega_1^2}}$ where $\theta = \tan^{-1}\dfrac{\omega_1}{\alpha}$
10	Chap. 14 13-6		$\dfrac{s}{(s-s_1)(s-s_2)}$		$\dfrac{1}{s_1-s_2}(s_1 e^{s_1 t} - s_2 e^{s_2 t})$
11			$\dfrac{1}{(s-s_1)(s-s_2)(s-s_3)}$		$\dfrac{e^{s_1 t}}{(s_1-s_2)(s_1-s_3)} + \dfrac{e^{s_2 t}}{(s_2-s_1)(s_2-s_3)} + \dfrac{e^{s_3 t}}{(s_3-s_1)(s_3-s_2)}$
12	Chap. 13		$\dfrac{1}{s^2}$		t
13	Chap. 16 1a 2-4		1		δ

ELECTRIC
NETWORKS

ELECTRIC NETWORKS

HUGH HILDRETH SKILLING

Professor of Electrical Engineering
Stanford University

$$s_{1,2} = \frac{-b \pm \sqrt{b^2 - 4ac}}{2a}$$

JOHN WILEY & SONS

New York   London   Sydney   Toronto

Library of Congress Cataloging in Publication Data

Skilling, Hugh Hildreth, 1905–
    Electric networks.

    Bibliography: p. 457
    1. Electric networks. I. Title.
TK454.2.S58 1974   621.319′2   73–14870
ISBN O-471-79420-1

Printed in the United States of America

10 9 8 7 6 5 4

PREFACE

Some readers like network analysis because it is a fascinating study. Others value networks as the basis of our electrical profession. Surely both are right, and a networks book has the pleasant responsibility of presenting the attractive ideas of the subject in a sequence designed to give the coverage that the real world requires. Now that the donkey work of routine calculations can be left to willing electronic computers, you and I can give our attention to these matters of concept and content.

This book, *Electric Networks*, is designed primarily to be a textbook for a first course, with careful explanations, numerical examples, and homework problems for the students. New concepts are built upon familiar ideas as chapter follows chapter. The plan is highly suitable for independent readers, too.

As a text, the book can be used for a semester, or two quarters, or a year, depending on the preference of the instructor and the number of weekly class meetings. It is shorter than my earlier book on *Electrical Engineering Circuits*, yet in some ways it goes farther, as I shall explain.

This book is mainly, of course, a presentation of linear networks. The conceptual plan is briefly as follows. We find in the first chapter that electrical currents in networks are described by differential equations. When voltages and currents are alternating, it is helpful to use complex algebra in these differential equations. Phasors and transforms represent alternating voltages and currents, and lead naturally to admittances, impedances, and transmittances. These network functions, first seen in terms of frequency, are readily extended to be functions of the complex variable s, and the *network function* is then perceived as determining both natural and forced components of a response. Thus the frequency domain is established.

Attention then goes to the time domain. The differential equations of the introductory chapters are now solved in the classical manner. Forced and natural responses are found by operating on time functions. State-variable equations are formulated and solved without reference to frequency or transformation. When formal solution becomes laborious, numerical methods are introduced, and the excellence of computer solution, analog or digital, is shown.

Returning to the complex-frequency domain, the general loop and node equations are formulated, and the simplicity gained by matrix notation is striking.

The groundwork has now been laid for presentation of a number of essential network theorems. Also, we are now prepared for two-port networks, and the two-port concepts that might so easily be confusing are seen to proceed without trouble.

Network functions in s require only the concept of Fourier series to bring the analysis of any periodic wave within our reach. From this, by a clear argument of plausibility, the Fourier integral gives us the very great ability to deal with inputs that are not cyclic, not repetitive, but discrete. Finally, the Laplace transformation is used, showing something of its power and beauty. This is the course of our conceptual plan.

Experienced teachers will recognize pedagogical aids woven into the book. There are simple examples to show the "go" of everything. The proverb about a picture being worth a thousand words is not forgotten. The book offers concepts to interest or challenge students at various levels. Ideas developed in early chapters are used again. An occasional glance ahead provides motivation, and this is heightened by references to practical use. Each chapter is followed by numerous problems for student solution. To fit the needs and wishes expressed by students, and by teachers too, answers to most of the problems are given in Appendix 4. The use of high-speed electronic computers for solving some of the home-work problems can be specified by the instructor, or not, as he prefers; some of the numerical problems are highly suitable for computer solution.

Regarding prerequisites, there are not many. In physics, a general familiarity with electricity is wanted, but not much more; even Ohm's law is stated in Chapter 1 although some previous acquaintance with it is assumed. In mathematics, ordinary algebra is of course expected, but complex algebra is introduced in Chapter 3. Differentiation and integration of such simple functions as the sine, the cosine, and the exponential are used throughout the book. In the later chapters, relating to Fourier and Laplace transformation, calculus is used in the proofs at about the level of integration by parts, but the actual utility of the transform method requires only the most modest calculus.

Technical terms are explained as they arise. Almost all technical definitions are standard, and reference is made to the IEEE Dictionary of *Electrical and Electronics Terms*. Units are those of the standard mks system. Symbols are more or less the customary ones, and are explained as needed. Abbreviations are used sparingly, to be easier for the readers.

A theorem in Chapter 10 says that matching a line to a load gives a stronger, clearer signal. So, too, does matching a book to a student. That is why this book, called *Electric Networks,* and the previous book on *Electrical Engineering Circuits,* by the same author and the same publisher, are both offered at the same time. You will not want to use both, of course; you will select the better match. Do you, or your students, prefer the more compact organization of this book, or the more comprehensive discussions of the other? Do you like the faster pace of this new book, with more subject matter though somewhat fewer pages? Or do you value the "professional chapters" of *Electrical Engineering Circuits?* Do you want the work on symmetrical components, the elementary filter design, and such? Or do you like the expanded treatment of Laplace theory and methods?

I am happy to have this opportunity to extend my thanks to readers of *Electrical Engineering Circuits* who have brought or sent me messages from all six continents. Every suggestion or correction has been used in new printings, or saved for a new edition. I hope that readers of this new book, too, will send me corrections and suggestions.

Some months ago the publisher asked a number of the users of *Electrical Engineering Circuits* about new material for this new book. Would they want something on state variables? Unanimously the answer was yes. The result is Chapter 7 of this book. Indeed, equations of state appear on the very first page of the book, though they are not stressed until later.

Replies to the publisher's other questions were divided, so this book is presented in such a way that electronic computers can be used or not, signal flow graphs are optional, and other techniques of computation, too, are left to the judgment of the teacher. It is my personal opinion that the "desk-top" computer, small and relatively inexpensive but quick and programmable, has a great future.

If you have seen any of my other books you know how they are prepared. Ideas come from my colleagues; I thank them. My students, working from a first draft, tell me what is clear and what is foggy. Also, they check examples, work problems, and verify the answers. Finally Hazel, my wife, writes the finished manuscript; she will not let me call her co-author, but each manuscript that bears my name is more and more her work. My friends with John Wiley & Sons then produce this book. I am grateful to them all.

Stanford, California   *Hugh Hildreth Skilling*

CONTENTS

CHAPTER 3

COMPLEX ALGEBRA   36

CHAPTER 4

THE FREQUENCY DOMAIN   55

CHAPTER 5

NETWORK FUNCTIONS AND THE s PLANE   102

CHAPTER 6

THE TIME DOMAIN   129

CHAPTER 7

STATE-VARIABLE EQUATIONS   161

CHAPTER 8

LOOP AND NODE EQUATIONS   182

CHAPTER 9

MATRIX SOLUTIONS 204

CHAPTER 13

THE FOURIER SERIES AND INTEGRAL   309

CHAPTER 14

THE LAPLACE TRANSFORMATION

CHAPTER 15

LAPLACE APPLICATIONS AND THEOREMS

CHAPTER 16

TRANSLATION, CONVOLUTION, AND THE IMPULSE FUNCTION 416

CHAPTER 17

THREE-PHASE SYSTEMS 434

ELECTRIC NETWORKS

CHAPTER

1

DEVICES
AND MODELS

1. EQUATIONS

Network equations can be formulated in several different ways. The equations of state of a system are especially adapted to solution by electronic digital computers. State equations consist of a set of first-order differential equations, and they may look like this for the network of Figure 1a:

$$\frac{di_1}{dt} = -\frac{R_{12}}{L_1}i_1 + \frac{R_{12}}{L_1}i_2 + \frac{1}{L_1}v_1 \tag{1-1}$$

$$\frac{di_2}{dt} = \frac{R_{12}}{L_2}i_1 - \frac{R_2 + R_{12}}{L_2}i_2 \tag{1-2}$$

Here the state variables are the two currents i_1 and i_2. When state variables are known the *state* of the system is known; interpreting this generalization

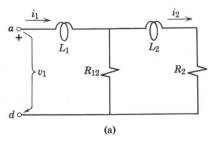

(a)

FIGURE 1a   Network for equations 1-1 and 1-2.

in terms of our particular network it means that when these two currents are known we have full information about what is going on in the network and in particular all voltages and currents from which we can determine the distribution of energy. Each of our two inductances can store energy and we must know their currents to know the state of energy distribution; the resistances store no energy, and in this particular example there are no capacitances—if there were capacitances in the network we might use their voltages as other state variables.

Suppose in this example we are given the voltage v_1 of a source to be connected to the network at terminals a and d, and that we want to know the value of each current i_1 and i_2 thereafter. That is, we are to find $i_1(t)$ and $i_2(t)$ given $v_1(t)$ by solving equations 1-1 and 1-2. But before this can be done it is necessary to know i_1 and i_2 at some particular instant from which our solution can begin.

Let us suppose that a switch is closed to introduce the source voltage v_1 into the network of Figure 1a; we shall arbitrarily call the instant of closing the switch our reference time at which $t = 0$, and we shall suppose that the two currents are known at this initial instant to have specific values that we may call $i_1(0)$ and $i_2(0)$. These *initial* values of the currents may or may not be zero. Since the source voltage v_1 is introduced into the network of Figure 1a by closing a switch, no doubt $i_1(0) = 0$, but $i_2(0)$ might be a current circulating through L_2, R_2, and R_{12} with a value that need not be zero at $t = 0$. In any case, whether zero or not, the initial values of the variables are known.

Then a solution for the set of differential equations can be found by any of several methods. A digital computer offers a rapid means of solution for numerical values, but its use may be expensive and is sometimes inconvenient as well as being limited, and it is best to know several ways to solve such a problem. Laplace transformation offers one of the most convenient attacks.

Equations 1-1 and 1-2 give state equations as an example, but the user may prefer loop equations or node equations, or he may work by combining the impedances of elements in series and parallel. These and other techniques are given in the chapters to come. In general it is best to use a means that is matched to the problem, and not "use a pile driver to drive a tack."†

Electrical problems are extremely varied. Electronic computers operate on discrete signals produced at a rate of perhaps a million a second. By contrast, a single pulse of electric current in a superconducting ring of lead has been seen to flow undiminished for months. Television signals from the

---

† Frazier; see bibliography, page 457.

moon, radio from Mars or Venus, or radar reflections from the sun, may give less than a millionth of a volt in our receiving apparatus, while electrical power engineers talk in terms of millions of volts on their transmission lines. Yet over all this range of time and intensity, of continuous waves and discrete pulses, the designers work with the same volts and amperes, ohms, henrys, farads, and watts. See Table 1-1.

TABLE 1-1   Unit Prefixes[a]

giga- $= 10^9$ (as in gigantic)
mega- $= 10^6$ (as in megaphone)
kilo- $= 10^3$ (as in kilowatt)
milli- $= 10^{-3}$ (as in milliampere)
micro- $= 10^{-6}$ (as in microfilm)
nano- $= 10^{-9}$ (meaning dwarfish)
pico- $= 10^{-12}$ (meaning very small)

[a] In naming electrical quantities, prefixes are often used to denote powers of 10. Thus it is well known that the kilowatt is 1000 watts, and the milliampere is 0.001 ampere.

2. MODELS

The first step of analysis is to idealize an actual network, keeping the essential characteristics of the network in what we call its *model*, and ignoring complications that make little or no difference in the desired result. Thus Figure 1a is a diagram of the model of a network, not a picture of the network itself. (With the same idea in mind, such a figure is sometimes called a schematic diagram.) In the model, for instance, R_2 is a resistor with pure resistance, and its shape, weight, color, and other attributes are ignored. Also the model omits the small but unavoidable capacitances existing between different parts of the network, and some unavoidable resistance in the branch containing L_1. When this model was made and the diagram was drawn, all these things were considered to be insignificant. Figure 1a might be called a graphical model, and equations 1-1 and 1-2 a mathematical model of the particular network that we are considering.

The following properties are attributed to the ideal elements of models (see Figure 2a):

The voltage v_R across *resistance* R carrying current i is

$$v_R = Ri \qquad (2\text{-}1)$$

R is usually (but not always) assumed to be constant.

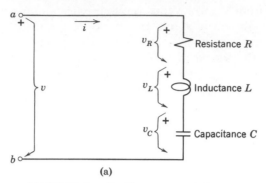

(a)

FIGURE 2a   To illustrate symbols.

The voltage v_L across *inductance* L carrying current i is the rate of change of the product, $d(Li)/dt$. If inductance is constant with time, this relation becomes

$$v_L = L\frac{di}{dt} \tag{2-2}$$

The current i entering *capacitance* C when v_C is the voltage across the capacitor is $d(Cv_C)/dt$. If capacitance is constant with time this relation becomes

$$i = C\frac{dv_C}{dt} \tag{2-3}$$

A number of other elements are also used in models and these will be described as need arises. Indeed, we already see the need for *sources*, and it is going to be helpful to distinguish between independent sources and dependent or controlled sources. An *independent* source provides a voltage, or perhaps a current, that is not dependent on what is happening in the network; thus an independent voltage source provides a voltage at its terminals that is not affected by current through the source, or by voltage or current elsewhere in the network. If a large 12-volt storage battery were providing a fraction of an ampere to the instrument-panel light of an automobile, the battery might well be modeled as an independent 12-volt source. It is entirely possible for the voltage of an independent source to vary with *time* (though not as a function of current or voltage); if the current to be drawn is small, an electric outlet in a house may be modeled as an independent but time-varying source of voltage that is alternating sinusoidally with time.

There may also be independent *current* sources in models. The output of certain types of rectifiers is sometimes a current that varies so little with load that it is quite well modeled as an independent current source. See Figure 2b.

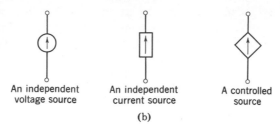

An independent
voltage source

An independent
current source

A controlled
source

(b)

FIGURE 2b Symbols for sources. For each, the magnitude must be given.

A *controlled* or dependent source has a voltage (or current) that is controlled by (and is often proportional to) a voltage or current somewhere in the network of which it is a part. Thus, as a simple example, the output current of a transistor model may be determined by the input voltage to the transistor.

Other elements are often included in models also. There are mutual inductance, possibly mutual capacitance and, occasionally, gyrators. Models of transistors and diodes provide means of analysis for networks with elements that are inherently nonlinear.[†] Some of these will be discussed in later chapters as needed.

Whether a model is a good model depends on whether it will predict accurately the behavior of the actual system.[‡] Has the physical network been idealized with proper judgment? Does the model act in essential ways like the real circuit? Of course, the only way to determine actual physical behavior of any element is by experiment and observation.

3. EXPERIMENTS

Measurement of resistance was reported by Georg Ohm in 1827 in his book entitled *The Galvanic Circuit Investigated Mathematically.* Current in a metallic conductor of electricity was found to be proportional to the voltage between the terminals. This is the physical fact observed by Ohm, and since this is in agreement with the mathematical model expressed in our equation 2-1, it follows that our mathematical model (with R a constant that is appropriately measured in *ohms*) is a good model of the physical circuit investigated by Ohm. However much Ohm's book may have been ridiculed at the time of publication,[§] his results are now so widely accepted that *Ohm's law* is often used by many who know little else about electricity.

---

† See Linvill, J. G., and Linvill and Gibbons, in the bibliography.
‡ See "An Operational View," Skilling (6) in the bibliography.
§ For this and other stories, see the author's *Exploring Electricity*, Skilling (2) in the bibliography.

An interesting tendency of electricity to remain in motion once it starts to flow, a kind of momentum, was discovered in 1832 at practically the same time by both Joseph Henry at Princeton University and Michael Faraday in London. This property of self inductance is introduced into the mathematical model by equation 2-2 and the unit of inductance is called the *henry*. The mathematical model with a constant value of L corresponds very closely to physical measurements on circuits and networks that do not involve a ferromagnetic material such as iron.

That electrical energy can be stored in capacitance has been known since the accident that startled Professor Muschenbroeck (or possibly Mr. Cuneous) in Leyden in 1746. Benjamin Franklin studied the Leyden jar in the same century, and Michael Faraday in the next, and our network model contains capacitance C in equation 2-3 that is measured in *farads*. Again the model corresponds to observed physical behavior with remarkable accuracy under most circumstances.

4. KIRCHHOFF'S LAWS

Our mathematical models also use a pair of relations that were determined experimentally by Kirchhoff and that have been confirmed by countless others since his time. One has to do with current, the other with voltage.

Kirchhoff's current law says that the sum of all electric currents flowing toward a junction is zero. Thus, in Figure 4a,

$$i_1 + i_2 + i_3 + i_4 + i_5 = 0 \tag{4-1}$$

This equation implies, of course, that if some of the currents, being *toward* the junction, are positive, others must be *away* from the junction and so be

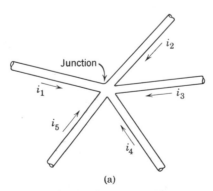

(a)

FIGURE 4a   **The sum of all electric currents flowing toward a junction is zero.**

negative. This equation of *continuity* can be made more general when generalization is needed.†

Kirchhoff's voltage law says that the sum of all voltages around a closed circuit is zero. Figure 4b shows a model of part of a network consisting of resistors and a battery; the battery is modeled as a source. The network might contain other elements with inductance and capacitance as well as resistance; it would still be found at every instant that the sum of voltages around such a loop as a–b–c–d–a is zero. That is,

$$v_{ab} + v_{bc} + v_{cd} + v_{da} = 0 \tag{4-2}$$

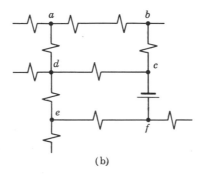

(b)

FIGURE 4b The sum of all voltages around a closed circuit is zero.

The loop may contain a source, as loop c–d–e–f–c. It may be more extensive as loop a–b–c–f–e–d–a. In all cases the sum of the voltages is zero at every instant.

The closed loop of Kirchhoff's law need not even be composed of conducting elements. Thus we can consider a loop a–b–c–a and write

$$v_{ab} + v_{bc} + v_{ca} = 0 \tag{4-3}$$

where v_{ca} is voltage from point c to point a. This leads to the concept of voltage being a potential difference and each point having a *potential*.‡ Noting that $v_{ac} = -v_{ca}$, equation 4-3 gives

$$v_{ac} = v_{ab} + v_{bc} \tag{4-4}$$

It is well to realize that this experimental relation, though simple, is not self-evident.

---

† Skilling (1), *Fundamentals of Electric Waves.*
‡ The generalization of this is the integral form of one of Maxwell's equations. See Skilling (1).

5. ENERGY AND POWER

Another experimental observation is that the power (in watts) supplied electrically to a resistor at a pair of terminals is equal to the product of the voltage between those terminals (in volts) and the current at the terminals (in amperes):

$$p = vi \qquad (5\text{-}1)$$

This can be established with a voltmeter, an ammeter, and a measurement of power obtained by observing the rate of heating, or it can be known as follows.

When voltage v is applied to a resistor there is current i, and the electric charge q entering the resistor in time t is $q = it$. There is current because the voltage produces an electric field e within the resistor; in a uniform resistor of length x, $v = ex$. The field exerts a force $f = qe$ on the electrons that carry the charge q. The electrons move along the resistor, and the work done on them while traveling its length x is $w = fx = qex = itex = itv$. This work produces heat as the electrons collide with molecules. Power p is the rate of doing work, and $p = w/t = iv$. This is equation 5-1.† This development is based on the experimental observation that $f = qe$.

Thus the power supplying heat to electrical resistance can be shown by either of two experimental means to be vi, and the electrical energy dissipated as heat is $\int vi\, dt$.

Purely inductive elements or purely capacitive elements do not become hot when voltage is applied at their terminals even though current flows and $\int vi\, dt$ is not zero. Electrical energy is not dissipated. This is interpreted to mean that energy is stored in a *reactive* element (inductance or capacitance), but not dissipated and the stored energy is returned to the electric circuit at some later time. See Sections 14 and 15 of Chapter 2.

6. DEVICES

A physical device that exhibits mainly resistance is called a *resistor*. One having mainly inductance or capacitance is an *inductor* or a *capacitor* (the term *condenser* is still heard, though obsolete). Resistors, inductors, and capacitors are all quite useful, but they are not equally common because of cost of manufacture.

The cost of a device depends on such matters as the current, the frequency,

---

† This development is given in more detail in Section 11, Chapter 1, of the author's *Electrical Engineering Circuits*, second edition. See Skilling (7).

and the type of manufacture. For a power system that operates with large current and low frequency, capacitance is exceedingly expensive and is used as little as possible. Communication systems, on the other hand, often use small currents and may operate with audio frequencies (below 20 kHz†) or at radio frequency (which might be typically 1000 kHz). Resistors, inductors, and capacitors are all practical in sizes required at audio and radio frequencies, often with a ferromagnetic core of some type for the inductors.

The radio-frequency use of *integrated* circuits employing semiconductor properties of materials and photographic techniques of manufacture has resulted in both capacitors and amplifiers (transistor units) being cheap; resistors are a bit more difficult, and inductors are almost impossible. As a result the designer of integrated circuits does everything he can with capacitance and amplification, using resistance where necessary, but avoiding inductance by every possible expedient. Since *integrated* circuits are highly practical when a large number of identical units are to be manufactured to operate at high frequency with small current and low power, it is distinctly interesting that their design is not the same as the design of systems composed of individual *discrete* elements. The ease of obtaining amplification and the difficulty of including inductance are having a profound effect.

7. NETWORKS

An actual physical network results when resistors, capacitors, inductors, transformers, generators and batteries, wires and cables, possibly transmission lines or wave guides and other devices are connected together in various combinations. For analysis, the network is idealized as a model containing elements of pure resistance, capacitance, inductance, sources, and other idealized elements as needed (Figure 7c).

Equations are written to describe the network (the system of equations is the mathematical model) and the equations are solved to find the behavior of the network. Network equations are written by combining the *element* equations 2-1, 2-2, and 2-3, which tell us that $v_R = Ri$, $v_L = L(di/dt)$, and $i = C(dv_C/dt)$, together with equations describing sources and other

---

† The hertz is a unit of frequency meaning one cycle per second. The kilohertz, abbreviated kHz, is 1000 Hz; the megahertz or MHz is 1000 kHz and so on. It would be logical to speak of the frequency of the ordinary house lighting circuit as 60 Hz, but power engineers prefer the more familiar 60 cycles per second or 60 c/s, and even refer to "60-cycle" current. This book will use both terms; remember, when Hz is used, that cycles per second is meant, and when c/s is printed, Hz is intended. Actually, the unit of radians per second (r/s) is much more needed in this theoretical treatment than either c/s or Hz.

elements, with Kirchhoff's two laws (equations 4-1 and 4-2) which define the *connections*.

Elements are often connected in series or in parallel. Elements are said to be in *series* if they carry the same current as in Figure 7a. Kirchhoff's current law says that the two elements in Figure 7a must carry the same current because there is no third connection to node *b*, and Kirchhoff's voltage law requires that the voltage from node *a* to *c* be the voltage from node *a* to *b* plus the voltage from node *b* to *c*.

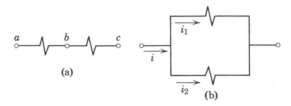

(a)

(b)

FIGURES 7a,b   Series and parallel elements.

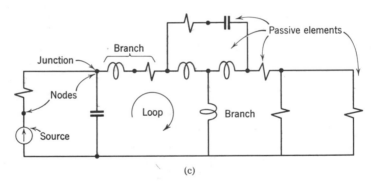

(c)

FIGURE 7c   Parts of a network.

Elements in *parallel* have the same voltage between terminals. Thus the same voltage is applied to the two elements of Figure 7b. Also, the terminal current is equal to the sum of the two element currents: $i = i_1 + i_2$.

An adequate set of element equations and connection equations will uniquely define a network. Note that no solution of a network should be attempted until all physical information about elements and connections has been put into the mathematics, for without a complete mathematical statement of the situation no answer can be expected; mathematics is a system of logic and will not supply missing facts.

8. DEFINITIONS

The following definitions will be needed repeatedly throughout our study of networks.†

An *element* of a model has already been discussed. A number of elements are connected into a network in Figure 7c. An equation relating current and voltage can be written for each element.

A *node* is a point in a network at which two or more elements are joined; see Figure 7c. If three or more elements are connected together at a node, that node is called a *junction*. (The IEEE Standard Dictionary gives node and junction as synonymous.)

A *branch* of a network extends from one junction to another, and may consist of one element or several elements in series. Thus there is a node at each end of every element and a junction at each end of every branch.

A *loop* or *mesh* (or, if the term is used carefully, a *circuit*) is a single closed path for current within a network.

A *passive* element contains no source of power. An *active* element delivers power to the network.

A *linear* element is independent of current and voltage. A linear network or system is made up of linear elements. Mathematically, if a disturbance $x_1(t)$ gives a response $y_1(t)$, and a disturbance $x_2(t)$ gives a response $y_2(t)$, then if a disturbance $Ax_1(t) + Bx_2(t)$, in which A and B are constants, gives a response $Ay_1(t) + By_2(t)$ the disturbed system is linear.

This mathematical definition is precise, but for many of our present purposes it is enough to say that a linear resistance, inductance, or capacitance must be independent of the current or voltage of that element. The value of the parameter may change with time (in which case it is said to be time-varying) or with some other independent variable, but not with current or voltage. For instance, if a coil has an iron core so that the inductance is less when the current is large, the inductance is not constant and the equation of an accurate mathematical model is not linear.

9. SUMMARY

This chapter starts with a summary of information that is probably familiar to the reader from earlier work.

For analysis, physical systems are replaced by their idealized *models*.

---

† Standard definitions are offered by the Institute of Electrical and Electronics Engineers in the *IEEE Standard Dictionary of Electrical and Electronics Terms*, published by John Wiley & Sons, New York, 1972. See bibliography, Appendix 2. Though authors do not always find it possible or expedient to follow standard definitions exactly, the usage of this book is quite close to the standard, slight deviations being noted.

The elements of which models are comprised may contain *resistance*, *inductance*, and *capacitance*, and therefore have (by definition) the properties given in equations 2-1, 2-2, and 2-3.

Models also contain *sources*. There are sources of independent voltage and sources with independent current; there are also controlled voltage sources and controlled current sources.

Experiment with actual physical *devices* shows that their behavior can usually be approximated, and often with amazing precision, by a combination of these ideal elements of models.

Experiment relates electrical quantities to *energy* and *power* as expressed in equation 5-1.

Resistors, inductors, and *capacitors* are highly practical devices for many purposes, though certain conditions of use or manufacture may make one or another prohibitively expensive.

Element equations together with *connection equations* based on Kirchhoff's laws give the mathematical model of a network.

A number of network terms are *defined*.

PROBLEMS

The number following each problem gives the most advanced section of the chapter that may be needed for the solution. For answers to problems marked with asterisks, see Appendix 4.

1-1. Combine equations 1-1 and 1-2 into a single equation of second order in i_1, eliminating i_2 and its derivatives. Use the following numerical values; $L_1 = 1$, $L_2 = 1$, $R_{12} = 2$, $R_2 = 3$. §1*

1-2. Combine equations 1-1 and 1-2 into a single equation of second order in i_2, eliminating i_1 and its derivatives. Use the following numerical values: $L_1 = 1$, $L_2 = 1$, $R_{12} = 2$, $R_2 = 3$. §1*

1-3. It is given that $v_L = d(Li)/dt$. This derivative expands to give equation 2-2 if L is constant. What is the comparable expansion if L (as well as i) is a function of time? Mention some situation in which L is variable. §2

1-4. It is given that current to capacitance is, in terms of voltage, $i = C(dv_C/dt)$. Give capacitor voltage in terms of current. Discuss the meaning of the constant of integration that appears in the answer, and the value to be given to it. §2

1-5. A floor lamp with a 300-watt bulb is plugged into a wall outlet. The owner, wanting to know how much electric current will be required, models this lamp as a pure resistance element supplied by an independent voltage source of 120 volts. List six or eight facts about the lamp that certainly exist but that make no difference in determining the current required, and six or eight others that make a difference so slight that the owner considers them negligible. (There is no sharp distinction between these categories.)   §6

1-6. To determine how much current it will require, an automobile headlight is modeled by the owner as a certain number of ohms of resistance supplied from a direct-current source of 12 volts. List six or eight characteristics of the devices involved that clearly make no difference in the operation, and six or eight others that are considered by the owner to make negligible difference.   §6

1-7. A circuit contains an element with inductance of 0.1 henry and resistance of 5 ohms. Current through the element is a triangular wave, shown in the figure as wave *B*. Find voltage across the element, plotting a curve of voltage as a function of time.   §7*

Note: In plotting results for this and succeeding problems of this chapter, the drawing may be freehand, but the work should be neat and done to scale. Scales should be indicated, and values of interest specified. It is not necessary to have a mathematical formula for each part of a curve, but each should be marked "straight," "parabolic," and so forth.

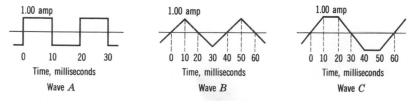

Problems 1-7 through 1-10.

1-8. Repeat Problem 1-7 for wave *C*. Discuss the probability of having current of the form of wave *A* in this circuit, considering wave *A* a limiting case of wave *C* as the sloping sides become steeper.   §7*

1-9. A coil in a circuit is idealized to an inductance of 0.1 henry and a resistance of 10 ohms in series. (*a*) Plot the voltage across this coil when the current is wave *B*. (*b*) Repeat for wave *C*.   §7*

1-10. A circuit has capacitance and resistance in series; the values are 20 microfarads and 500 ohms, respectively. (*a*) Plot voltage across the circuit terminals when the current is wave *A*. (*b*) Repeat for wave *C*.   §7*

1-11. (*a*) A source of 10 milliamperes (10^{-2} ampere) constant current is connected to a capacitor of $C = 10^{-3}$ farad capacitance at time $t = 0$. Find the voltage on the capacitor and the stored energy at $t = 1$ second. (*b*) A source of 1 volt constant voltage is connected to an inductive element with $L = 10^{-3}$ henry at time $t = 0$. Find the current through the inductance and the stored energy at $t = 0.1$ second.   §7*

1-12. The current in the figure (p. 14) flows into a 100-microfarad capacitor. Draw v_C, the voltage across the capacitor, letting $v_C = 0$ when the current begins at $t = 0$.   §7

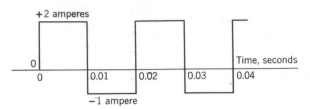

Problem 1-12.

1-13. In a cathode-ray tube the jet of electrons carries current of 5 microamperes $(5 \cdot 10^{-6})$. The electrons are accelerated by an electric field with a potential difference of 1500 volts. In what length of time is 1 watt-second of work done on the electron jet by the electric field? §5*

1-14. Show that the following equation is linear (R and L being constant) according to the mathematical definition of Section 8:

$$Ri + L\frac{di}{dt} = v(t) \qquad \text{§8}$$

1-15. Show that the following equation is linear (R, L, and C being constant) according to the mathematical definition of Section 8:

$$RI + L\frac{di}{dt} + \frac{1}{C}\int_0^t i\,dt = v(t) \qquad \text{§8}$$

1-16. Show that the following equation (which might apply to a crystalline semiconductor such as Thyrite) is not linear according to the mathematical definition of Section 8:

$$i = Kv^3 \qquad \text{§8}$$

1-17. The energy of a particle with a small electric charge is often measured in *electron-volts*. One electron-volt is the energy gained (or lost) by an electron while moving through a potential difference of one volt. How many electron-volts equal one joule of energy? The electric charge on an electron is approximately -1.602×10^{-19} coulomb, as may be found in any electrical handbook.

§5

CHAPTER
2

ALTERNATING CURRENT

1. CURRENTS AND COMPONENTS

Electric current to actuate the receiver of a telephone, or a radio or television speaker, has a wave form similar to that of the sound it is to reproduce. A single note of a musical instrument gives current that varies periodically; the simplest wave, perhaps from one pipe of an organ, can be described mathematically as a sinusoidally varying function of time, and to produce an audible sound its frequency will be something less than 20 kHz.

The incoming signal from the antenna of a radio or television set alternates at a much higher frequency. This radio-frequency current is modulated by the information that it bears, but it is approximately sinusoidal and it can usually be analysed as if it were exactly so.

Power to an electric light, to a home refrigerator, or to the air-compressor motor in a gasoline station is almost exactly sinusoidal in form, with constant voltage, and frequency that is quite precisely 60 cycles per second (60 Hz) in the United States, or 50 in many other countries.

Current that does not vary with time but remains constant is called direct current (or dc). For analysis, direct current can be treated as alternating current with zero frequency.

Electric current may be periodic and yet not have the smooth form of a sinusoidal wave. The note of a violin gives an audio-frequency wave that is far from sinusoidal, or a square-wave generator gives an output that is periodic but almost discontinuous at each half-cycle. Any such periodic but nonsinusoidal function can be analysed, however, by Fourier series, into the sum of a direct-current component, a fundamental-frequency component, and harmonic components with frequencies that are multiples of the fundamental. Thus all periodic forms can be treated as sums of sinusoidal forms.

There are also nonperiodic forms and discrete signals. Here we employ analysis by Fourier or Laplace integrals. Consider a current that is non-repetitive, such as the result of closing a switch, or as an extreme but simple example consider a single discrete pulse of the kind used to operate digital electronic computers. Any such nonrepetitive or discrete form can be analysed into the sum of sinusoidally varying components. There may be an infinity of components, each of infinitesimal size, but each component is sinusoidal (or similar) and network analysis proceeds as it does for any sinusoidal current.

Thus any current whatever can, if desired, be treated as the sum of sinusoidal components, and thus the overriding importance of the sinusoidal form is clearly seen. The purpose of this foregoing section is to explain the great attention given to the sinusoidal form in all study of networks.

2. STEADY ALTERNATING CURRENT

A mathematical expression for sinusoidal alternating current as in Figure 2a can be written:

$$i = I_m \cos \omega t \tag{2-1}$$

At any time t, the current is i. I_m is the maximum current, a constant; ω is a constant related to frequency.

Figure 2a shows i as a function of t. One cycle is the time from positive crest to positive crest. The value of ωt at the end of one complete cycle is 2π, at which point the cosine function begins to repeat. Hence $\omega t_1 = 2\pi$, and t_1,

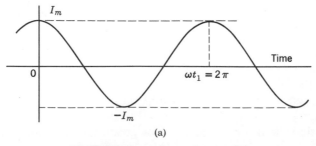

(a)

FIGURE 2a   Alternating current.

the time of one cycle, is $2\pi/\omega$. From this, the number of cycles in 1 second is $\omega/2\pi$. We call the number of cycles per second the *frequency f*, and

$$f = \frac{\omega}{2\pi} \text{ cycles per second (or hertz)} \tag{2-2}$$

We call ω the *radian frequency*, and

$$\omega = 2\pi f \text{ radians per second} \tag{2-3}$$

Equation 2-1 expresses current as a cosine function. It would be equally possible to use the sine function $i = I_m \sin \omega t$. We prefer the cosine form for convenience. When it is necessary to use a fully general form of sinusoidal function, we shall write

$$i = I_m \cos(\omega t + \theta) \tag{2-4}$$

Then by adjusting θ the curve of Figure 2a can be slid sideways as desired; θ is called a *phase angle*.

3. ALTERNATING CURRENT THROUGH RESISTANCE

Figure 3a shows part of a network that contains resistance. By equation 2-1 of Chapter 1, the resistance, voltage, and current are related by

$$R = \frac{v}{i} \tag{3-1}$$

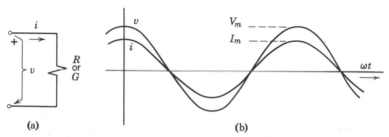

(a)   **(b)**

FIGURES 3a,b   Voltage and current of resistance.

We also speak of the *conductance G*. If an element of a network has pure resistance (without inductance or capacitance), the conductance of that element is the reciprocal of the resistance and

$$G = \frac{1}{R} = \frac{i}{v} \tag{3-2}$$

A more general relation between R and G will be given in Section 13 of Chapter 4.

With alternating current as in equation 2-1, voltage across resistance is

$$v = Ri = R(I_m \cos \omega t) = V_m \cos \omega t \qquad (3\text{-}3)$$

in which

$$V_m = RI_m \qquad (3\text{-}4)$$

V_m is the maximum value of the voltage wave. Figure 3b shows i and v as functions of time;[†] both are cosine waves, and at every instant v is proportional to i. The crest values, V_m and I_m, are indicated.

Similar equations can be written in terms of conductance:

$$i = Gv = G(V_m \cos \omega t) = I_m \cos \omega t \qquad (3\text{-}5)$$

in which

$$I_m = GV_m \qquad (3\text{-}6)$$

These equations apply, of course, to the same element, and the same curves illustrate the relation.

4. ALTERNATING CURRENT THROUGH INDUCTANCE

When an element contains only inductance, the terminal voltage is given by equation 2-2 of Chapter 1 as:

$$v = L\frac{di}{dt} \qquad (4\text{-}1)$$

With sinusoidally alternating current as in equation 2-1,

$$v = L\frac{d}{dt}(I_m \cos \omega t) = LI_m\frac{d}{dt}\cos \omega t = -\omega LI_m \sin \omega t \qquad (4\text{-}2)$$

Current being a cosine function of time, the voltage turns out, following the differentiation in equation 4-2, to be a negative sine function. Current and voltage are plotted in Figure 4a. It will be seen that both are sinusoidal waves, but that one is displaced sideward relative to the other. By a trigonometric change in equation 4-2, voltage can be written

$$v = -\omega LI_m \sin \omega t = \omega LI_m \cos(\omega t + \pi/2) \qquad (4\text{-}3)$$

[†] In Figure 3b and others following, the independent variable is indicated as ωt rather than t. This makes no essential difference, ω being merely a constant. It is done for convenience. Values of ωt are *angle*, and it is customary and more convenient to speak of short intervals along the horizontal axis as intervals of angle rather than time. Thus, in Figure 4a the angle $\pi/2$ is marked; if the axis were calibrated in t rather than ωt, this interval would have to be $\pi/2\omega$.

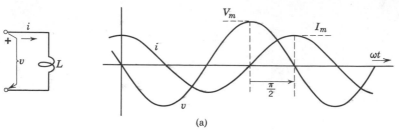

FIGURE 4a Voltage and current of inductance.

The maximum value of the voltage wave, V_m, is seen from equation 4-3 to be

$$V_m = \omega L I_m \qquad (4\text{-}4)$$

That is, V_m is related to I_m by the factor ωL. Similar expressions involving ωL appear so often that a special name and symbol are given to this quantity. It is called *inductive reactance* and is designated X. Thus

$$X = \omega L = 2\pi f L \qquad (4\text{-}5)$$

Equation 4-4 can then be written in terms of X:

$$V_m = X I_m \qquad (4\text{-}6)$$

The maximum value of the current wave in Figure 4a is reached at a later time than the maximum value of the voltage wave. The current wave is said to *lag* the voltage wave, and the amount of lag is $\frac{1}{4}$ cycle. From a different point of view, one may say that the voltage *leads* the current wave by $\frac{1}{4}$ cycle.

5. ALTERNATING CURRENT TO CAPACITANCE

When an element contains only capacitance, the current is given by equation 2-3 of Chapter 1 to be

$$i = C\frac{dv}{dt} \qquad (5\text{-}1)$$

If the voltage is sinusoidally alternating so that $v = V_m \cos \omega t$, current is

$$i = C\frac{d}{dt}(V_m \cos \omega t) = -\omega C V_m \sin \omega t = \omega C V_m \cos(\omega t + \pi/2) \quad (5\text{-}2)$$

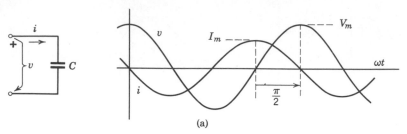

FIGURE 5a   Voltage and current of capacitance.

Current and voltage waves are shown in Figure 5a. Both waves are sinusoidal. The wave of current through a pure capacitance *leads* the wave of voltage by $\frac{1}{4}$ cycle ($\pi/2$ radians or 90 degrees). The maximum value of the current wave is, from equation 5-2,

$$I_m = \omega C V_m \qquad (5\text{-}3)$$

Maximum current and voltage of a capacitor are thus related by the quantity ωC. We find it convenient to give a name to this quantity also; we call it *capacitive susceptance*. The symbol for susceptance is B. In terms of susceptance, then, equation 5-3 is written

$$I_m = B V_m \qquad (5\text{-}4)$$

where

$$B = \omega C \qquad (5\text{-}5)$$

6. DIRECT CURRENT

Direct current need not be considered separately. The same methods, the same laws, the same theorems apply for direct as for alternating current. It is sometimes said that direct current can be treated as alternating current of zero frequency. It is more precise to say that the behavior of direct current is the limit approached by alternating current as the frequency approaches zero. Computations of direct current involve only the resistances of circuits. Inductance is ineffective; an unvarying current does not produce voltage across inductance. Capacitance is equivalent to an open circuit, for where there is a capacitor no steady current can pass. Only resistance remains for computation. When ac methods are known, dc methods will be obvious.

7. THEOREM FOR LINEAR NETWORKS

It is of great importance that sinusoidally alternating current to any element having constant resistance, inductance, or capacitance is accompanied by a voltage between the terminals of the element that is also sinusoidally alternating, with the same frequency. This has been shown in the three preceding sections. Now the sum or difference of any two sinusoidal waves of the same frequency is another sinusoidal wave, and therefore it follows that *if the current or voltage at any part of any (linear) network is sinusoidal, the currents and voltages at every part of the network are sinusoidal with the same frequency.*

This statement assumes that (1) the network is composed of linear elements having constant resistance, inductance, and capacitance, and (2) voltages have been applied and currents have been flowing for so long a time that transient disturbances have died away.†

8. EFFECTIVE ALTERNATING CURRENT

The meaning of a unit of direct current, an ampere, is well known. It is a measure of the rate of passage of electricity. An ampere is defined by the National Bureau of Standards in terms of the force produced on another electric current. The concept is simple, and there is no question about the nature of the ampere of direct current.

A unit of alternating current is not so easily conceived. In a circuit carrying alternating current, the current is actually varying from instant to instant as indicated by the wave of Figure 8a. How is the amount of the current to be specified?

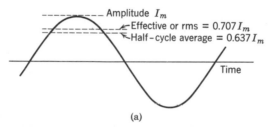

(a)

FIGURE 8a   Effective and average values.

One way to specify the amount of alternating current is to give the maximum instantaneous value I_m, the amplitude of the wave. Yet surely it would be misleading to say that the number of amperes of alternating current is equal to I_m, for a wave with crests reaching I_m only momentarily each half-cycle is in no way equivalent to a steady current that is continuously equal to the same value.

---

† See Section 8 of Chapter 1 for a precise mathematical definition of *linear*.

In the above generalization, currents and voltages are functions of frequency. In certain exceptional networks (not simple connections in series and parallel, but including some lattice or Maxwell-bridge networks), there can be frequencies at which the current of a branch passes through zero; such a null frequency may be excluded from the generalization.

Some kind of an average value of the current wave might well be suggested as a measure. An average of a sinusoidal current wave throughout a complete cycle would be meaningless, for such an average is zero. The current is negative as much as it is positive. However, the average current during any one positive half-cycle can be computed; it is $2I_m/\pi = 0.637I_m$. Is this a desirable measure of alternating current?

To answer this question it is necessary to consider the purpose of measuring alternating current. What use is to be made of the measurement? The answer is: We want to find what the current is able to do. This must usually be expressed in terms of *power*. Whether the alternating current is to operate a telephone receiver, a loudspeaker, or an electric motor, to heat an electric light or a vacuum-tube filament, we are primarily concerned with its ability to do work. Also, power is a measure of the difficulty of producing the electricity, and therefore determines how much we must pay for it.

We want to be able to say of an alternating current that it has an effective value, to be expressed as a certain number of amperes, meaning that the power available from this alternating current is the same as the power available from an equal number of amperes of direct current. Before this definition can be completed, it will be necessary to discuss power.

9. POWER

Power is defined as the rate of doing work; it is energy per unit time; electrical power is measured in watts, and 1 watt is 1 joule per second.

Electrical power is voltage times current† as in equation 5-1 of Chapter 1:

$$p = vi \tag{9-1}$$

This power comes from the source, perhaps a generator or battery. Electromagnetic energy follows the circuit and reappears as heat energy in a resistor, or as the energy of mechanical motion in a motor or loud-speaker, as luminescence in a fluorescent lamp, as chemical energy in a charged battery, or perhaps in some other nonelectrical form.‡

Let us think specifically of a resistor. The voltage between its terminals is v

---

† In general, power is proportional to voltage times current. To be *equal*, as in equation 9-1, the units must be a consistent set. Watts, volts, and amperes are a consistent set. (But kilowatts, volts, and amperes are not.)

‡ To be precise it is necessary to consider that the part of the circuit in which we are interested is within an imaginary closed surface that passes through the circuit terminals. The steady-state average electromagnetic energy entering through this surface is transformed to nonelectromagnetic forms, such as heat, and so forth. (The Poynting field is applicable; see Skilling (1), or any book on electromagnetic theory.) If the net power is outward, there are generators or other sources within the closed surface. See bibliography; *Institute of Electrical and Electronics Engineers, Standard Dictionary*, "power, instantaneous," and "Poynting's vector."

and current through the resistor is i at a particular instant. It is immaterial whether the current is constant or changing; the power at that particular instant is $p = vi$. By obvious substitutions in this expression, power can be written in terms of current only, or voltage only:

$$p = vi = (iR)(i) = i^2R \qquad (9\text{-}2)$$

$$p = vi = v(v/R) = v^2/R \qquad (9\text{-}3)$$

If current is alternating, these equations apply at each instant of the alternating wave. Figure 9a shows a wave of current through a resistor and the corresponding wave of voltage across the terminals. By multiplying these two together at each instant the curve of instantaneous power p is obtained. It will be seen that the power varies between $v_m i_m$ and zero.

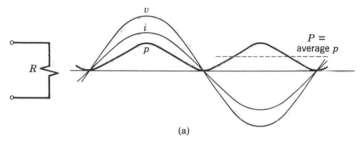

(a)

FIGURE 9a   Instantaneous and average power in resistance.

10. AVERAGE POWER

But we have little practical use for the curve of instantaneous power; what we really need is the *average* power. If instantaneous power is

$$p = i^2R \qquad (10\text{-}1)$$

then average power (which we shall call P) is

$$P = \text{Average } p = \text{Average } (i^2R) = (\text{Average } i^2) \cdot R \qquad (10\text{-}2)$$

This expression for average power P brings into prominence the average of the current squared. Note that the average of i^2 is not at all the same thing as the square of the average of i. Figure 10a shows a cycle of an ac wave as a solid curve. A dashed curve shows i^2. The value of i^2 is positive at every instant, for the square is positive whether current is positive or negative. The average of i^2 through the cycle is shown by a horizontal dashed line. The average of i through a cycle, on the other hand, is zero, and the square of the average of i is zero.

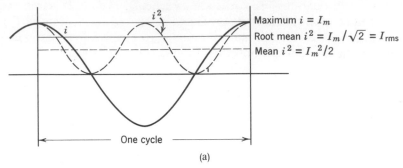

Maximum $i = I_m$
Root mean $i^2 = I_m/\sqrt{2} = I_{rms}$
Mean $i^2 = I_m^2/2$

One cycle

(a)

FIGURE 10a   **Instantaneous and mean values of** $i = I_m \cos \omega t$. **(Drawn for** $I_m = 1$.**)**

The mean-square value of voltage is important, also. Equation 10-2 was based on equation 9-2 for power in terms of current. Using equation 9-3 instead, and expressing power to a resistor in terms of voltage, we obtain

$$P = \text{Average } p = \text{Average } (v^2/R) = (\text{Average } v^2)/R \qquad (10\text{-}3)$$

Here average power is expressed in terms of the mean-square voltage. Note that equations 10-2 and 10-3 apply to current and voltage of *any periodic form*.

11. ROOT-MEAN-SQUARE VALUES

Now we are prepared to specify an *effective* value for a current of any periodic form. The power delivered by such current is, from equation 10-2,

$$P = I^2 R \qquad (11\text{-}1)$$

where P means average power (during one cycle) and I^2 means the average of i^2 during the cycle. That is,

$$I^2 = \text{Average } i^2 \qquad (11\text{-}2)$$

From this it follows, of course, that

$$I = \sqrt{\text{Average } i^2} \qquad (11\text{-}3)$$

This effective value is valid for any periodic current. It is commonly called the root-mean-square value, abbreviated *rms*. An italic capital I is the symbol used for rms current, and sometimes it is written I_{rms} for emphasis. The unit is the *ampere*.

A similar measure of effective voltage is required, and the root-mean-square voltage is seen from equation 10-3 to be the necessary quantity. Average power (through a cycle) is

$$P = V^2/R \qquad (11\text{-}4)$$

provided V^2 means

$$V^2 = \text{Average } v^2 \qquad (11\text{-}5)$$

through the cycle, and hence

$$V = \sqrt{\text{Average } v^2} \qquad (11\text{-}6)$$

An italic capital V is used, and the root-mean-square derivation of this effective value can be emphasized by writing the symbol V_{rms} : the unit is *volts*.

In equations 11-3 and 11-6 we have effective values for current and voltage that are obtained by considering power conveyed to a load. Two other important and useful relations involving rms values are easily obtained. From equation 2-1 of Chapter 1 (Ohm's law) $v = iR$ and hence $v^2 = i^2R^2$. Therefore

$$(\text{Average } v^2) = (\text{Average } i^2)\, R^2 \qquad (11\text{-}7)$$

and, taking the square roots,

$$\sqrt{(\text{Average } v^2)} = \sqrt{(\text{Average } i^2)} \cdot R \qquad (11\text{-}8)$$

This gives a form of Ohm's law involving rms values applied to a resistance:

$$V = IR \quad \text{or} \quad V_{\text{rms}} = I_{\text{rms}} R \qquad (11\text{-}9)$$

Now we are able to write, combining equations 11-1 and 11-9,

$$P = I^2R = (IR)I = VI \quad \text{or} \quad P_{\text{av}} = V_{\text{rms}} I_{\text{rms}} \qquad (11\text{-}10)$$

The equations 11-1, 11-4, 11-9, and 11-10 relate rms values of V and I when average power P is conveyed by a periodic current to a load of constant resistance R. Note that the current may be of any periodic form (or even direct current). Sinusoidally alternating current is a special case.

12. RMS RELATIONS FOR SINUSOIDAL WAVES

In Figure 9a, a curve of instantaneous power is shown; it has the appearance of a displaced double-frequency sinusoid. To see that this is indeed its true shape, consider that it is a curve of $p = i^2R$. Let $i = I_m \cos \omega t$. By a common trigonometric identity:

$$p = i^2R = I_m{}^2(\cos \omega t)^2 R = I_m{}^2 R \cdot \tfrac{1}{2}(\cos 2\omega t + 1) \qquad (12\text{-}1)$$

This is the expression for instantaneous power shown by the curve of Figure 9a.

Average power P can now be found by inspection. Since the average of $\cos 2\omega t$ through any integer number of cycles is zero, that term vanishes from the average, leaving as an expression for the average power:

$$P = \frac{I_m^2}{2} R = \left(\frac{I_m}{\sqrt{2}}\right)^2 R \qquad (12\text{-}2)$$

Comparison with equation 11-1 shows that the quantity in parentheses is the effective or rms value of a sinusoidal wave. That is, for a sinusoidal wave only,

$$I_{rms} = \frac{I_m}{\sqrt{2}} \quad \text{or} \quad I_m = \sqrt{2}\, I_{rms} \qquad (12\text{-}3)$$

See Figure 10a.

This same result is reached more formally as follows: Assume $i = I_m \cos \omega t$ as a typical sinusoidal wave. The square is $i^2 = I_m^2(\cos \omega t)^2$, and I_{rms}^2, the average of i^2, is found by integrating with respect to time for 1 cycle and dividing by the time of 1 cycle. The time of 1 cycle, T, is $2\pi/\omega$. Hence

$$I_{rms}^2 = \frac{1}{T}\int_0^T i^2\, dt = \frac{\omega}{2\pi}\int_0^{2\pi/\omega} I_m^2(\cos \omega t)^2\, dt$$
$$= \frac{I_m^2 \omega}{2\pi}\left(\frac{\pi}{\omega}\right) = \frac{I_m^2}{2} \qquad (12\text{-}4)$$

$$I_{rms} = \frac{I_m}{\sqrt{2}} \qquad (12\text{-}5)$$

That is,

$$\text{Rms value} = \frac{1}{\sqrt{2}}(\text{Maximum value}) = 0.707\ (\text{Maximum value}) \qquad (12\text{-}6)$$

For a wave of any form the ratio of maximum to rms value is called the *crest*, *peak*, or *amplitude factor*. Equation 12-6 shows that for a sinuoidal wave the crest factor is $\sqrt{2} = 1.414$. This value will become very familiar, but care must be used not to apply it to waves of nonsinusoidal form—a common error.

13. PRACTICAL USE OF RMS VALUES

Effective or rms values of voltage and current are so commonly used that they are always understood to be implied when a numerical value is given for alternating voltage or current. If an alternating current is stated to be 10 amperes, it is understood that 10 is its rms value; if this current wave is sinusoidal its maximum value is 14.1 amperes. A 120-volt circuit is understood to supply an rms value of 120 volts. All ordinary voltmeters and ammeters for alternating current are calibrated to indicate rms values.

Rms values of voltage and current are almost always used in computation, also. This practice is so common, indeed, that it is usual to omit subscripts from V_{rms} and I_{rms} in all ordinary computations, writing merely V and I to mean effective rms values.

The foregoing discussion of average power has stipulated that the average is to be taken through 1 cycle or some integer number of cycles. This is the only way that P can have exact meaning. In practice, however, we are usually interested in average power over an unspecified length of time. Except for extremely short intervals of time, this does not lead to appreciable difficulty. If the practical average of power is obtained over a time equal to some hundreds of cycles, the possible error from inclusion of the power of a fraction of a cycle at the end of the interval is quite negligible.

Physical measurements, as with a wattmeter, are always made over a time of many cycles, and no difficulty is encountered. In computation, however, the exact meaning of P must be remembered. It is possible that the rms value of a half-cycle or a quarter-cycle may be the same as that of a full cycle. This can result from symmetry, as in the cosine wave of Figure 10a, and it is often a help in computation. But basically the rms value is to be obtained from some integer number of cycles.

14. POWER TO INDUCTANCE

Section 9 discusses power to resistance, the instantaneous power being $p = vi$. Since inductance, as defined in equation 2-2 of Chapter 1, can carry current when there is voltage between its terminals, does this imply that power is being delivered to it? The interpretation is that inductance does accept power but does not dissipate it; power received by inductance is stored in the magnetic field of the inductance and is later returned to the electric network.

Mathematically, power to inductance at any instant is

$$p = vi = iL\frac{di}{dt} \tag{14-1}$$

and energy W received during the time from t_1 to t_2 is

$$W = \int_{t_1}^{t_2} iL\frac{di}{dt}\,dt = L\int_{i_1}^{i_2} i\,di = \frac{1}{2}L(i_2{}^2 - i_1{}^2) \tag{14-2}$$

where i_1 and i_2 are currents at t_1 and t_2 respectively. If current is periodic, and i_1 and i_2 are currents to the inductance at the beginning and end of the period, they will necessarily be equal so that $i_1 = i_2$ and W in equation 14-2 is zero; that is, the energy received by inductance during any complete period is zero, and hence the *average* power during any complete period or during any integer number of periods is zero.

Energy received by inductance is stored. If current at t_1 equals zero, and current at t_2 equals i, then by equation 14-2,

$$W = \tfrac{1}{2}Li^2 \tag{14-3}$$

and this is considered to be the energy stored in inductance L when carrying current i. When current is zero, stored energy is zero. While current increases, stored energy increases, and power delivered *to* the inductance is positive. In equation 14-1, while i and di/dt are both positive (or else both are negative), power, being positive, goes into the inductance. But as current decreases, if i is positive but di/dt is negative (or, if i is negative, while di/dt is positive), power is negative. Power is then coming *from* the inductance where it has been held in storage.

Figures 14a and 14b show these relations using sinusoidal current for illustration. Power to inductance is shown by the heavier line in Figure 14a,

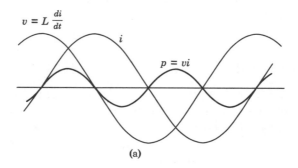

(a)

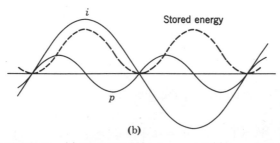

(b)

FIGURE 14   (a) Power to inductance, and (b) stored energy.

first positive and then negative as energy flows toward and then from the inductance. Stored energy is shown by the dash line in Figure 14b, proportional to the integral of power and proportional to the square of current.

Notice that current in an inductive element and voltage across the element are both sinusoidal in Figure 14a, and the current wave lags one-quarter cycle behind the voltage wave. This illustrates the phase relation derived in equation 4-3.

It should perhaps be emphasized that this section deals with the inductance of a network model as defined by equation 2-2 of Chapter 1. However, in the physical network there may be *magnetic saturation*; if so, inductance is still an energy-storage property but L is a function of i, and the network is therefore *nonlinear*. There may be *magnetic hysteresis*; then the model must provide energy dissipation as well as energy storage in its inductive elements.

15. POWER TO CAPACITANCE

Capacitance, as defined by equation 2-3 of Chapter 1, is likewise an element of a network model that can store energy but does not dissipate it. Power to capacitance at any instant is

$$p = vi = vC\frac{dv}{dt} \tag{15-1}$$

and energy W received during the time t_1 to t_2, as voltage changes from v_1 to v_2 is

$$W = \int_{t_1}^{t_2} vC\frac{dv}{dt}\, dt = C\int_{v_1}^{v_2} v\, dv = \frac{1}{2}C(v_2{}^2 - v_1{}^2) \tag{15-2}$$

If voltage at the end of an interval is equal to voltage at the beginning of the interval, as at the end of a cycle of periodic voltage, the energy received by capacitance during that interval is zero, and hence the *average* power is zero.

Energy received by capacitance is interpreted as being stored in the electric field. If voltage at time t_1 is zero, and voltage at time t_2 is v, then equation 15-2 gives

$$W = \tfrac{1}{2}Cv^2 \tag{15-3}$$

as the energy stored in capacitance C when its terminal voltage is v.

Figure 15a shows these relations using sinusoidal voltage for illustration. Power to capacitance and stored energy are both shown. Energy that is stored during one quarter-cycle of voltage is released and returned to the network during the following quarter-cycle. Stored energy is proportional to the integral of power input, and also proportional to the square of the voltage.

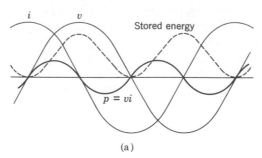

(a)

FIGURE 15a   Power and energy to capacitance.

Notice that current to a capacitive element and voltage across the element are both sinusoidal in Figure 15a and the current wave *leads* one quarter-cycle ahead of the voltage wave. This illustrates the phase relation derived in equation 5-2.

It is interesting, and will later be of great importance in connection with resonance phenomena, that the energy stored in capacitance is greatest at the crest of the voltage wave when energy stored in inductance is zero, and it is zero at the crest of the current wave when energy stored in inductance is greatest. If a network is so arranged that stored energy can shuttle back and forth between capacitance and inductance as sinusoidal current flows from one to the other at a particular frequency, the network is said to be resonant, and the frequency is the *resonant frequency*.

16. PHASORS

Graphically, an alternating quantity is conveniently represented by the projection of a rotating line on a fixed axis. We shall call the rotating line a *phasor*. It is common knowledge that if a line, such as the one marked I_m in Figure 16a, rotates, like the spoke of a wheel, with constant velocity about the origin O, the length of its projection on the horizontal axis varies sinusoidally with time. If I_m is the length of the line, the length of its projection on the horizontal axis is $I_m \cos \omega t$, where ω is the angular velocity of the line, assumed counterclockwise, and t is the time that has elapsed since an instant when the rotating line was itself in a horizontal position.

A phasor can thus represent a current $I_m \cos \omega t$ or a voltage $V_m \cos \omega t$. To represent a voltage such as $V_m \cos(\omega t + \varphi)$, it is necessary only to use another phasor advanced by an angle φ. This phasor has an angle $\omega t + \varphi$, as shown in Figure 16b.

Current and voltage are both represented by the rotating lines of Figure 16c. Both lines are rotating relative to the axes, and as they turn and

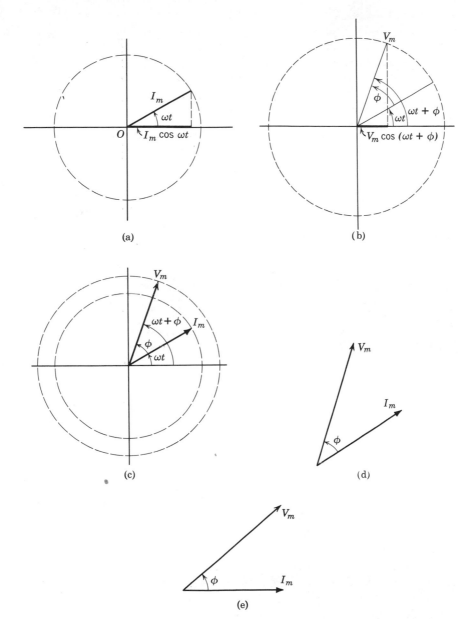

FIGURE 16 **(a)** A rotating line. **(b)** Phase angle. **(c)** Voltage and current in a phasor diagram. **(d)** The rotating lines of Figure 16c. **(e)** The rotating lines at the instant $t = 0$.

continue to turn indefinitely, one line will always lead the other by the phase angle φ. The two radial lines tell all that there is to know about these quantities; the rest of the diagram can be left out with no loss of information, and Figure 16d shows all that is essential of Figure 16c. Even the orientation of Figure 16d is not significant, for although two rotating lines must be drawn at the correct angle to one another, they should be visualized as rotating together at synchronous speed about their common origin, and any picture of them is merely a snapshot at one instant. Any other snapshot would be equally valid.

Since there must be a choice, let us elect to take our snapshot at the instant at which the angle of the current phasor is zero. Figure 16e results, and in such an orientation I_m is said to be the *reference phasor*. Always, when drawing a phasor diagram, the first step is to select some one phasor as reference, having zero angle (or any angle desired). All other phasors must then be drawn with the correct *relative* angles.

17. MATHEMATICAL FORMULATION

Mathematically, an alternating current or voltage can be represented by a trigonometric expression (as it has been in this chapter) or by expressions containing exponentials with complex exponents. These three equations all represent the same voltage:

$$v = V_m \cos \omega t \tag{17-1}$$

or

$$v = V_m \frac{e^{j\omega t} + e^{-j\omega t}}{2} \tag{17-2}$$

or

$$v = \mathscr{R}e\, V_m e^{j\omega t} \tag{17-3}$$

where $\mathscr{R}e$ means "the real component of."

Trigonometric functions are awkward to deal with, but exponentials can be multiplied and divided, differentiated and integrated quite easily. Indeed, equation 17-3 is so much the simplest form for expressing alternating quantities that it is used by practically everyone.

The use of complex exponentials is so common and so helpful that a few pages will now be devoted to the mathematics of the subject.

18. SUMMARY

Electrical quantities that vary sinusoidally are of great practical importance. Moreover, it will soon be seen that *any* current or voltage can be treated as the *sum of sinusoidal components*.

The relations of alternating current and voltage are determined for *resistance* or conductance, *inductance*, and *capacitance*. Direct current is a special case of alternating current.

A *theorem* is given regarding the sinusoidal form of steady current in all parts of linear networks.

A *unit* of alternating current is defined to have the same ability to do work as a unit of direct current. Consideration of power to a resistive load leads to the *root-mean-square* definition of unit current and voltage, for either sinusoidal or non-sinusoidal waves.

It is shown that *average* power to a *reactive* load is *zero*. Energy is *stored* in inductance or capacitance but is not dissipated.

Alternating currents and voltages are conveniently represented in graphical form by *phasors*.

Mathematically, *exponential* expression is more convenient than trigonometric.

PROBLEMS

2-1. A theorem for linear networks is given in Section 7. Explain why this theorem is true when applied to simple circuits and networks. (A demonstration for all networks will be possible after Chapter 9.) §7

2-2. Verify by integration that the average of a sinusoidal current wave through a positive half-cycle is $2I_m/\pi$. §8

2-3. Find the rms value of the wave shown. §11*

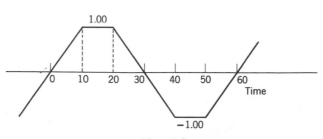

Problem 2-3.

2-4. Find the rms value of the square wave shown as Wave *A* with Problem 1-7, page 13. Also find the rms value of the triangular wave marked Wave *B* in the same figure. §11*

2-5. Two waves are shown, a half-cycle of each. Find the effective (rms) current of each wave. Compare the two. (Figure on page 34.) §11*

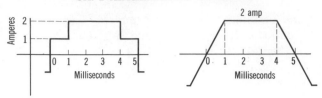

Problem 2-5.

2-6. Find the rms value of the wave shown. It is symmetrical, and one quarter-cycle of the wave is

$$i = I_m\sqrt{t}.$$ §11

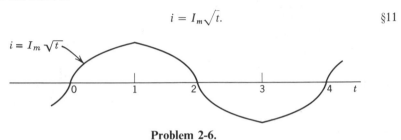

Problem 2-6.

2-7. For each of the three waves shown, find the rms value, the half-cycle average value, and the crest factor. In wave c i varies as t^2 from $t = 0$ to $t = T/2$, at which point $i = K$, and is then symmetrical (odd symmetry about $t = T/2$).   §12*

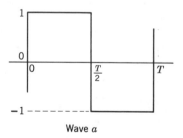

Wave a

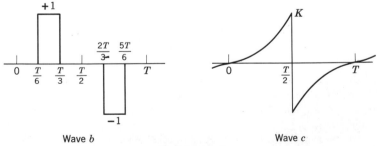

Wave b Wave c

Problem 2-7.

2-8. An electric heater (pure resistance) gives out 1 kilowatt of heat when the applied voltage is alternating, sine wave, with an rms value of 120 volts. It is desired to produce the same heat from the same heater with an applied square wave of voltage (having the shape of Wave *a* of Problem 2-7). What voltage must be used? (Give the maximum value, the flat top, of the square wave of voltage.) §11

2-9. For the offset square current wave shown with Problem 1-12, page 14, find (*a*) average current, (*b*) average of current rectified with a full-wave rectifier (that is, the average absolute value), and (*c*) rms current. §11*

2-10. The output voltage of a full-wave rectifier is a succession of sinusoidal half-waves as shown in Figure 9c of Chapter 13, page 326. Find the rms value of this output voltage, the maximum being 1. §11*

2-11. The output voltage of a half-wave rectifier retains only the positive sinusoidal half-waves as shown in Figure 9a of Chapter 13, page 326. Find the rms value of this output voltage, the maximum being 1. §11

2-12. Alternating voltage, sine wave, with an rms value of 115 volts is applied to a load, and there results a current of square-wave form. (*a*) Can you multiply the rms value of voltage by the rms value of current to get the average power to the load, using equation 11-10? Tell why or why not. (*b*) What can you say about the load? §11*

CHAPTER
3

COMPLEX ALGEBRA

1. THE NEED FOR IMAGINARY NUMBERS

Addition of positive integers, such as 2 and 5, can be conceived in terms of physical things, such as 2 apples and 5 apples. Multiplication is repeated addition. Subtraction is the inverse of addition, and division the inverse of multiplication.

To include in our system positive numbers that are not integers, a correspondence can be established between all positive numbers and points on a line, such as the line in Figure 1a; the line begins at the point zero and extends indefinitely to the right. Definitions of algebraic operations must now be extended to include the positive numbers that are *not* integers. We desire that these operations be defined in such a way that the new definitions include the definitions already used for integers, as positive integers are special cases of positive numbers. We therefore define addition, subtraction, multiplication, division, powers, and roots, and the definitions of these operations are, of course, well known to us. When these operations are performed on positive numbers, the results are other positive numbers, all of which can be located on the line—except when a larger number is subtracted from a smaller number. The subtraction of a larger number from a smaller number yields a new kind of number for which there is no place on our line. We decide to call this a *negative* number, we distinguish it by writing a symbol in front of the number (a short horizontal line, for example, -2, -6), and we make a place for negative numbers in the diagram by extending

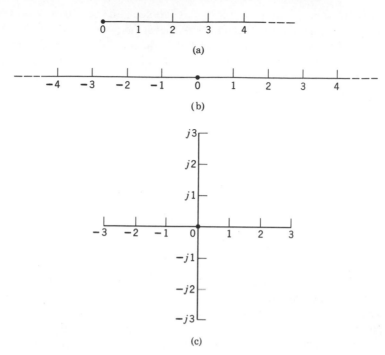

FIGURE   1 (a) Positive numbers.   (b) Real numbers.   (c) Complex numbers.

our line to the left of the point 0 as in Figure 1b. We call the positive numbers together with the negative numbers *real* numbers.†

It is now necessary to devise new definitions for multiplication of these negative numbers. To have a simple and consistent system of rules for multiplying all real numbers, we shall decide that the product of two negative numbers is positive and the product of a positive and a negative number is negative.

The result of any algebraic operation on any real number is another real number—except when the square root of a negative number is attempted. No real number multiplied by itself will give a negative number because of the way multiplication is defined, so the square root of a negative number does not fit in our system.

We therefore invent another new kind of number. It is a number whose square is a negative number, by definition, and we decide to call it an

---

† This presentation, although given with extreme simplicity and touching only the points needed for our use, follows the broad outline of a sound mathematical treatment. I am indebted to Dr. George Polya for reviewing the mathematics.

imaginary number. (This is an unfortunate choice of name, for this number is no more imaginary than other numbers; nevertheless, that is its name.) We distinguish it by writing a symbol in front of the number; the mathematician uses the letter i for this purpose (for example, $i2$, $i4$) but the electrical engineer uses j ($j2$, $j4$) to avoid confusion with the symbol i for current. We make a place for these new numbers in our diagram by extending a line upward from the point 0, and also downward for negative imaginary numbers, as in Figure 1c.

2. THE COMPLEX PLANE

If two real numbers are added, the sum is another real number. If two imaginary numbers are added, the sum is an imaginary number. If a real number and an imaginary number are added, the result is called a *complex* number. Thus if we add a real 2 and an imaginary 1, the sum is the complex number $(2 + j1)$. Using the symbol Z to represent† this complex number, $Z = (2 + j1)$. Note that real and imaginary numbers are special cases of complex numbers: a real number is a complex number with zero for its imaginary component (for example, $3 + j0$), and an imaginary number is a complex number with no real component.

A complex number can be represented in Figure 1c by two points, one on the line of reals and one on the line of imaginaries, but it is more convenient to adopt the convention of representing it by a single point so chosen that its projection on the line of reals gives the real component of the complex number, and its projection on the line of imaginaries gives the imaginary component of the complex number. In this way point Z in Figure 2a represents the complex number $(2 + j1)$. All such points, representing all complex numbers, fill the whole plane of Figure 2a.

The plane used in this way to represent complex numbers is called the *complex plane*; the two lines are called the real and imaginary axes. It is specified, for convenience, that the axes are to be at right angles to each other, that the divisions along the axes are to be equal, giving uniform number scales, and that both axes are to have the same scale. It is important, and is sometimes overlooked, that the scale on the axis of imaginaries must be the same as the scale on the axis of reals.

Having arrived at complex numbers, we have completed our job of inventing a consistent number system. No other kinds of numbers are needed, for we shall see that the result of any algebraic operation on any complex number is another complex number, and there is a place for it in the complex plane.

---

† As a matter of notation, it is customary to use boldface italic type for the symbols that represent complex numbers, as $Z = x + jy$, or $W = u + jv$.

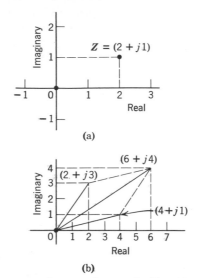

FIGURE 2 (a) A complex number, and (b) addition of complex numbers.

3. DEFINITIONS

It is necessary to extend some of the definitions of real algebra in order to apply them to complex numbers. We can, of course, define operations on complex quantities in any way we wish (or at least the man who first invented complex quantities had that privilege). We shall be guided by the following principles. We shall adopt definitions that are consistent with the well-known definitions applying to real numbers and that reduce to these well-known definitions when the complex quantities have zero imaginary components, and we shall adopt definitions that are useful and convenient. We first need a definition of equality.

Equality. Two complex numbers are equal if their real components are equal and their imaginary components are also equal. Thus if $Z = x + jy$ and $W = u + jv$, we can say that $Z = W$ if and only if $x = u$ and $y = v$.

Addition and Subtraction. The sum of two complex numbers has a real component equal to the sum of the real components and an imaginary component equal to the sum of the imaginary components. Thus $(2 + j3) + (4 + j1) = (6 + j4)$. To subtract, we change the signs of the components of the subtrahend and add.

The parallelogram method used for adding and subtracting vectors applies to the addition and subtraction of quantities in the complex plane, as in Figure 2b.

Multiplication.   The product of a real and an imaginary number is imaginary. Thus $2(j3) = j6$. The product of two positive imaginary numbers is real and negative: $(j2)(j3) = -6$, and $(j3)(-j4) = -(j3)(j4) = +12$. Special emphasis may well be given to the fact that $(j1)(j1) = -1$. This is true by definition, and it provides the basis for all complex algebra. Written in another way,

$$j1 = \sqrt{-1} \tag{3-1}$$

Complex numbers are multiplied by the ordinary rules of algebra supplemented by these foregoing rules for imaginary numbers. As an example:

$$(2 + j3)(4 + j1) = 2 \cdot 4 + 2 \cdot j1 + j3 \cdot 4 + j3 \cdot j1 = 8 + j2 + j12 - 3$$
$$= 5 + j14$$

These quantities are shown in Figure 3a. More generally,

$$(a + jb)(c + jd) = (ac - bd) + j(ad + bc) \tag{3-2}$$

Division.   By way of illustration, let us consider division of $(5 + j10)$ by $(2 + j1)$:

$$\frac{5 + j10}{2 + j1} \tag{3-3}$$

The first step is to multiply both numerator and denominator by $(2 - j1)$, using equation 3-2:

$$\frac{(5 + j10)(2 - j1)}{(2 + j1)(2 - j1)} = \frac{(10 + 10) + j(-5 + 20)}{(4 + 1) + j(-2 + 2)} = \frac{20 + j15}{5} \tag{3-4}$$

As a result of this mathematical sleight of hand (which is called *rationalization*), we find that the denominator is now a real number and can be divided into each component of the numerator to give $4 + j3$. That is:

$$\frac{5 + j10}{2 + j1} = 4 + j3 \tag{3-5}$$

This process can be generalized by writing letters instead of numbers:

$$\frac{a + jb}{c + jd} = \frac{(a + jb)(c - jd)}{(c + jd)(c - jd)} = \frac{(ac + bd) + j(bc - ad)}{c^2 + d^2}$$
$$= \frac{ac + bd}{c^2 + d^2} + j\frac{bc - ad}{c^2 + d^2} \tag{3-6}$$

Conjugate.   The process of making the denominator a real number, as exemplified in equations 3-4 and 3-6, is called rationalization of the quotient. It is done by multiplying the denominator by its conjugate.

By definition, the conjugate of a complex number is another complex number with the same real component and with an opposite imaginary component. An asterisk is used to indicate the conjugate; thus, if A is a complex number, A^* is its conjugate.† To write the definition in symbols,

$$\text{if } A = a + jb \quad \text{then} \quad A^* = a - jb \tag{3-7}$$

Figure 3b shows a pair of conjugate complex quantities. Obviously,

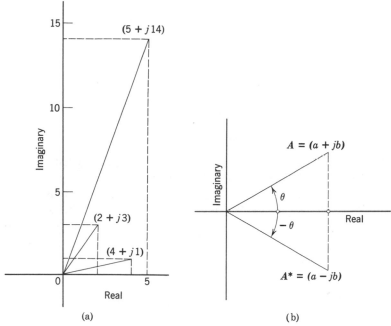

FIGURE 3   (a)   **Multiplication of complex numbers.**   (b)   **A conjugate pair.**

conjugates always come in pairs; each is the conjugate of the other. The product of the conjugates is always real:

$$A \cdot A^* = (a + jb)(a - jb) = (a^2 + b^2) + j(-ab + ab) = a^2 + b^2 \tag{3-8}$$

It is also true that the sum of a pair of conjugates is always real, but this fact is of no use to us at the moment.

Powers.   Squares and higher powers of complex quantities are obtained by repeated multiplication. The square of a real number is real and positive.

---

† Some authors use $\bar{A}$ for the conjugate.

The square of an imaginary number is real and negative. The square of a complex number is, in general, complex.

Powers of $j1$ are of particular interest. By definition, $j1 = \sqrt{-1}$. Hence

$$
\begin{aligned}
(j1)^2 &= -1 & (jx)^2 &= -x^2 \\
(j1)^3 &= -j1 & (jx)^3 &= -jx^3 \\
(j1)^4 &= +1 & (jx)^4 &= x^4 \\
(j1)^5 &= j1 & (jx)^5 &= jx^5
\end{aligned}
\tag{3-9}
$$

The imaginary unit $j1$ is often abbreviated to j. This notation makes j look like an algebraic quantity, a number, instead of a symbol to label imaginary numbers. It is as if the quantity -1 were written just $-$, and unfortunately it is likely to be confusing.[†]

A further discussion of powers and roots will come more appropriately after the exponential form of the complex quantity has been developed.

4. TRIGONOMETRIC FORM

It will be noticed that in Figures 2b, 3a, and 3b lines have been drawn from the origin to points in the complex plane. These lines are not necessary, but they are a convenient means of indicating the specified points. Also, the lengths and angles of such radial lines are significant quantities.

A complex quantity $A = (a + jb)$ is indicated in Figure 4a. A radial line is drawn from the origin to the point $(a + jb)$. If the length of this radial line is called r, and its angle with the axis of reals is θ, then $a = r \cos \theta$ and $b = r \sin \theta$, and

$$
A = a + jb = r \cos \theta + jr \sin \theta = r(\cos \theta + j \sin \theta)
\tag{4-1}
$$

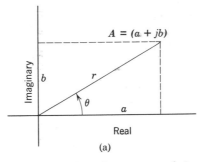

(a)

FIGURE 4a   Components of A.

---

† If this notation is not clearly understood some queer results can appear. See for instance H. M. Bacon and H. H. Skilling in *Electrical Engineering*, Vol. 54, pages 460–462, 1935.

Thus a complex quantity A can be expressed in terms of a magnitude r and an angle θ. The magnitude and the angle are, of course, both real numbers.

As a matter of notation it is convenient and customary to write a complex quantity in what is called the *polar* form as $A = r\underline{/\theta}$, meaning "A equals r at an angle of θ." This is just another way to write a complex quantity, and means neither more nor less than the *rectangular* form $A = a + jb$. If r and θ are known, a and b can be found; or if a and b are known, r and θ can be found:

$$a = r \cos \theta \qquad r = \sqrt{a^2 + b^2}$$

$$b = r \sin \theta \qquad \theta = \tan^{-1}\frac{b}{a} \tag{4-2}$$

5. EULER'S THEOREM

Equation 4-1 expresses a complex quantity in algebraic form in terms of a and b, and in trigonometric form in terms of r and θ. A third useful expression is the exponential form.

The exponential function is, by definition, equal to the limit approached by an infinite series:

$$e^x = 1 + x + \frac{x^2}{2!} + \frac{x^3}{3!} + \cdots \tag{5-1}$$

The trigonometric functions can be defined in terms of the exponential function. The cosine function is

$$\cos \theta = \tfrac{1}{2}(e^{j\theta} + e^{-j\theta}) \tag{5-2}$$

and the sine function is

$$\sin \theta = \frac{1}{j2}(e^{j\theta} - e^{-j\theta}) \tag{5-3}$$

Adding these equations, after multiplying the latter by $j1$, gives

$$\cos \theta + j \sin \theta = \tfrac{1}{2}(e^{j\theta} + e^{-j\theta} + e^{j\theta} - e^{-j\theta})$$
$$= e^{j\theta} \tag{5-4}$$

This equation, $e^{j\theta} = \cos \theta + j \sin \theta$, is known as Euler's theorem. See Figure 5a.

When Euler's theorem, equation 5-4, is introduced into the trigonometric form of complex quantity, equation 4-1, it gives an exponential form of expression. Any complex number can be written

$$A = a + jb = r(\cos \theta + j \sin \theta) = re^{j\theta} \tag{5-5}$$

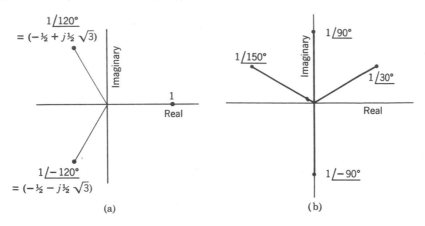

FIGURE 7 (a) The cube roots of 1. (b) The quantity $1/90°$ and its cube roots.

The three cube roots of 1 look a little less formidable when written in the polar form as $1/0°$, $1/120°$, and $1/-120°$. Moreover, in this form it is easy to see that they are indeed cube roots of 1, for when any one of them is raised to the third power, 1 results. Thus

$$(1/120°)^3 = (1)^3/(120) \cdot 3 = 1/360° \tag{7-3}$$

and $1/360°$ is identical with $1/0°$. Similarly, $(1/-120°)^3$

$$= 1/-360° \text{ and this, also, is the same as } 1/0°.$$

It is equally correct to say that $1/240°$ is a cube root of 1, for $(1/240°)^3 = 1/720°$, and this again is indistinguishable from $1/0°$. However, there are still only three cube roots of 1, for $1/240°$ is identical with $1/-120°$, which we counted as one of the three roots. See Figure 7a. There are unlimited ways of expressing the cube roots, but in a diagram, or when written in the rectangular form, there are only three distinct values.

In a similar manner there are three cube roots of 8. They are $2/0°$, $2/120°$, and $2/-120°$.

Let us consider the cube roots of the quantity $8/90°$. Equation 7-2 gives us one root: $2/30°$. The other two cube roots have the same magnitude and are spaced 120 degrees apart. They are $2/(30 + 120)°$ and $2/(30 - 120)°$, as may be proved by cubing. See Figure 7b.

Similarly, the fourth roots of any complex quantity are equal in magnitude and are spaced 90 degrees apart in the complex plane. The fifth roots are spaced 72 degrees apart, and so on.†

These operations on complex quantities will now be illustrated by a few numerical examples.

8. EXAMPLES

Example 1. Find the product of A and B if $A = 1.5e^{j\pi/6}$ and $B = 1.2e^{j\pi/4}$.

$$A \cdot B = (1.5)(1.2)e^{j(\pi/6 + \pi/4)} = 1.8e^{j5\pi/12} \tag{8-1}$$

It is usually more convenient to work in degrees than radians, so the problem may be repeated in different notation:

$$A \cdot B = (1.5)(1.2)\underline{/30° + 45°} = 1.8\underline{/75°} \tag{8-2}$$

Note that an algebraic form such as $e^{j75°}$ is undesirable, if not meaningless, when the definition of the exponential as an infinite series in equation 5-1 is considered.

Example 2. Find the product of $j1$ and $5\underline{/30°}$.

$$(j1)(5\underline{/30°}) = (1\underline{/90°})(5\underline{/30°}) = 5\underline{/120°} \tag{8-3}$$

Basically, j is a symbol to mark an imaginary quantity. However, it is frequently referred to as an *operator that rotates a line* in the complex plane by 90 degrees. Thus, in equation 8-3, $j(5\underline{/30°}) = 5\underline{/(30 + 90)°}$. See Figure 8a.

Example 3. Divide:

$$\frac{10 + j5}{10\underline{/45°}} = \frac{11.18\underline{/26.6°}}{10\underline{/45°}} = 1.118\underline{/-18.4°} \tag{8-4}$$

Example 4. Find the square root of $-0.90 + j1.20$. First compute that $-0.90 + j1.20 = 1.50\underline{/126.8°}$ as in Figure 8b. Then

$$\sqrt{1.50\underline{/126.8°}} = \sqrt{1.50}\underline{/126.8°/2} = 1.225\underline{/63.4°} \tag{8-5}$$

This gives one of the square roots of $-0.90 + j1.20$. There should be another, and we see that the other is found by interpreting $-0.90 + j1.20$ not as $1.50\underline{/126.8°}$ but rather as $1.50\underline{/-233.2°}$. See Figure 8b for this value

---

† The ambiguity of sign of the square root, the multiplicity of higher roots, and many paradoxes relating to complex quantities arise from the fact that if $e^{jx} = a + jb$, x is multivalued. By Euler's theorem, $e^{jx} = e^{j(x + 2\pi)}$.

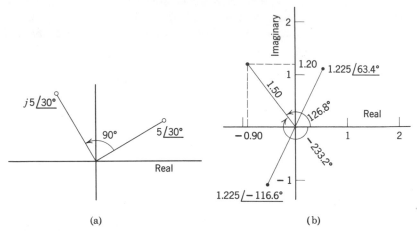

(a)   (b)

FIGURE 8   **(a)**   **The symbol j as an operator of rotation.**   **(b)**   **Square roots.**

also, noting that the one point in the complex plane can be interpreted either way. Then

$$\sqrt{1.50\underline{/-233.2°}} = 1.225\underline{/-116.6°}$$

This is the other square root of $-0.90 + j1.20$. One square root is the negative of the other.

9. A CIRCLE AND A SPIRAL

Figure 6a shows a radial line, $Ae^{j\alpha}$. Such a line can be made to rotate about the origin of the complex plane by letting its angle be proportional to time. Thus if t is time, $Ae^{j\omega t}$ describes a line of length A that is rotating about the origin with angular velocity ω. The tip of the line traces a complete circle each time the angle ωt increases by 2π.

Similarly, an exponential quantity $Ae^{(\alpha + j\omega)t}$ with a *complex* exponent describes a line that has variable length as it rotates in the complex plane, and its tip traces a *spiral*. Figure 9a shows the spiral traced by the exponential $e^{(-\alpha + j\omega)t}$. Let

$$s_2 = -\alpha + j\omega \tag{9-1}$$

with α and ω real.† Then

$$e^{s_2 t} = e^{(-\alpha + j\omega)t} = e^{-\alpha t}e^{j\omega t} \tag{9-2}$$

---

† The letter s is given the subscript 2 for reasons that are not now evident but will appear in the next section.

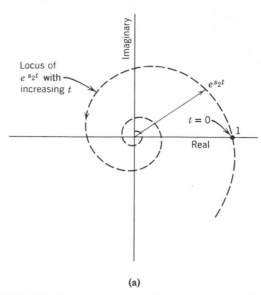

(a)

FIGURE 9a   An exponential spiral; s_2 is complex.

It is assumed in drawing Figure 9a that ω is positive and the line therefore rotates in the positive (counterclockwise) direction with time, and that α is also a positive quantity which makes the line grow shorter as it rotates. If α were a negative quantity the curve would spiral outward instead of inward with increasing t.

It is of course evident that the circle traced by a line of constant length is the special case of this exponential spiral with $\alpha = 0$. The exponent of $e^{s_2 t}$ is then purely imaginary and the rotating line remains constant in length.

10. APPLICATIONS TO NETWORK THEORY

Sinusoidal Wave.   Some marked advantages that result from the use of complex algebra in alternating-current theory were first pointed out by Steinmetz.† A number of conveniences can now be shown.

A steadily alternating voltage or current such as $v = V_m \cos \omega t$ can be described as the real component of an exponential:

$$v = \mathscr{R}e\, V_m e^{j\omega t} \tag{10-1}$$

where $\mathscr{R}e$ means "The real component of ...". The equivalence is shown by Euler's theorem; since V_m is real, equation 10-1 gives

$$\mathscr{R}e\, V_m e^{j\omega t} = V_m\, \mathscr{R}e\, (\cos \omega t + j \sin \omega t) = V_m \cos \omega t \tag{10-2}$$

† C. P. Steinmetz, 1893; see bibliography.

Equation 10-1 is also the mathematical expression of a *phasor*. In Section 16 of Chapter 2 a phasor is described as a rotating line, and the projection of the phasor on a horizontal axis represents a sinusoidally alternating voltage or current. Equation 10-1 says the same in mathematical terms. It was mentioned in Chapter 2 that there may be a phase angle associated with a phasor. Such a phase angle is put into the mathematical form by writing

$$v = \mathscr{R}e\, V_m e^{j(\omega t + \varphi)} \tag{10-3}$$

φ being the phase angle.

Perhaps it should be emphasized that a voltage or a current is a real quantity. Although complex exponentials are used in their representation, the value of current or voltage is the *real* component of the complex quantity.

Damped Cosine.   When a source is suddenly introduced into a network, as by closing a switch, there may be a component of natural current in the network with the form

$$i = K_1 e^{s_1 t} + K_2 e^{s_2 t} \tag{10-4}$$

where s_1 and s_2 are conjugate complex quantities such that

$$s_1 = -\alpha - j\omega \tag{10-5}$$

$$s_2 = -\alpha + j\omega \tag{10-6}$$

Let us assume that K_1 and K_2 are real and equal, and write

$$K_1 = K_2 = K \tag{10-7}$$

Then i of equation 10-4 can be seen from the discussion of Section 9 to be the sum of two rotating lines that trace spirals in the complex plane. Figure 10a shows two such lines with their spiral loci; they turn in opposite directions and each diminishes as it turns. At the instant $t = 0$ each line lies on the axis of reals and (because of equation 10-7) each has then the length K. As time passes the tips of the lines move along their respective spiral loci and the sum of the two lines remains always real. Mathematically,

$$\begin{aligned} i &= K e^{s_1 t} + K e^{s_2 t} = K e^{-\alpha t}\left(e^{j\omega t} + e^{-j\omega t}\right) \\ &= 2K e^{-\alpha t}\cos \omega t \end{aligned} \tag{10-8}$$

This is a damped cosine function with initial amplitude $2K$. It is always a real quantity.

Damped Sine.   A result that is even more often of practical value is obtained by making a different assumption regarding the coefficients. Instead of

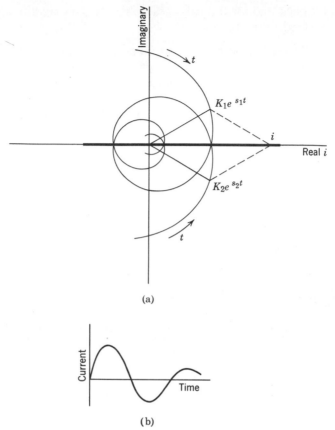

(a)

(b)

FIGURE 10 (a) Spiral loci of two complex exponential components. (b) A damped sine.

letting them be real and equal, let us assume them to be imaginary and of equal magnitude but opposite sign. Thus we can let

$$K_1 = jk \quad \text{and} \quad K_2 = -jk \tag{10-9}$$

When these values are used in equation 10-4:

$$i = jke^{s_1 t} - jke^{s_2 t} = jke^{-\alpha t}(e^{-j\omega t} - e^{+j\omega t})$$
$$= 2ke^{-\alpha t} \sin \omega t \tag{10-10}$$

This is a damped sine function as in Figure 10b. Its initial value, when $t = 0$, is zero, for at this instant the two rotating lines that add to give the function are both vertical, one up and one down, equal in magnitude. At later instants the two rotating lines add to give a quantity such as i in Figure 10a that oscillates back and forth on the heavy line of the diagram, always on the horizontal axis of reals.

Damped General Sinusoid.   This sine function of equation 10-10 differs only in phase from the cosine function of equation 10-8. In Figure 10a, the rotating lines are initially on the horizontal axis for the cosine function, and initially on the vertical axis for the sine function. A more general statement that includes both of these phase relations as special cases results if we assume that the two coefficients are neither real nor imaginary, but conjugate complex. That is, let

$$K_1 = K + jk \qquad K_2 = K - jk \qquad (10\text{-}11)$$

Figure 10a still applies. The sum of the exponentials can now be expressed as

$$i = K_1 e^{s_1 t} + K_2 e^{s_2 t} = 2e^{-\alpha t}(K \cos \omega t + k \sin \omega t) \qquad (10\text{-}12)$$

or alternatively as

$$i = 2\sqrt{K^2 + k^2}\, e^{-\alpha t} \cos(\omega t - \tan^{-1} k/K) \qquad (10\text{-}13)$$

This more general form becomes the damped cosine function of equation 10-8 if $k = 0$ and the damped sine function of equation 10-10 if $K = 0$. This form also gives the undamped cosine function $i = 2K \cos \omega t$ if both k and α are zero, and the undamped sine function $i = 2k \sin \omega t$ if both K and α are zero. All of these various expressions will be used in coming chapters.

11. SUMMARY

As algebraic operations on positive numbers require negative numbers, so operations on real numbers require numbers that are *imaginary and complex.*

Operations on complex numbers are defined. A complex number can be expressed in rectangular or polar coordinates or, by *Euler's theorem*, in exponential form. Multiplication and division, powers and roots, are best computed in exponential form.

An exponential with *imaginary* exponent proportional to time t is represented graphically by a *line rotating* about the origin of the complex plane, with its tip tracing a circle. If the exponent is *complex*, the line becomes shorter (or longer) as it rotates, and its tip traces a *spiral.*

The following relations which are here derived will be found useful in later chapters:

A *sinusoidal function* such as the voltage $v = V_m \cos(\omega t + \varphi)$, which can be represented graphically by a *phasor*, can be represented mathematically by $v = \mathscr{R}e\, V_m e^{j(\omega t + \varphi)}$.

A *damped* sinusoidal function can be written exponentially; for instance,

$i = I_0 e^{-\alpha t} \sin \omega t = I_0 e^{-\alpha t}(e^{j\omega t} - e^{-j\omega t})/2j$. This mathematical expression can be represented graphically as the sum of two rotating lines with *spiral* loci. This example can be generalized.

Problems

3-1. Perform the following operations. For each, show the complex quantities in a sketch of the complex plane. §3

    a. $(3 + j4) + (3 + j4)^*$ *d.* $\dfrac{6 + j22}{22 + j6}$

    b. $(3 + j4)[(3 + j4)^*]$ *e.* $\dfrac{7 + j13}{(7 + j13)^*}$

    c. $\dfrac{6 + j8}{4 + j3}$

3-2. Perform the following operations. Draw sketches as for Problem 3-1. §3

    a. $(6 + j8) + (6 + j8)^*$ *d.* $\dfrac{10 + j15}{15 + j10}$

    b. $(6 + j8)[(6 + j8)^*]$ *e.* $\dfrac{6 + j8}{(6 + j8)^*}$

    c. $\dfrac{3 + j4}{8 + j6}$

3-3. Prove that $e^{j\pi/2} = j1$, or that it does not. This is to be a strictly algebraic proof with no reference to lines, angles, or planes. §5*

3-4. Find the real and imaginary components of $(10 - j5)/(-3 + jx)$. §6

3-5. (*a*) Find the product $(4.86 + j7.35)(1.24 - j0.83)$ using the formula of equation 3-2. (*b*) Find the same product by changing each complex quantity to polar form and applying the formula of equation 6-1. Show that the result of (*b*) is the same as that of (*a*). Which is easier?

    Also (*c*) find the product $(9.33\underline{/62°})(1.48\underline{/-22°})$, multiplying in polar form. (*d*) Find the same product by changing each complex quantity to rectangular form and applying equation 3-2. Show that the result of (*d*) is the same as that of (*c*). §6

3-6. (*a*) Perform the division $(2.93 + j1.08)/(1.32 - j1.25)$ using the formula of equation 3-6. (*b*) Perform the same division by changing each complex quantity to polar form and applying the formula of equation 6-2. Show that the two results are the same. Which method is easier? §6

3-7. (*a*) How many different complex fourth roots has a quantity? (*b*) Find and plot all values of $\sqrt[4]{5 + j8.66}$. (*c*) Find and plot all values of $\sqrt[5]{14.14 - j14.14}$. §7*

3-8. Find all different complex values of (a) $\sqrt[7]{-100}$; (b) $\sqrt[4]{j35}$. Tell, in (a) and (b), which roots are conjugate to each other. §7

3-9. Given $Z = 64\underline{/60°} = 32 + j55.5$, find, in polar and rectangular form, Z^*, $1/Z$, $-Z$, $\sqrt[3]{Z}$, magnitude of Z. Show each in a sketch of the complex plane. §7

3-10. (a) Give the real and imaginary components of A if $A = (5 + j5)^3$. (b) Give in polar form all different complex values for W if $W = \sqrt[5]{j32}$. Show the computed values in a sketch of the complex plane. Tell which are conjugate to each other. §7*

3-11. Find the real and imaginary components of $\ln(Ae^{j\alpha})$. (A and α are real; ln means the natural logarithm.) §7*

3-12. Find three distinct cube roots of $\ln(-3)$. From a table, $\ln 3 = 1.09861$. §7*

3-13. (a) Find all different complex values of $\sqrt[6]{64}$ and (b) of $\sqrt[4]{j100}$. Plot these roots, and tell which are conjugate to each other. §7

3-14. Given $A = 10e^{j25t}$, find $\mathscr{R}e\, A$ and $\mathscr{I}m\, A$, the real and imaginary components of A. §9*

3-15. Derive the trigonometric relation that is used to change equation 10-12 to equation 10-13, starting with the trigonometric identities to be found in any table or handbook. §10

CHAPTER
4

THE FREQUENCY DOMAIN

1. THE TRIGONOMETRIC FORM

We might have tried, in Chapter 1, to solve the network equations 1-1 and 1-2 which we formulated as follows:

$$\frac{di_1}{dt} = -\frac{R_{12}}{L_1} i_1 + \frac{R_{12}}{L_1} i_2 + \frac{1}{L_1} v_1 \qquad (1\text{-}1)$$

$$\frac{di_2}{dt} = \frac{R_{12}}{L_2} i_1 - \frac{R_2 + R_{12}}{L_2} i_2 \qquad (1\text{-}2)$$

We might have specified that the applied voltage was sinusoidal and let $v_1 = V_m \cos \omega t$; we should then have known by the theorem of Chapter 2, Section 7, that the currents would also be sinusoidal and might be written $i_1 = I_{1m} \cos(\omega t + \theta_1)$ and $i_2 = I_{2m} \cos(\omega t + \theta_2)$ the coefficients being as yet undetermined. Substitution of these expressions for v_1, i_1, and i_2 into equations 1-1 and 1-2 gives a pair of trigonometric equations that could no doubt have been solved; the differentiation is easy, and then trigonometric substitution makes possible the elimination of one current to permit solving for the other. But this looks like a fairly lengthy operation, and we have fortunately already discovered a quicker method. Let us not express the applied voltage as a function of $\cos \omega t$, but rather as a function of $\mathscr{R}e\, e^{j\omega t}$, which is exactly equivalent. A few paragraphs of preparation will be necessary first, introducing the idea of complex transforms.

55

2. THE EXPONENTIAL FORM

A sinusoidally alternating voltage can be written in either of the following forms:

$$v = V_m \cos(\omega t + \theta_1)$$
$$= V_m \, \mathscr{R}e \, e^{j(\omega t + \theta_1)} = \mathscr{R}e \, V_m \, e^{j\theta_1} e^{j\omega t} \tag{2-1}$$

We shall now, for convenience of notation, write

$$V_m e^{j\theta_1} = \boldsymbol{V_m} \tag{2-2}$$

where $\boldsymbol{V_m}$, printed as a boldface italic letter, is called the *transform* of the voltage.† The transform is, in general, a complex quantity. With this notation the voltage can be written

$$v = \mathscr{R}e \, \boldsymbol{V_m} \, e^{j\omega t} \tag{2-3}$$

In an exactly similar manner a sinusoidally alternating current can be written

$$i = I_m \cos(\omega t + \theta_2)$$
$$= I_m \, \mathscr{R}e \, e^{j(\omega t + \theta_2)} = \mathscr{R}e \, I_m e^{j\theta} e^{j\omega t} \tag{2-4}$$

and defining a current transform as

$$I_m e^{j\theta_2} = \boldsymbol{I_m} \tag{2-5}$$

we obtain

$$i = \mathscr{R}e \, \boldsymbol{I_m} e^{j\omega t} \tag{2-6}$$

3. TRANSFORMS

Resistance. Voltage across resistance is $v_R = Ri$; this is equation 2-1 of Chapter 1. If voltage and current are sinusoidal we can rewrite this equation, using equations 2-3 and 2-6 for v_R and i, to say that across resistance:

$$\mathscr{R}e \, \boldsymbol{V_m} e^{j\omega t} = R \, \mathscr{R}e \, \boldsymbol{I_m} e^{j\omega t} \tag{3-1}$$

---

† *Transform* is here used for the complex quantity, *phasor* for the rotating line that graphically represents it. The IEEE definition uses the word phasor for both the complex quantity and the rotating line.

In this equation R and ω are real constants.† This equation 3-1 must be true through all time t (such an equation is sometimes called an identity), and the real components of equation 3-1 can be equal at all values of t only if their respective complex quantities are themselves equal. Therefore:

$$V_m e^{j\omega t} = R I_m e^{j\omega t} \tag{3-2}$$

Each side of the equation is now divided by $e^{j\omega t}$, leaving the transform equation:

$$V_m = R I_m \tag{3-3}$$

This looks like Ohm's law, but it is more. It implies that both voltage and current are sinusoidally alternating at the same frequency, and V_m and I_m are complex quantities involving both magnitude and phase. Note that magnitudes of voltage and current are related by the constant R, and that the phase angle between them is zero.

Inductance. By equation 2-2 of Chapter 1, voltage across inductance is $v_L = L\, di/dt$. From this we obtain

$$\begin{aligned} \mathscr{R}e\, V_m e^{j\omega t} &= L \frac{d}{dt} \mathscr{R}e\, I_m e^{j\omega t} \\ &= \mathscr{R}e\, j\omega L I_m e^{j\omega t} \end{aligned} \tag{3-4}$$

Again we have an identity that must be valid for all values of time. With ω and L constant, the identity of real components in equation 3-4 can be correct only if the corresponding complex quantities are equal, so

$$V_m e^{j\omega t} = j\omega L I_m e^{j\omega t} \tag{3-5}$$

When each side of this equation is divided by $e^{j\omega t}$ we have the relation of transforms:

$$V_m = j\omega L I_m \tag{3-6}$$

Thus, with pure inductance, the transforms of voltage and current are complex quantities that are related in magnitude by ωL (which is called the

---

† Rules for operations involving only real components are readily derived from the definitions of complex operations in Section 3 of Chapter 3.

1. Addition and subtraction: $\mathscr{R}e\, A + \mathscr{R}e\, B = \mathscr{R}e(A + B)$.

2. Multiplication and division by a real number: If N is real, $N\, \mathscr{R}e\, A = \mathscr{R}e\, NA$.

3. Multiplication and division by a complex number: $(\mathscr{R}e\, A)(\mathscr{R}e\, B) \neq \mathscr{R}e\, AB$. It is apparent from equation 3-2 of Chapter 3 that the product of the real components is *not* equal to the real component of the product.

4. Differentiation with respect to a real variable (such as time) is the limiting value of division by a real number. Therefore $(d/dt)\mathscr{R}e\, A = \mathscr{R}e(d/dt)A$.

5. Integration with respect to a real variable (such as time) is the limiting value of a summation of products of a complex and a real number. Therefore $\int \mathscr{R}e\, A\, dt = \mathscr{R}e \int A\, dt$.

inductive reactance X) and in angle by j signifying a 90-degree phase difference. Correspondingly, the phasor of current *lags* 90 degrees behind the
phasor of voltage as in Figure 3a, and the wave of current, as in Figure 4a of
Chapter 2, is delayed a quarter-cycle behind the wave of voltage.

Capacitance.   Current to capacitance, by equation 2-3 of Chapter 1, is
$i = C \, dv_C/dt$. Using the exponential notation of equations 2-6 and 2-3,

$$\mathscr{R}e \, I_m e^{j\omega t} = C \frac{d}{dt} \mathscr{R}e \, V_m e^{j\omega t} = \mathscr{R}e \, j\omega C V_m e^{j\omega t} \qquad (3\text{-}7)$$

Taking note that this must be valid for all t, and that C and ω are real
constants, we equate

$$I_m e^{j\omega t} = j\omega C V_m e^{j\omega t} \qquad (3\text{-}8)$$

Each side is then divided by the exponential to obtain the transform equation that applies to capacitance:

$$I_m = j\omega C V_m \qquad (3\text{-}9)$$

Thus the transforms of current to capacitance and of voltage across the
capacitance are related in magnitude by ωC (the capacitive susceptance B)
and in angle by a 90-degree phase difference; with capacitance, the phasor of
current *leads* the voltage phasor by 90 degrees, as in Figure 3b, and the
sinusoidal wave of current is a quarter-cycle ahead of the wave of voltage as
in Figure 5a of Chapter 2.

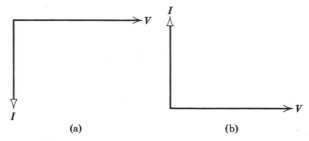

(a)   (b)

**FIGURE 3   (a)   Current through inductance lags behind voltage by 90
degrees.   (b)   Current to capacitance leads voltage by 90 degrees.**

4. ADMITTANCE AND IMPEDANCE

The *admittance* of any two-terminal element, branch, or network is the ratio
of the transform of entering current to the transform of terminal voltage.
That is,

$$Y = \frac{I_m}{V_m} \qquad (4\text{-}1)$$

Also, by definition, *impedance* is the reciprocal of admittance, or the ratio of the transform of voltage to that of current:

$$Z = \frac{1}{Y} = \frac{V_m}{I_m} \qquad (4\text{-}2)$$

Let us think of a network within the box of Figure 4a. This is, of course, a model, idealized from the physical network. Impedance and admittance at the terminals are suggested in the diagram. The value of the terminal impedance (or admittance) can be determined from what is within the box, and this determination will be the subject of much of the rest of this chapter.

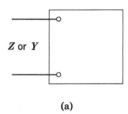

(a)

FIGURE 4a   A two-terminal network.

As an example that is almost absurdly simple, let us consider that within the box, between the two terminals shown, there is nothing but an element with resistance R. Then the terminal impedance $Z = R$, and the terminal admittance $Y = 1/R = G$.

Other single elements give similarly simple values of terminal impedance and admittance. If there were within the box nothing but an element of pure inductance, $Z = j\omega L$ or $Y = 1/j\omega L$. If the element were pure capacitance, $Y = j\omega C$ or $Z = 1/j\omega C$.

5. IMPEDANCE OF ELEMENTS IN SERIES

But suppose the box contains several elements in series. Then the input impedance at the terminals of the box is the sum of the individual impedances. That is, if a number of elements with impedances that we may call Z_1, Z_2, Z_3, and Z_4 are connected in series, as in Figure 5a, they can be replaced by a single equivalent impedance Z if

$$Z = Z_1 + Z_2 + Z_3 + Z_4 \qquad (5\text{-}1)$$

Proof: Voltage across the elements in series is the sum of the voltages across the individual elements; therefore the transform of the total voltage is equal to the sum of the transforms $Z_1 I + Z_2 I + Z_3 I + Z_4 I$. The transform of voltage across a single equivalent Z will be ZI, and it must be equal to the sum of the transforms. Equating and dividing by I, equation 5-1 results.

Example 1. The resistance of a coil of wire is 1.50 ohms, and its inductance is 5.30 millihenrys. See Figure 5b: (*a*) What is its impedance to 60-cycle current? (*b*) If current through the coil is $i = 5.66 \cos \omega t$ amperes, what is the terminal voltage?

Solution: (*a*) Impedance, a complex quantity, is (at 60 Hz)

$$Z = R + j\omega L = 1.50 + j(2\pi)(60)(5.30)10^{-3} = 1.50 + j2.00 \qquad (5\text{-}2)$$

Impedance is often more useful in the polar form:

$$Z = \sqrt{(1.50)^2 + (2.00)^2} \; \underline{/\tan^{-1}(\omega L/R)} = 2.50\underline{/53.1^\circ} \qquad (5\text{-}3)$$

(*b*) The transform of the given current is complex with magnitude 5.66; regarding angle, a reference phasor is to be selected, so let us choose this current as reference and give it the arbitrary angle of zero. Thus

$$I_m = 5.66\underline{/0^\circ} \qquad (5\text{-}4)$$

Then from equation 4-2,

$$V_m = ZI_m = (2.50\underline{/53.1^\circ})(5.66\underline{/0^\circ}) = 14.14\underline{/53.1^\circ} \qquad (5\text{-}5)$$

This is the transform of terminal voltage. For many purposes this can be considered the answer to our problem, because the expression for the actual wave of voltage is now so easily written. The actual wave is

$$\begin{aligned} v &= 14.14 \; \mathscr{R}e \; e^{j(\omega t + 53.1\pi/180)} \\ &= 14.14 \cos(\omega t + 53.1\pi/180) \end{aligned} \qquad (5\text{-}6)$$

in which $\omega = 2\pi(60) = 377$ radians per second (for a 60-cycle-per-second current).

Computation in this example has all been done in terms of I_m, the maximum value or amplitude of a sinusoidal current, and V_m, the maximum value or amplitude of the voltage. We have used the equation

$$V_m = ZI_m \qquad (5\text{-}7)$$

For practical purposes, however, values given for alternating current and voltage are ordinarily the rms or effective values, not the maximum values. The rms values are less by $\sqrt{2}$. Let us therefore define $V = V_m/\sqrt{2}$ and $I = I_m/\sqrt{2}$, and call V and I the transforms with rms magnitudes. We can then divide each side of equation 5-7 by $\sqrt{2}$ and write

$$V = ZI \qquad (5\text{-}8)$$

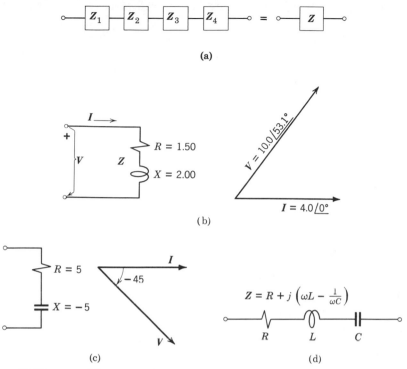

FIGURE 5 (a) Impedance is added for elements in series. (b) Resistance and inductance in series. (c) Resistance and capacitance in series. (d) R, L, and C in series.

This equation is so familiar to engineers that it is often called "Ohm's law for alternating currents."

The use of V_m and I_m in equation 5-6 is consistent with later chapters of this book; it is the notation adopted by many authors, and it is accepted by the IEEE standards. However, V and I are nearly always used for practical circuit calculations and all ordinary voltmeters and ammeters indicate these rms values. We shall surely do well to be familiar with both notations.

Let us now repeat the latter part of this example using transforms with rms magnitudes. Putting the numerical values of our problem into equation 5-8 we have, in rms quantities,

$$V = ZI = (2.50\underline{/53.1})\frac{5.66\underline{/0°}}{\sqrt{2}}$$

$$= (2.50\underline{/53.1°})(4.00\underline{/0°}) = 10.0\underline{/53.1°}$$

(5-9)

In Figure 5b these rms lengths of the phasors are given.

Example 2. A circuit, as in Figure 5c, has resistance and capacitance in series. $R = 5$ ohms, $C = 10^{-4}$ farads or 100 microfarads, and $\omega = 2000$ rn/sec or 318 Hz. Current of 5.0 amperes (rms) is sinusoidal. Find terminal voltage.

Solution:

$$Z = R + \frac{1}{j\omega C} = 5 + \frac{10}{j2} = 5 - j5 = 7.07\underline{/-45°} \qquad (5\text{-}10)$$

$$V = ZI = (7.07)(8.0)\underline{/-45°} = 56.6\underline{/-45°} \qquad (5\text{-}11)$$

We do not carry this solution as far as finding the instantaneous voltage v, but stop with V. This is quite usual, and indeed it is common practice for V and I to be called voltage and current, simply omitting the obvious words "transforms of."

Example 3. Impedance is to be found for a circuit consisting of $R = 8.00$ ohms, $L = 2.38$ millihenrys, and $C = 14.14$ microfarads in series (Figure 5d). Frequency is 500 Hz.

Solution:

$$Z = R + j\omega L + \frac{1}{j\omega C}$$

$$= 8.00 + j\left(2\pi \cdot 500 \cdot 2.38 \cdot 10^{-3} - \frac{10^6}{2\pi \cdot 500 \cdot 14.14}\right)$$

$$= 8.00 + j(7.50 - 22.5) = 8.00 - j15.0 = 17.0\underline{/-61.9°}$$

The impedance of this series circuit is predominantly capacitive. Note that the angle of an inductive impedance is positive, that of a capacitive impedance negative.

6. VOLTAGE DISTRIBUTION

The voltage across each of a number of elements in series is proportional to the impedance. Thus in Figure 6a the transform of current is equal to any of the following ratios:

$$I = \frac{V_1}{Z_1} = \frac{V_2}{Z_2} = \frac{V_3}{Z_3} = \frac{V_4}{Z_4} = \frac{V}{Z} \qquad (6\text{-}1)$$

from which

$$V_1 = \frac{Z_1}{Z}V \qquad V_2 = \frac{Z_2}{Z}V \quad \text{etc.} \qquad (6\text{-}2)$$

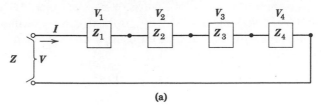

(a)

FIGURE 6a Voltage distribution among series elements.

The relation of equations 6-2, so apparent that it hardly needs demonstration, is sometimes called the potentiometer equation or the voltage-divider equation.

7. ADMITTANCE OF BRANCHES IN PARALLEL

Branches of a network are said to be connected in parallel if the same voltage is applied to each, as in Figure 7a.

If a number of branches with admittances that we may call Y_1, Y_2, Y_3, and Y_4 are connected in parallel, they can be replaced by a single equivalent element with admittance Y if

$$Y = Y_1 + Y_2 + Y_3 + Y_4 \qquad (7\text{-}1)$$

Proof: Current through the branches in parallel is the sum of the currents in the individual branches; therefore the transform of the total current is equal to the sum of the transforms $Y_1 V + Y_2 V + Y_3 V + Y_4 V$. The transform of current through a single equivalent admittance is YV, and it must be equal to the sum of the transforms. Equating and dividing by V, equation 7-1 results.

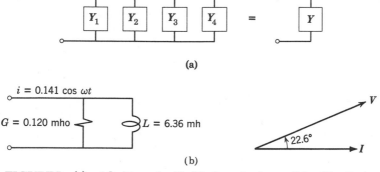

(a)

(b)

FIGURE 7 (a) Admittance is added for branches in parallel. (b) Resistance and inductance in parallel.

Example.   Current to the circuit of Figure 7b is to be $i = 0.141 \cos \omega t$. The frequency is 500 Hz, making $\omega = 3141$. Find the necessary terminal voltage (magnitude and phase angle), values of inductance and conductance being as indicated.

Solution: To find the voltage we first find Y:

$$Y = G + \frac{1}{j\omega L} = 0.120 + \frac{1}{j(3141)(6.36)10^{-3}}$$

$$= 0.120 - j0.050 = 0.130\underline{/-22.6°}$$

Since

$$i = 0.141 \cos \omega t \qquad I_m = 0.141\underline{/0°} \qquad I = 0.100\underline{/0°}$$

Then

$$V = \frac{I}{Y} = \frac{0.100\underline{/0°}}{0.130\underline{/-22.6°}} = 0.769\underline{/22.6°}$$

This is the transform V, and its time function v is

$$v = \sqrt{2}\,(0.769)\cos(\omega t + 22.6°)$$

(Note that the phase angle is expressed in degrees, and we must therefore understand that ωt is in degrees also.)

   This v is the voltage that was to be determined. The solution is carried out to the end, to find the actual time function of voltage, instead of stopping as is more usual with the computation of the transform $V = 0.769\underline{/22.6°}$.

8. CURRENT DISTRIBUTION

The current through each of a number of branches in parallel is proportional to the admittance. Thus in Figure 8a the transform of voltage is any of the following:

$$V = \frac{I_1}{Y_1} = \frac{I_2}{Y_2} = \frac{I_3}{Y_3} = \frac{I_4}{Y_4} = \frac{I}{Y} \tag{8-1}$$

Then the division of the total current among the branches is

$$I_1 = \frac{Y_1}{Y}I \qquad I_2 = \frac{Y_2}{Y}I \quad \text{etc.} \tag{8-2}$$

   The analogy to equation 6-2 is obvious. Such analogy between *voltage* and *current*, *impedance* and *admittance*, and *series* and *parallel* is called duality.

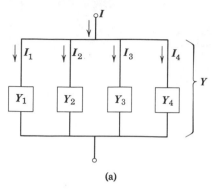

(a)

FIGURE 8a   Current distribution among parallel branches.

9. IMPEDANCE OF BRANCHES IN PARALLEL

Since a branch with admittance Y_1 has impedance Z_1 such that $Y_1 = 1/Z_1$, equation 7-1 for branches in parallel can be written

$$\frac{1}{Z} = \frac{1}{Z_1} + \frac{1}{Z_2} + \frac{1}{Z_3} + \frac{1}{Z_4} \qquad (9\text{-}1)$$

Solving for the impedance of the four branches in parallel,

$$Z = \frac{1}{\dfrac{1}{Z_1} + \dfrac{1}{Z_2} + \dfrac{1}{Z_3} + \dfrac{1}{Z_4}} \qquad (9\text{-}2)$$

$$= \frac{Z_1 Z_2 Z_3 Z_4}{Z_1 Z_2 Z_3 + Z_1 Z_2 Z_4 + Z_1 Z_3 Z_4 + Z_2 Z_3 Z_4}$$

Much the most important special case is for the impedance of *two* elements in parallel:

$$Z = \frac{Z_1 Z_2}{Z_1 + Z_2} \qquad (9\text{-}3)$$

Figure 9a shows this combination of impedances as an alternative to the sum of admittances for dealing with branches in parallel. An analogous combination that can be used to find the equivalent admittance of elements in series is shown in Figure 9b, but this combination is not nearly so often used.

Besides finding the equivalent impedance of two branches in parallel by

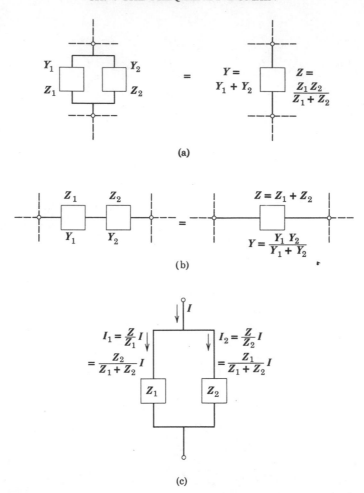

FIGURE 9 **(a)** **Impedance of branches in parallel.** **(b)** **Admittance of elements in series.** **(c)** **Current division between two parallel branches.**

equation 9-3, it is often useful to determine the *current distribution* between these two branches in terms of their impedances. Thus equation 8-2 becomes

$$I_1 = \frac{Z}{Z_1}I \qquad I_2 = \frac{Z}{Z_2}I \quad \text{etc.} \tag{9-4}$$

This is the general formula. A special case that applies when there are only two branches in parallel is illustrated in Figure 9c. Combining equation 9-4 with equation 9-3, we obtain

$$I_1 = \frac{Z}{Z_1} I = \frac{Z_2}{Z_1 + Z_2} I$$

$$I_2 = \frac{Z}{Z_2} I = \frac{Z_1}{Z_1 + Z_2} I$$

(9-5)

Both equations 9-3 and 9-5 are important enough to be memorized; it will save time in the end.

10. SERIES-PARALLEL NETWORKS

When the elements of a network are not all in series, or all in parallel, but present some combination of series and parallel arrangement, they can be handled in groups.

Example. Find the impedance of the network of Figure 10a.

Solution: The first step is to find an equivalent for the two branches in parallel. (The two parallel branches will be identified by subscripts *bce* and *bde*; the single element equivalent to the two in parallel by *be*.) By equation 5-1

$$Z_{bce} = Z_{bc} + Z_{ce} = 10.0 + j0.1\omega$$

Also

$$Z_{bde} = \frac{1}{j10\omega10^{-6}} = -j\frac{10^5}{\omega}$$

By equation 9-3, the impedance of the two branches in parallel is

$$Z_{be} = \frac{(10.0 + j0.1\omega)(-j10^5/\omega)}{(10.0 + j0.1\omega) - (j10^5/\omega)}$$

Multiplication in the numerator and addition in the denominator give

$$Z_{be} = \frac{10^4 - j10^6/\omega}{10.0 + j(0.1\omega - 10^5/\omega)}$$

This is the impedance from *b* to *e*. The series resistance between *a* and *b* must now be added:

$$Z_{ae} = 10.0 + \frac{10^4 - j10^6/\omega}{10.0 + j(0.1\omega - 10^5/\omega)}$$

(10-1)

Here we have the terminal impedance Z_{ae} and since it is a function of frequency it can be (and often is) written $Z_{ae}(j\omega)$. Z_{ae} has a real or resistive

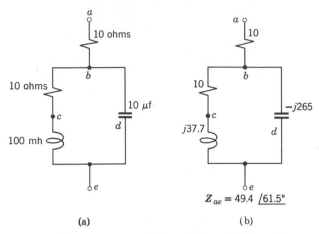

(a)   (b)

FIGURE 10   (a)   A series-parallel example.   (b)   Impedances of elements.

component, and an imaginary or reactive component, and both components vary with frequency (see equation 10-2). It is easy to compute the numerical value of the terminal impedance Z_{ae} at any given frequency; thus at 60 cycles per second it is $49.4/61.5°$ but at a different frequency it would have another value.

Current and voltage distribution are according to relations already derived. By way of example, let us assume an applied terminal voltage of 100 volts rms, alternating at 60 cycles per second. Taking this applied voltage as reference,

$$V_{ae} = 100/0°$$

Putting $\omega = 377$ radians per second into equation 10-1,

$$Z_{ae} = 10.0 + \frac{10^4 - j2650}{10.0 + j(37.7 - 265)}$$

$$= 10.0 + 45.5/72.7° = 49.4/61.5° \qquad (10\text{-}2)$$

$$I_{ae} = \frac{V_{ae}}{Z_{ae}} = \frac{100/0°}{49.4/61.5°} = 2.02/\overset{\frown}{-61.5°}$$

Computing the impedances of the elements by letting $\omega = 377$, the values given in Figure 10b are obtained. Current then is found to divide between branches as follows:

$$I_{bce} = \frac{(-j265)(2.02/-61.5°)}{(10.0 + j37.7) - j265} = \frac{(-j265)(2.02/-61.5°)}{10.0 - j227}$$

$$= \frac{535/-151.5°}{227/-87.5°} = 2.36/-64.0°$$

$$I_{bde} = \frac{(39.0/75.1°)(2.02/-61.5°)}{227/-87.5°} = 0.347/101.1°$$

Whenever a division of current or of voltage is computed, there is available an obvious check on the results by adding the component currents. This opportunity to catch mistakes is much too valuable to be overlooked.

$$I_{ae} = I_{bce} + I_{bde} = 2.36/-64.0° + 0.347/101.1°$$
$$= (1.035 - j2.12) + (-0.0667 + j0.340)$$
$$= 0.968 - j1.78 = 2.03/-61.5°$$

Comparison with the original value of I_{ae} shows reasonable agreement for slide-rule computation.

We now have found current in each branch of the network. It remains to find voltage distribution. Let us find V_{ab} and V_{be} (since the resistance and reactance of the inductive branch are probably the constants of a single coil, there is no physical significance to the potential of point c and it will be disregarded). Then

$$V_{ab} = Z_{ab}I_{ae} = 10(2.02/-61.5°) = 20.2/-61.5°$$

Similarly, using a value from equation 10-2,

$$V_{be} = Z_{be}I_{ae} = (45.5/72.7°)(2.02/-61.5°)$$
$$= 92.2/11.2°$$

Here, again, a valuable check on some of the algebra is found by adding the values of V_{ab} and V_{be} to obtain, hopefully, $100/0°$.

11. A PHASOR DIAGRAM

Drawing a phasor diagram of the quantities involved should be part of every circuit analysis. If drafting of the diagrams is done accurately, this constitutes a graphical analysis of the circuit and nothing more is needed. However, it is usually easier to do the analysis numerically with a slide rule or a desk-type computer, drawing only freehand sketches of phasor diagrams. These sketches should *never* be omitted (unless the problem is so simple that you can see the diagram mentally without having to use pencil and paper), for they not only help in understanding what is going on, but they also are tremendously time-saving in making obvious any gross mistake in the arithmetic such as reversal of sign or wrongly placed decimal point.

Let us therefore draw a phasor or transform diagram of the currents and voltages in the example just completed. This would most profitably have been done as the solution proceeded. We should first have drawn the reference transform V_{ae}, as in Figure 11a. This can have any convenient length to represent 100 volts (its rms value), and the scale of all voltage transforms of the diagram is decided when this choice of length is made.

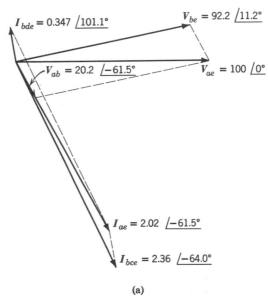

$I_{bde} = 0.347 \;\underline{/101.1°}$

$V_{be} = 92.2 \;\underline{/11.2°}$

$V_{ab} = 20.2 \;\underline{/-61.5°}$

$V_{ae} = 100 \;\underline{/0°}$

$I_{ae} = 2.02 \;\underline{/-61.5°}$

$I_{bce} = 2.36 \;\underline{/-64.0°}$

(a)

FIGURE 11a   Transform or phasor diagram for the network of Figure 10b.

The first computation gives I_{ae}, which is then drawn at the appropriate negative angle of 61.5 degrees. The scale that determines the length of this transform can also be chosen merely for convenience, since this is the first current in the diagram. I_{bce} and I_{bde} are next drawn to the same scale; they should (and do) add to equal I_{ae}, as indicated by dashed lines in the figure.

V_{ab} and V_{be} are next computed and plotted. V_{ab} is naturally colinear with I_{ae}, for it is voltage across a resistance; if they were not colinear it would be too bad. V_{be} should be in quadrature with I_{bde}, the capacitor current—actually an error of 0.1 degree has crept in, but this is detected from the arithmetic and not from the diagram. Also, I_{bce} through the inductive branch should lag 75.1 degrees behind voltage V_{be}, but the same slight error appears here. Finally, V_{ab} and V_{be} should add to give V_{ae}, which they do. Thus all the simple relations that are obvious from the circuit diagram are seen to be satisfied, and we can feel sure that nothing has gone radically wrong.

12. COMPONENTS OF Z AND Y

It is customary to denote the real and imaginary parts of the impedance Z by the letters R and X, thus:

$$Z = R + jX \tag{12-1}$$

R and X may or may not be the resistance and reactance of a single branch of a circuit. Thus if Z is the impedance of a coil of wire, R and X are the resistance and reactance of that coil. But if Z is the impedance of the network of Figure 10a as given in equation 10-1, R is not the resistance of any particular element or group of elements, nor is X a particular reactance. Rather, Z is the ratio of input voltage to input current, and R and X are its components; often, as in this network, R and X are both functions of frequency.

In a similar manner, Y is the terminal admittance of a network. By definition it is the reciprocal of Z, and its real and imaginary components are denoted G and B:

$$Y = \frac{1}{Z} = G + jB \tag{12-2}$$

In general, the values of G and B are affected by all the parts of a network and cannot be referred to any single element.

13. RELATIONS OF COMPONENTS

Since $Y = 1/Z$, the components of Y must be related to the components of Z. That is, we can find G and B in terms of R and X, and vice versa.

The relation is shown by writing

$$Y = G + jB = \frac{1}{Z} = \frac{1}{R + jX} = \frac{R - jX}{R^2 + X^2} \tag{13-1}$$

In any complex equation the real components are equal and the imaginary components are equal, so equation 13-1 gives

$$G = \frac{R}{R^2 + X^2} \quad \text{and} \quad B = \frac{-X}{R^2 + X^2} \tag{13-2}$$

Thus the components of admittance can be found if the components of impedance are known. Note that G *is not the reciprocal of* R unless X is zero. Similarly B is not the negative reciprocal of X unless R is zero.

In a similar manner, R and X can be found from G and B:

$$Z = R + jX = \frac{1}{Y} = \frac{1}{G + jB} = \frac{G - jB}{G^2 + B^2} \qquad (13\text{-}3)$$

Hence

$$R = \frac{G}{G^2 + B^2} \quad \text{and} \quad X = \frac{-B}{G^2 + B^2} \qquad (13\text{-}4)$$

In equations 13-2 and 13-4, the denominators involving squares are always positive. R and G are therefore always of the same sign, and for passive networks they are both positive. X and B, on the contrary, are of opposite sign; if X is positive, B is negative, and vice versa. Thus if the impedance of a network has a positive angle its admittance has a negative angle, and vice versa.

14. THE LADDER METHOD

A particularly easy solution is possible if a network has the form of a ladder, and this is quite common. There are no new principles in the ladder method; it can be applied to a direct-current problem, using voltages, currents, and resistances, or to an alternating-current problem, using transforms of voltages, transforms of currents, and complex impedances. The method is illustrated by the ladder network of Figure 14a.

It is desired to find current I in the 10-ohm load. This could be done by combining the network impedances in series and in parallel until the source current was obtained, and then computing voltage and current distributions until the load current finally emerged. Fortunately, however, Kirchhoff's two laws can be applied to this problem in an easier way.

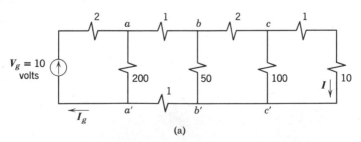

(a)

FIGURE 14a   A ladder network; the elements are pure resistance with values, as shown, in ohms. Load current is to be found.

Let current in the 10-ohm load equal I. Voltage from c to c' is then $11I$. Current through the 100-ohm element is $11I/100 = 0.11I$, and current from b to c is $I + 0.11I = 1.11I$. Continuing,

$$V_{bc} = (1.11I)(2) = 2.22I$$

$$V_{bb'} = 11I + 2.22I = 13.22I$$

$$I_{bb'} = \frac{13.22I}{50} = 0.264I$$

$$I_{ab} = I_{bc} + I_{bb'} = 1.11I + 0.264I = 1.374I$$

This current flows through both of the 1-ohm resistors, between ab and $a'b'$, so

$$V_{aa'} = (1.374I)(2) + 13.22I = 15.97I$$

$$I_{aa'} = \frac{15.97I}{200} = 0.080I$$

$$I_g = 1.374I + 0.080I = 1.454I$$

$$V_g = (1.454I)(2) + 15.97I = 18.88I$$

The load current, the answer to our problem, is therefore

$$I = \frac{V_g}{18.88} = \frac{10}{18.88} = 0.530 \text{ ampere}$$

We now substitute 0.530 for I in all the expressions for current and voltage in the above solution. Thus

$$V_{aa'} = (15.97)(0.530) = 8.45 \text{ volts}$$

This method, although commonly described in books on circuit analysis, seems to have no generally accepted name. It will therefore be referred to as the *ladder method* because it is so plainly suitable to a network of that form, and because the ladder network is such an important one in both communication and power systems.

The essence of the ladder method is to find a key spot at which current may be assumed, making solution for other voltages and currents easy.

15. A LADDER EXAMPLE

This method is not restricted to obvious ladder networks. It can well be applied, for instance, to the same network that was treated earlier in this

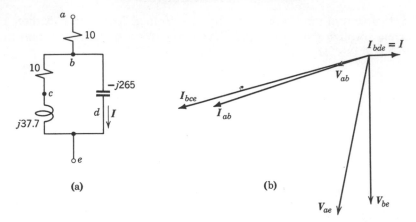

FIGURES 15a,b   A network, with 60-cycle impedances given in ohms, and its transform diagram.

chapter, shown again in Figure 15a. Let us find, by the ladder method, the 60-cycle input impedance. Assume the capacitor current to be I. Then

$$V_{be} = Z_{bde}I = -j265I$$

We start at once sketching the transform diagram of Figure 15b, and we shall continue to add each current and voltage as it is computed.

$$I_{bce} = \frac{V_{be}}{Z_{bce}} = \frac{-j265I}{10 + j37.7} = \frac{265I\underline{/-90°}}{39.0\underline{/75.15°}} = 6.80I\underline{/-165.15°}$$

$$I_{ab} = I(1 + 6.80\underline{/-165.15°}) = (1 - 6.60 - j1.74)I$$

$$= (-5.60 - j1.74)I = 5.85I\underline{/-162.7°}$$

$$V_{ab} = 10I_{ab} = (-56.0 - j17.4)I$$

$$V_{ac} = V_{ab} + V_{be} = (-56.0 - j17.4 - j265)I = (-56.0 - j282.4)I$$

$$= 288I\underline{/-101.2°}$$

Each of the currents and voltages has now been computed. Each has been added to the transform diagram. Since the diagram appears reasonable, we are encouraged to believe that no great mistakes have been made. Impedance is the object of the computation, so we divide input voltage by input current and obtain:

$$Z_{ae} = \frac{288I\underline{/-101.2°}}{5.85I\underline{/-162.7°}} = 49.3\underline{/61.5°}$$

This result may be compared with the impedance obtained by parallel and series combinations in equation 10-2. If we wish to find current and voltage distribution in the network as a result of some given applied voltage, it is now the work of a moment to apply the appropriate ratio to each of the above values. It will be observed that the transform diagram is oriented differently from that of Figure 11a for the same network; this is due to our assumption of $I = I_{bde}$ as reference, V_{ae} having been the reference transform in the earlier solution.

It is not unreasonable to apply the ladder method of analysis to the network of Figure 15a, for this network can be redrawn in ladder form if desired. A good many networks that do not look like ladders can be changed to ladder form, either by simply redrawing or perhaps by redrawing and making one or two simplifications at the same time.

16. AN EXAMPLE OF RESONANCE

Referring again to Figure 10a, it would be interesting to know whether there is any frequency at which the impedance of this parallel-series combination of elements is a pure resistance instead of being the inductive impedance we found at 60 Hz. Study of the transform diagram of Figure 11a leads us to suspect that this might be possible: at a frequency higher than 60 Hz the capacitor current I_{bde} would become greater and the current I_{bce} through the inductive branch would become less; their sum, the total current, I_{ae}, should at some frequency be in phase with the applied voltage V_{ae}. The equivalent impedance Z_{ae} would then be purely real.

How can we compute the frequency that will make Z_{ae} real? For Z_{ae} to be real, the imaginary part of Z_{ae} in equation 10-1 must be zero. To find the imaginary part of Z_{ae} the fraction in that equation must be rationalized. Using the method of equation 3-6 of Chapter 3, numerator and denominator of the fraction are multiplied by the conjugate of the denominator, giving

$$\frac{[10^4 - j10^6/\omega][10.0 - j(0.1\omega - 10^5/\omega)]}{(10.0)^2 + (0.1\omega - 10^5/\omega)^2} \tag{16-1}$$

The denominator is now real, so if the imaginary part of the numerator can be made to equal zero the whole expression for Z_{ae} will be purely real. We therefore perform the indicated multiplication in the numerator; we retain only the imaginary terms and set them equal to zero:

$$10^4\left(0.1\omega - \frac{10^5}{\omega}\right) + \left(\frac{10^6}{\omega}\right)10.0 = 0$$

$$10^3\omega - \frac{10^9}{\omega} + \frac{10^7}{\omega} = 0 \tag{16-2}$$

Solving for frequency,

$$\omega^2 - 10^6 + 10^4 = 0 \qquad \omega = 995$$

$$f = \frac{995}{2\pi} = 158 \text{ Hz} \tag{16-3}$$

This tells us that the equivalent impedance is a pure resistance if the frequency is 158 Hz. Let us compute the equivalent impedance at this frequency.

We use equation 10-1 again, letting $\omega = 995$. Then

$$Z_{ae} = 10.0 + \frac{10,000 - j1005}{10.0 + j(99.5 - 100.5)} = 10.0 + \frac{10,000 - j1005}{10.0 - j1.0} \tag{16-4}$$

$$= 10.0 + 1000 = 1010 \text{ ohms}$$

It will be seen that at this frequency of 158 Hz the input impedance to the two parallel branches of the circuit is pure resistance with the surprisingly high value of 1000 ohms.

An interesting relation appears when we compare the reactance of the inductive branch of the circuit with that of the capacitive branch at this particular frequency of 158 Hz. For the inductance, $X = \omega L = 99.5$ ohms. For the capacitance, $X = -1/\omega C = -100.5$ ohms. Thus the reactances are very nearly equal and opposite at this frequency. The magnitudes of impedance of the two branches are even more nearly the same: impedance of the inductive branch is $Z_{bce} = 10 + j99.5 = 100.0 \underline{/84.26°}$.

Current Division. With 100 volts applied (see Figure 16a), current I_{ae} entering the terminals is

$$I_{ae} = \frac{100 \underline{/0°}}{1010 \underline{/0°}} = 0.0990 \underline{/0°}$$

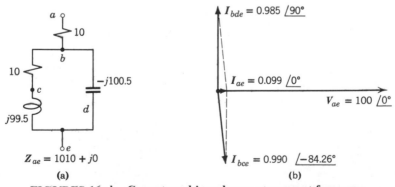

Z_{ae} = 1010 + j0
(a)

(b)

FIGURES 16a,b   Currents and impedances at resonant frequency.

This very small current, in phase with applied voltage, is shown in Figure 16b. The voltage drop through the 10-ohm resistor between a and b is

$$V_{ab} = (0.0990/\underline{0°})(10/\underline{0°}) = 0.990/\underline{0°}$$

Subtracting this from the applied voltage, 100 volts, gives voltage across the parallel branches:

$$V_{be} = 100/\underline{0°} - 0.990/\underline{0°} = 99.0/\underline{0°}$$

Current in each of the parallel branches is now found from this voltage:

$$I_{bce} = \frac{99.0/\underline{0°}}{100.0/\underline{84.26°}} = 0.990/\underline{-84.26°} = 0.099 - j0.985$$

$$I_{bde} = \frac{99.0/\underline{0°}}{100.5/\underline{-90°}} = 0.985/\underline{+90°} = 0.0 + j0.985$$

(Here, again, we have a check: adding these two gives $0.099 + j0$, which is seen to be the correct value of I_{ae}.)

Phasors for the two branch currents are drawn in Figure 16b. They are almost equal in magnitude, but since one leads the applied voltage and the other lags they are nearly opposite to each other. They can be thought to represent a current, a fairly large current of nearly an ampere, that circulates around the resonant loop, while only a small current (less than one-tenth as much) enters the resonant loop from the external circuit. Current in the resonant loop is almost self-supporting as stored energy surges back and forth from capacitance to inductance. Energy flowing in from the external circuit is only the amount needed to supply loss in the 10-ohm resistance.

Indeed, if this resonant loop were cut loose from the external circuit, perhaps by opening a switch at a, while current was circulating in the loop, this surging current would continue to oscillate back and forth, though with constantly diminishing amplitude. The oscillations of current would grow less and less as energy stored in inductance and capacitance was gradually consumed by the resistance of the circuit.

This suggests that current of a certain form—a certain frequency and a certain rate of decay—can continue in a circuit without input. This leads to a concept of natural current and of poles and zeros which will be used in the next chapter.

17. THE PRINCIPLE OF RESONANCE

In the network of the foregoing section there is parallel resonance at the frequency at which the two *parallel* branches, one inductive and the other capacitive, have the same magnitude of susceptance. Another equally useful type of resonance appears in a network with capacitance and inductance in *series* as in Figure 17a. Resonance in the *parallel* arrangement can give extremely *high* impedance (at the resonant frequency) between input terminals; *series* resonance can give extremely *low* impedance between terminals. Both types of resonance can be very sensitive to frequency if resistive loss is low, and this frequency sensitivity is the reason for their great practical importance. Resonant circuits are used for *tuning* or *filtering* in communication circuits, as in radios or telephone systems, to select a desired frequency while blocking other unwanted frequencies.

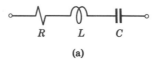

$$R \qquad L \qquad C$$

(a)

FIGURE 17a   A series-resonant circuit.

By definition, *resonance* occurs at the frequency at which terminal current and terminal voltage of a reactive network are in phase with each other.† A complicated network with several reactive branches can have several resonant frequencies.

18. SERIES RESONANCE

For the series-resonant circuit of Figure 17a, an expression for impedance is written in the usual manner. There are three elements in series:

$$Z = R + j\omega L + \frac{1}{j\omega C} = R + j\left(\omega L - \frac{1}{\omega C}\right) \qquad (18\text{-}1)$$

The reactance terms are in parentheses. Inductive reactance is positive, capacitive reactance is negative. It is evident that there is some frequency, some value of ω, at which these individual reactances will be equal and opposite, leaving the total reactance of the circuit zero. This, by definition, is the frequency of *resonance*. If the frequency of resonance is called f_0, and $2\pi f_0$ is designated ω_0, it follows that at resonance

$$\omega_0 L - \frac{1}{\omega_0 C} = 0 \quad \text{or} \quad \omega_0 L = \frac{1}{\omega_0 C} \qquad (18\text{-}2)$$

† There are other definitions of resonance but they are hardly distinguishable in low-loss systems; this, the most common and probably the most useful, will be accepted in this book.

From this,

$$\omega_0{}^2 = \frac{1}{LC} \quad \text{or} \quad \omega_0 = \frac{1}{\sqrt{LC}} \tag{18-3}$$

and the frequency of resonance is given by

$$f_0 = \frac{1}{2\pi\sqrt{LC}} \tag{18-4}$$

The characteristic features of resonance are much more striking if the circuit resistance is small. We shall therefore be interested in circuits in which the reactance of either of the reactive elements (either $\omega_0 L$ or $1/\omega_0 C$) is at least several times as great as R. This ratio will be called Q, the quality factor of the circuit.

Figure 18a shows transforms of voltage and current in the series resonant branch of Figure 17a at the frequency of resonance, and at frequencies a little lower and a little higher. In drawing these diagrams it is assumed that Q is only about 4; in practical use a Q of 40 might be more likely but the diagrams would be more difficult.

Figure 18b shows the various reactances and impedances as functions of frequency. The inductance L has reactance ωL, shown by a straight line through the origin. The capacitance C has reactance $-1/\omega C$, a hyperbolic curve at the bottom of the diagram. The sum of these two, marked X and extending from lower left to upper right, is the total reactance of the branch. The resistance of the branch is R, assumed small and constant.

The frequency of resonance is that at which $X = 0$, where the curve for X crosses the axis. At this frequency the total impedance of the resonant branch is merely R (equation 18-1), and a curve for the magnitude of impedance, marked $|Z|$, passes through R at ω_0. Above resonance, at frequencies higher than ω_0, the total impedance of the branch is approximated by the inductive reactance, and at low frequencies by the magnitude of the capacitive reactance; equation 18-1 tells us that $|Z| = \sqrt{R^2 + (\omega L - 1/\omega C)^2}$.

In this diagram, resistance is assumed constant, independent of frequency. In fact the best model of resistance in a typical resonant branch probably has a value of resistance that rises slightly with frequency, thereby taking into account such additional losses as those due to skin effect in the conductor, proximity effect in the coil, and perhaps iron losses in a metal core through the inductive coil. Practically, the amount of resistance should be correct at the resonant frequency, where it is the whole impedance of the branch, but any variation of the resistance makes little difference to the magnitude of the impedance at frequencies that are much higher or lower than ω_0.

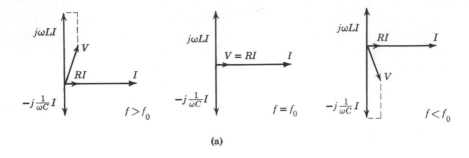

(a)

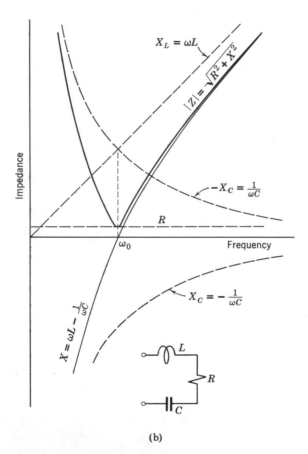

(b)

FIGURE 18 (a) Transform diagrams at frequencies near resonance.
(b) Impedance of a series-resonant circuit and its three components.

19. ADMITTANCE NEAR RESONANCE

Since nearly all the important information about the behavior of the resonant circuit is contained in the rounded tip at the bottom of the V-shaped curve, it is much more useful to draw the reciprocal quantity, admittance, as in Figure 19a. This makes it possible to show more clearly

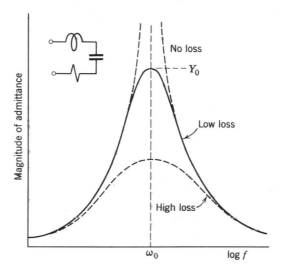

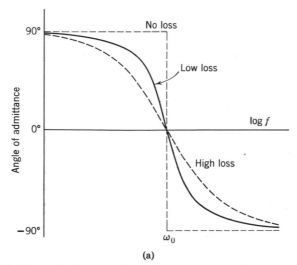

(a)

FIGURE 19a   Admittance of a series-resonant circuit; magnitude and angle.

the effect of the resistance of the branch at the resonant frequency, an effect almost imperceptible at the bottom of the V in Figure 18b.

Since the admittance is vanishingly small at frequencies far from resonance (still assuming low loss), it is unnecessary to show in our diagram the entire frequency scale all the way to zero; instead we draw Figure 19a with an expanded frequency scale so that only a small part of the total frequency range is included—perhaps 10% each way from the resonant frequency. A logarithmic frequency scale is used; this is not important (although it is rather customary when plotting frequency functions), but it has the minor advantage of making the resonance curve symmetrical about the vertical line ω_0. The angle as well as the magnitude of the admittance can be deduced from the transform diagrams. This is plotted as a separate curve in Figure 19a. The angle of I relative to V is positive at low frequency, as in Figure 18a, zero at resonant frequency, and negative at high frequency.

In Figure 19a the effect of high or low resistance—that is, of high or low loss in the circuit—is readily shown. The height of the resonance curve is determined by the circuit resistance alone; the crest value is $1/R$. Hence the high-loss curve is, as shown, flatter than the low-loss curve. If there were no loss at all $(R = 0)$, the resonance curve would be infinitely high. On the other hand, at frequencies a little away from resonance the loss makes slight difference and the three curves of Figure 19a tend to merge.

20. EQUATIONS OF RESONANCE

From equation 18-1,

$$Z = R + j\left(\omega L - \frac{1}{\omega C}\right) \tag{20-1}$$

We know from equation 18-3 that L, C, and resonant frequency are related by $C = 1/\omega_0^2 L$; using this to eliminate C from equation 20-1,

$$Z = R + j\left(\omega L - \frac{\omega_0^2 L}{\omega}\right) \tag{20-2}$$

We have previously referred to the quality factor Q. Let us now define it[†] (with the subscript 0 referring to conditions at resonant frequency) as

$$Q_0 = \frac{\omega_0 L}{R_0} \tag{20-3}$$

[†] The symbol Q has three different meanings in common use: quality factor, electric charge, reactive power. These three meanings are quite unrelated to each other. Since they are unlikely to be used in the same equation, there is little danger of confusion.

and write

$$Z = R + j\omega_0 L\left(\frac{\omega}{\omega_0} - \frac{\omega_0}{\omega}\right) = R_0\left[\frac{R}{R_0} + jQ_0\left(\frac{\omega}{\omega_0} - \frac{\omega_0}{\omega}\right)\right] \qquad (20\text{-}4)$$

This is a neat expression, but it is not well adapted to computation because ω, near resonance, is almost equal to ω_0, and in the parentheses we have the small difference of two relatively large quantities. This can be avoided by introducing a new symbol to represent, not frequency, but the difference between actual frequency and resonant frequency—the "detuning." Actually, it is not the detuning in radians per second that is most convenient but the ratio of this detuning to the resonant frequency—the "fractional detuning." This is called δ, and defined in terms of ω it is

$$\delta = \frac{\omega - \omega_0}{\omega_0} \qquad (20\text{-}5)$$

Simple algebra gives us:

$$1 + \delta = \frac{\omega}{\omega_0} \qquad (20\text{-}6)$$

Hence

$$\frac{\omega}{\omega_0} - \frac{\omega_0}{\omega} = 1 + \delta - \frac{1}{1 + \delta} = \delta\frac{2 + \delta}{1 + \delta} \qquad (20\text{-}7)$$

When this is introduced into equation 20-4, the result is

$$Z = R_0\left(\frac{R}{R_0} + jQ_0\delta\frac{2 + \delta}{1 + \delta}\right) \qquad (20\text{-}8)$$

To this point, it will be observed, there are no limiting assumptions and no approximations of any kind. This expression is as general as equation 18-1. The symbols may be tabulated:

Z is impedance at the terminals of a series-resonant circuit
R, L, C are the three circuit parameters
R_0 is resistance (effective) at resonant frequency
Q_0 is defined in equation 20-3
δ is defined in equation 20-5
ω is $2\pi f$, where f is frequency
ω_0 is $2\pi f_0$, where f_0 is resonant frequency

R, the resistance at ω, may or may not be the same as R_0, the resistance at ω_0 (at resonance).

For a high-Q circuit, certain approximations are now made. Through the frequency range near resonance, where the fractional detuning δ is small compared to 1, approximately

$$Z = R_0(1 + j2Q_0\delta) \tag{20-9}$$

Note that the impedance at resonance $(\delta = 0)$, by both the exact and approximate formulas, is R_0.

We now compute Y from this approximate expression for Z and get

$$Y = \frac{1}{Z} = \frac{Y_0}{1 + j2Q_0\delta} \tag{20-10}$$

where Y_0 is the admittance at resonance, the reciprocal of R_0. Actually the most useful form is the ratio:

$$\frac{Y}{Y_0} = \frac{1}{1 + j2Q_0\delta} \tag{20-11}$$

Although this is an approximation, it is quite satisfactory for all ordinary purposes if Q_0 is 20 or greater.†

Figure 19a shows the admittance to a series-resonant circuit as computed from equation 20-10. Sharpness of the crest depends on the Q of the circuit; a low-loss circuit has a high crest (low circuit resistance), and a high-loss circuit has a low crest.

The angle of the admittance (the angle between current and voltage, the power-factor angle) increases from zero at resonance to nearly 90 degrees as resonance is lost. Looking in at the terminals, the series circuit looks like low resistance at resonant frequency, high capacitive reactance at lower frequency, and high inductive reactance at higher frequency.

21. THE UNIVERSAL RESONANCE CURVE

Since the shape of the resonance curve is essentially the same for all circuits, being merely higher or lower, broader or narrower, for different resonant frequencies and amounts of loss, a single dimensionless curve can be used to represent resonance in all circuits. Following equation 20-11, Y/Y_0 is plotted in Figure 21a as a function of $Q_0\delta$. The result is the *universal resonance curve*. The curve marked "Total" is Y/Y_0 from equation 20-11. It has a maximum value of 1 at resonant frequency. The horizontal scale is frequency-off-resonance; it is the fractional detuning δ times Q_0. This product, δQ_0, sometimes called α, is the *relative* fractional detuning.

---

† If $Q_0 = 20$, the error in Y barely exceeds 1% of Y_0 for any δ, and is less for small δ. If Q_0 is 10, the error is somewhat more than doubled and δQ_0 must be less than 1 for 1% accuracy.

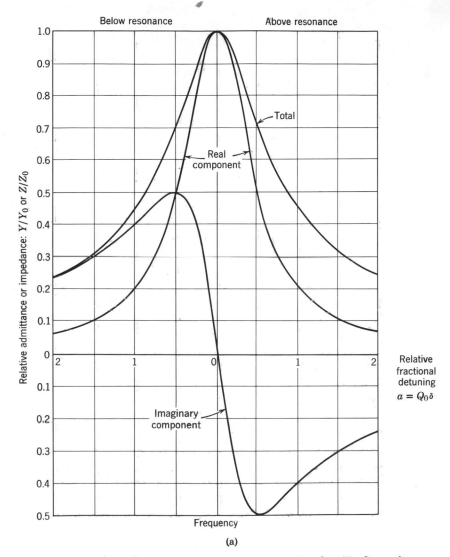

FIGURE 21a The universal resonance curve. Read Y/Y_0 for series resonance, or Z/Z_0 for parallel resonance.

Figure 21a shows curves of the real and imaginary components of admittance as well as the total admittance. Near resonance the admittance is nearly real. Far from resonance the admittance is mainly imaginary. Conductance and susceptance are equal in magnitude and the angle of admittance is ± 45 degrees where the two curves cross and $Q_0 \delta = \mp 1/2$.

The points at which conductance and susceptance are equal in magnitude are called the half-power points, or the 3-db points, or the 70% points. The reasons are briefly as follows. These points mark the frequencies at which power to a resonant circuit is reduced by half (from resonance), voltage remaining the same. This reduction to half power is approximately a reduction by 3 decibels, abbreviated 3 db. Also, these are the frequencies at which the magnitude of admittance is reduced by $\frac{1}{2}\sqrt{2}$, or approximately to 70%.

The range of frequencies between the two half-power points is sometimes called the bandwidth of the resonant circuit; Figure 21b. This lies between

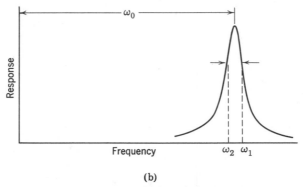

(b)

FIGURE 21b   Bandwidth. $Q_0 = \omega_0/(\omega_1 - \omega_2)$.

an upper frequency which we may call ω_1 at which (as we know from Figure 21a) $Q_0 \delta_1 = 1/2$, and a lower frequency ω_2 where $Q_0 \delta_2 = -1/2$. Then

$$Q_0(\delta_1 - \delta_2) = 1 \quad \text{and} \quad \delta_1 - \delta_2 = \frac{1}{Q_0} \qquad (21\text{-}1)$$

Since $\delta_1 - \delta_2 = (\omega_1 - \omega_2)/\omega_0$ we have:

$$\omega_1 - \omega_2 = \frac{\omega_0}{Q_0} \qquad (21\text{-}2)$$

Thus the quantity that is sometimes called, not very meaningfully, the bandwidth, is seen to be inversely proportional to Q_0, and a sharply selective circuit must have low loss.

22. DEFINITION OF Q

Q was defined for the series resonant circuit in equation 20-3, but a broader definition can be devised to include also other types of resonance. One such definition is in terms of the response of a circuit to detuning.

For any type of resonance to which the "universal" resonance curve of Figure 21a applies, we can use equation 21-2 to define Q:

$$Q_0 = \frac{\omega_0}{\omega_1 - \omega_2} \tag{22-1}$$

where ω_0 is the frequency of resonance and ω_1 and ω_2 are the half-power frequencies. This equation is useful for experimental determination of Q_0 as well as for providing a formal definition, since the frequency of resonance and the frequencies of 70% current response either side of resonance are easily measured.

An even broader definition relates the energy stored in the resonant system to the loss of energy in the system. The constant amount of energy stored in any resonant electrical system is oscillating between the magnetic field of inductance and the electric field of capacitance. (To include mechanical resonant systems, such as the pendulum of a clock, we can refer to the components of stored energy as kinetic and potential.) Energy dissipated per cycle is changed to heat. The definition, which can be shown to reduce to equation 22-1, is simply

$$Q_0 = \frac{\text{Energy stored}}{\text{Energy dissipated per cycle}} 2\pi \tag{22-2}$$

This is a nice definition from the theoretical point of view, but it does not lend itself to experimental determination of Q.

23. PARALLEL RESONANCE

Figures 23a, 23b, and 23c show three circuits with resonance between elements of inductance and capacitance in parallel. The first, Figure 23a, is the exact dual of the series resonant circuit. However, this arrangement is not usually a practical model, for an inductive device necessarily has at least a little resistance, and the model of Figure 23b results. But the simpler two-branch arrangement of Figure 23c is just as useful as the three-branch network for most purposes, so this is the circuit that is commonly used in practice to provide parallel resonance, often under the name of "tank circuit."

Whereas series resonance provides *low* impedance at one particular resonant frequency, parallel resonance provides *high* impedance at its resonant

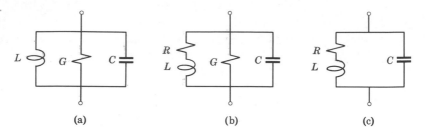

FIGURES 23a,b,c   Circuits with parallel resonance.

frequency. To show this, we write input admittance for the circuit of Figure 23c:

$$Y = \frac{1}{R + j\omega L} + j\omega C = \frac{\dfrac{RC}{L} + j\left(\omega C - \dfrac{1}{\omega L}\right)}{1 + \dfrac{R}{j\omega L}} \tag{23-1}$$

In a low-loss circuit (a circuit with high Q), near resonance, R is very much smaller than ωL, and, approximately,

$$Y = \frac{C}{L}\left[R + j\left(\omega L - \frac{1}{\omega C}\right)\right] \tag{23-2}$$

This equation has the same form as equation 20-1 for the series-resonant circuit, and when the same substitutions and approximations are made as in Section 20, equation 23-2 becomes

$$\frac{Z}{Z_0} = \frac{1}{1 + j2Q_0\delta} \tag{23-3}$$

But the right-hand member is identically the same as that of equation 20-11, the equation from which the "universal" resonance curve is plotted, so it follows that the "universal" curve can be used for both parallel resonance and series resonance; for the series resonant circuit it is a curve of Y/Y_0 whereas for parallel resonance it is Z/Z_0.

Neither equation 23-3 nor the universal curve can be given numerical values until Z_0, the impedance at resonance, is known. But this comes from equation 23-2 which gives, at resonance,

$$Y_0 = \frac{CR_0}{L} \quad \text{or} \quad Z_0 = \frac{L}{CR_0} \tag{23-4}$$

The value of Q of a two-branch resonant circuit is also needed, and it is found to be the same as for a series resonant circuit. Using either definition of Q from Section 22, the value for the two-branch resonant circuit is

$$Q_0 = \frac{\omega_0 L}{R_0} = \frac{1}{\omega_0 C R_0} \tag{23-5}$$

which is the same as Q_0 of equation 20-3 for the series circuit. (Note that in either circuit the three elements carry practically the same current at resonance, and it is to be expected that the ratio of loss to stored energy will be the same.)

Z_0 is given in equation 23-4, but it is perhaps more useful and meaningful in terms of Q. Thus

$$Z_0 = (\omega_0 L)Q_0 = \frac{1}{\omega_0 C}Q_0 = R_0 Q_0{}^2 \tag{23-6}$$

Notice that impedance is purely resistive at the resonant frequency, and that it is very large if Q_0 is high. A high value of Q_0 is to be expected in practical resonant circuits. Indeed most of the equations of resonance are approximations that are accurate only if Q_0 is at least 10 or 20.

24. POWER TO A TWO-TERMINAL NETWORK

One way of computing power to any two-terminal network is to determine the resistance of each branch, compute current in that branch, find the power consumption as I^2R, and add the power consumptions in all the various branches to get the total power consumption.

It is often helpful to be able to compute average power for an entire network, however, in terms of *terminal* voltage and current, and there are two basic ways of doing this. The first yields a result in terms of power and power factor. The second yields a result in terms of active power and reactive power.

Consider any linear network with a pair of input terminals as in Figure 24a. The input impedance at this pair of terminals is known to be $Z = Z/\underline{\varphi}$. Sinusoidal voltage is applied, and sinusoidal current flows into the network. Instantaneous current, voltage, and power are

$$i = I_m \cos \omega t \tag{24-1}$$

$$v = V_m \cos(\omega t + \varphi) \tag{24-2}$$

$$p = vi = V_m I_m[\cos(\omega t + \varphi)](\cos \omega t) \tag{24-3}$$

Figure 24b shows the waves of voltage and current, the phase angle φ being indicated. The figure must necessarily be drawn to show some specific

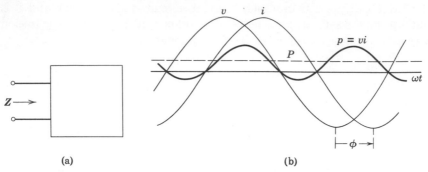

(a)   (b)

FIGURE 24 (a) Input impedance. (b) Instantaneous power p and average power P.

phase angle, so for purposes of illustration it is assumed that current lags voltage by something less than $\frac{1}{4}$ cycle. This implies an impedance of resistance and inductance. The discussion and the mathematics, however, are not limited to this particular kind of impedance; they are quite general.

The curve of instantaneous power p is the product of v and i. Current lags behind voltage in Figure 24b, so p is negative in part of each half-cycle. The average power P is shown. To find the amount of P, we work from equation 24-3. The trigonometric identity

$$\cos x \cos y = \tfrac{1}{2}[\cos(x - y) + \cos(x + y)] \tag{24-4}$$

is used to change equation 24-3 to

$$p = \frac{V_m I_m}{2}[\cos \varphi + \cos(2\omega t + \varphi)] \tag{24-5}$$

The average of equation 24-5 is the average power, P. The average of $\cos(2\omega t + \varphi)$ is zero, through 1 cycle or any integer number of cycles; this term, therefore, contributes nothing to the average of p. The remaining term of equation 24-5 is $\frac{1}{2}V_m I_m \cos \varphi$. The average power is therefore

$$P = \frac{V_m I_m}{2} \cos \varphi \tag{24-6}$$

As in other expressions for power, it is more convenient to use the effective rms values of voltage and current than the maximum values. Substituting $V_m = \sqrt{2}\,V_{\text{rms}}$ and $I_m = \sqrt{2}\,I_{\text{rms}}$,

$$P = V_{\text{rms}} I_{\text{rms}} \cos \varphi \tag{24-7}$$

This equation says that the power entering any network is the product of the effective values of terminal voltage and current and the cosine of the phase angle. It applies to sinusoidal voltage and current only.

The cosine of the phase angle is so often used that it is given a special name. For reasons that are obvious from equation 24-7 it is called the *power factor*. Thus, when current and voltage are sinusoidal, the power factor is defined as

$$F_p = \cos \varphi \qquad (24\text{-}8)$$

and†

$$P = V_{\text{rms}} I_{\text{rms}} F_p \qquad (24\text{-}9)$$

25. ACTIVE AND REACTIVE POWER

It is only when the impedance of a load is purely resistive that average power is as much as VI. Only with a resistive load is the current in the circuit fully engaged in conveying power from the generator to the load resistance. When reactance as well as resistance is present, a component of the current in the circuit is engaged in conveying the energy that is periodically stored in and discharged from the reactance. This stored energy, being shuttled to and from the magnetic field of an inductance or the electric field of a capacitance, adds to the current in the circuit but does not add to the *average* power.

From this point of view, the average power in a circuit is called *active* power, and the power that supplies energy storage in reactive elements (as in Figures 14b and 15a in Chapter 2) is called *reactive* power. Active power, as in equation 24-7, is

$$P = VI \cos \varphi \qquad (25\text{-}1)$$

and reactive power, designated Q, is

$$Q = VI \sin \varphi \qquad (25\text{-}2)$$

In both equations, V and I are *rms* values of terminal voltage and current, and φ is the phase angle by which current lags voltage.‡

The geometrical interpretation of these equations is helpful. In Figure 25a, the transforms V and I are shown. These both have rms magnitudes. To compute power, find the projection of I on V and multiply by V. Since the projection of I on V is $I \cos \varphi$, this method is obviously in accord with equation 25-1. Active power is thus considered to be the voltage times the in-phase or active component of current.

---

† More generally, with nonsinusoidal waves, for which equation 24-8 has no meaning, equation 24-9 is taken as the *definition* of power factor.
‡ The algebraic sign of reactive power, which was under debate for many years, seems now to have been settled by the "IEEE Standard Dictionary of Electrical and Electronic Terms."

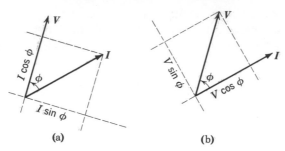

(a)   (b)

FIGURES 25a,b   Geometry of P and Q.

Similarly, *reactive* power is the voltage times the quadrature or reactive component of current. Q is the projection of I on a line normal to V, as in Figure 25a, multiplied by V.

Alternatively, the same V and I are shown in Figure 25b. Here the projection of V on I is found, giving $V \cos \varphi$. The projection of V on a line normal to I gives $V \sin \varphi$. These, when multiplied by I, give P and Q respectively. The final results are thus the same whether we find components of current in phase and in quadrature with voltage, or components of voltage in phase and in quadrature with current.

26. COMPLEX POWER

Consider, in Figure 25b, that V, $V \cos \varphi$, and $V \sin \varphi$ are each multiplied by I, the rms value of current. When the components of voltage are multiplied by current they become P and Q, respectively. We shall now define a new quantity which we shall call S, the *complex power*, of which P and Q are the components. P, Q, and S are shown in Figure 26a. P and Q are measured along real and imaginary axes, in what is called the complex power plane. Geometrically, Figure 26a is similar to Figure 25b rotated to make the current transform coincide with the axis of reals.

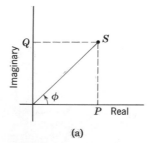

(a)

FIGURE 26a   Complex power.

By definition

$$S = P + jQ = VI \cos \varphi + jVI \sin \varphi$$
$$= VI(\cos \varphi + j \sin \varphi) = VIe^{j\varphi} = VI\underline{/\varphi} \qquad (26\text{-}1)$$

Here we have used Euler's formula to show that "complex power" is a complex quantity with magnitude equal to the product of terminal voltage and current (both rms) and with angle equal to the phase angle by which current lags voltage, as in Figure 26a.

The *magnitude* of S, equal to VI, is called *apparent power*. This magnitude is sometimes denoted U.

Equation 26-1 is not always the easiest form to use for computing S. When voltage and current are expressed as transforms† of the form $I = Ie^{j\theta}$ and $V = Ve^{j(\theta + \varphi)}$, the most convenient formula for S is

$$S = VI^* \qquad (26\text{-}2)$$

I^*, the conjugate of I, is $Ie^{-j\theta}$. To show that equation 26-2 is correct, according to the definition of S in equation 26-1, substitute for V and I^*:

$$S = VI^* = Ve^{j(\theta + \varphi)}Ie^{-j\theta} = VIe^{j\varphi} \qquad (26\text{-}3)$$

We shall now derive from equation 26-2 an expression for *power* that does not include either impedance or phase angle explicitly, an expression that is convenient for theoretical work. By equation 26-1,

$$S = P + jQ \qquad S^* = P - jQ$$

First adding, and then subtracting, these formulas,

$$S + S^* = 2P \qquad S - S^* = 2jQ$$

$$P = \frac{1}{2}(S + S^*) \qquad Q = \frac{1}{2j}(S - S^*)$$

Now equation 26-2 gives $S = VI^*$, and it follows that $S^* = V^*I$, so

$$P = \frac{1}{2}(VI^* + V^*I) \qquad Q = \frac{1}{2j}(VI^* - V^*I) \qquad (26\text{-}4)$$

(These voltage and current transforms have rms magnitudes.) The former of these expressions is often useful, the latter only rarely.

---

† It is important to notice that S is a complex number but not a transform. It does not represent a sinusoidally varying quantity, as do V and I. Z is another example of a complex quantity that does not represent a sine wave. Equation 26-5 shows a relation between S and Z.

Another useful form of equation 26-1 results when impedance is introduced. From the definition of impedance in equation 4-2, $Z = V/I = (V_{rms}/I_{rms})e^{j\varphi}$ (using the angle φ from Figure 26a). Then from equation 26-1:

$$S = VIe^{j\varphi} = I^2 \frac{V}{I} e^{j\varphi} = I^2 Z = I^2 R + jI^2 X \qquad (26\text{-}5)$$

Since $S = P + jQ$, we see that

$$P = I^2 R \quad \text{and} \quad Q = I^2 X \qquad (26\text{-}6)$$

This expression for P, of course, is already familiar, and we know it to be useful. The similar expression for Q will be useful in the same manner: that is, the reactive power Q to each reactive element can be computed from the current. *Notice particularly that I in equations 26-6 is the real rms value, not a complex number.*

Equation 26-6 tells us that, as a resistive load consumes *active* power, so an inductive load (with positive reactance) consumes *reactive* power. On the other hand, a capacitive load (having negative reactance) may be said to consume negative reactive power. If a line supplies two loads, one inductive and the other capacitive, the two loads together consume only the difference between their two reactive powers.

It may equally well be said that a capacitive load *produces* reactive power. Thus a capacitor, or an over-excited synchronous machine, will produce reactive power, while an inductive load, an induction motor, or an under-excited synchronous machine consume reactive power.

In any system there is conservation of reactive power just as there is conservation of active power. Whatever amount is consumed by one device must be produced by another.

The unit of power is the watt (kilowatt, kw; megawatt, Mw). Reactive power is measured in vars (kilovars, kvar; megavars, Mvar) from "*volt-amperes reactive.*" The unit of S and of its magnitude U is the volt-ampere (va; kilovolt-ampere, kva; megavolt-ampere, Mva). Of course, the watt and the var are dimensionally the same as the volt-ampere.

27. TRANSFORMATION OF OPERATIONS

The work of this chapter has been taken out of what is called the "time domain" by *transformation*. Basically the equation describing a branch that contains resistance, inductance, and capacitance in series is an integrodifferential equation:

$$v = Ri + L\frac{di}{dt} + \frac{1}{C}\int i \, dt \qquad (27\text{-}1)$$

But in a study that is specifically for sinusoidal quantities, the integrodifferential equation becomes merely algebraic, and in place of equation 27-1 we obtain (by a method due to Steinmetz) the much simpler

$$V = RI + j\omega LI + \frac{1}{j\omega C}I = ZI \qquad (27\text{-}2)$$

In equation 27-1, v and i are functions of time and are often written $v(t)$ and $i(t)$. This equation is in the *time domain*. In equation 27-2, their transforms are V and I. These are functions of frequency, often written $V(j\omega)$ and $I(j\omega)$, and this equation is in the *frequency domain*.

Transformation of the functions v and i is illustrated by these equations, and so also is transformation of *operations*. Following the methods of Section 3, these can be summarized in the following rule.

Write the differential equations for a network. Then write *transformed* equations that are similar except that V replaces v, I replaces i, $j\omega$ replaces d/dt, and $1/j\omega$ replaces $\int \cdots dt$.

In this chapter the frequency variable is $j\omega$. In the next chapter, s will be used, and equations will be written in terms of $V(s)$, $I(s)$, $Z(s)$, and $Y(s)$.

28. SUMMARY

The mathematics is easier if an alternating voltage or current is represented as the *real component* of an exponential function than it is if a trigonometric function is used.

A complex quantity such as $V_m = V_m e^{j\theta}$ gives, by its magnitude V_m, the amplitude of a wave of voltage and, by its angle θ, the phase angle of the wave. The complex quantity is called a *transform* and its graphical representation is a *phasor*. A similar current transform is I_m.

The magnitudes of V_m and I_m are the maximum values of the waves they represent. It is more convenient and much more common to use transforms with magnitudes equal to the *rms* rather than the maximum values of the waves. These are $V = (V_m/\sqrt{2})e^{j\theta}$, and a similar rms transform for current. Note that rms transforms and maximum transforms differ only by $\sqrt{2}$.

Impedance between the terminals of an element or a network is the ratio of the transform of voltage between terminals to the transform of the current entering at the terminals. Thus $Z = V_m/I_m$ or $Z = V/I$. Z, the impedance, is in general a complex quantity. Its reciprocal $Y = 1/Z = I_m/V_m = I/V$, is the *admittance*.

The impedance of elements or branches *in series* is the sum of their impedances. *Voltage divides* among elements in series in proportion to their impedances.

The admittance of elements *in parallel* is the sum of their admittances. *Current divides* among elements in parallel in proportion to their admittances.

Admittances and impedances at the terminals of more complicated *series-parallel* networks, and current and voltage distributions among their elements, are computed by repeated applications of the foregoing principles.

At any pair of terminals, G and B are the *components of* Y; R and X are the *components of* Z.

Phasor *diagrams*, with lines representing the complex transforms of currents and voltages, should be drawn as part of the solutions of all network problems. Visualization is helped and the chance of error is diminished. Examples are given.

The *ladder method* offers a convenient attack on certain common network problems.

Resonance provides a useful way to admit signals of one frequency while impeding others, or to block signals of one frequency while passing others. Resonance occurs at the frequency at which impedance (or admittance) of a reactive network is real. A low-loss (or high-Q) network can be *highly selective*.

The "*universal*" *resonance curve* is useful with good approximation for either series or parallel resonance. Q is defined.

Average *power* to a network can be computed in terms of rms voltage, rms current, and *power factor*, or in terms of *active and reactive power*. The latter are the components of *complex power*, which is related to impedance.

The introduction of transforms at the beginning of this chapter leads to the *transformation of operations*, and also to the concepts of *admittance* and *impedance*. This will be used continually in all the study of networks that is to come.

PROBLEMS

4-1. It is stated in the footnote to Section 3 that $\mathcal{R}e\,A + \mathcal{R}e\,B = \mathcal{R}e(A + B)$. Also $(\mathcal{R}e\,A)(\mathcal{R}e\,B) \neq \mathcal{R}e(AB)$. Prove these two statements from definitions of operations on complex quantities. §3

4-2. If we had chosen to express voltage as $v = V_m \sin(\omega t + \varphi)$ instead of using the cosine function as in equation 2-1, it would have been convenient to write $v = \mathcal{I}m\,V_m e^{j(\omega t + \varphi)}$ where $\mathcal{I}m$ means "the imaginary component of" as in $A = \mathcal{R}e\,A + j\,\mathcal{I}m\,A$. Using this expression for v instead of that of equation 2-1, find the impedance of a purely inductive element, and also find the impedance of a purely capacitive element. §4

4-3. A wave of voltage $v = 503 \cos(377t - 0.4\pi/180)$ is applied to a circuit consisting of $R = 30$ ohms, $L = 0.02$ henry, and $C = 100$ microfarads in series. Give the impedance Z of the circuit, the transforms V and I, and the wave of steady current i. §5*

4-4. The following R, L, and C are connected in series: $R = 8.00$ ohms, $L = 2.38$ millihenrys, $C = 14.14$ microfarads. (*a*) Find the impedance (magnitude and angle) at a frequency of 500 Hz. (*b*) Find the frequency at which the angle of impedance is zero. What is the magnitude of impedance at this frequency? §5

4-5. Voltage $v = 250 \cos(377t)$ is applied to a circuit of R and L in series. $R = 7.00$ ohms, $L = 63.7$ millihenrys. Find the transform of voltage, the complex impedance, the transform of current, and i. §5*

4-6. Repeat Problem 4-5, the circuit resistance being increased to 10.0 ohms. §5*

4-7. Four boxes, connected in parallel, have these individual impedances:

$$Z_1 = 0.80\underline{/0°} \qquad\qquad Z_3 = 0.685\underline{/-59.0°}$$
$$Z_2 = 0.894\underline{/-26.5°} \qquad Z_4 = 0.660\underline{/-9.46°}$$

(*a*) Find the admittances of each of the four boxes.
(*b*) Find the input *impedance* in polar form (magnitude and angle) to the network of the four boxes in parallel. §7

4-8. A coil with impedance $Z_1 = 75 + j38$ is connected in parallel with another coil with impedance $Z_2 = 40 + j17$. (*a*) Find the impedance of the two in parallel. (*b*) The total current through the coils in parallel is 10 amperes (rms), 60 Hz. Find the current in each coil. §8

4-9. A choke coil with 10 henrys inductance and negligible resistance is connected in parallel with a resistor with 2500 ohms resistance. Find the impedance (magnitude and angle) of the parallel combination at 60, 120, and 180 Hz. §9*

4-10. In the figure, $R = 600$ ohms, $f = 1000$ Hz. It is required that input impedance be $Z_{in} = 400 + jX_{in}$. What must C be? (X_{in} is not specified.) §9*

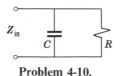

Problem 4-10.

4-11. As shown on page 98, inductance is added to the circuit of Problem 4-10, which otherwise remains the same. It is now required that the input impedance be $Z_{in} = 400 + j0$. What must L be?
  Note: This is a practical circuit for impedance matching, to make a 600-ohm load look like 400 ohms at the terminals. §10*

4-12. If 4.0 volts, 1000 Hz, is applied to the circuit of Problem 4-11, find the current in each element. Find the voltage across each element. Draw a phasor diagram. §10

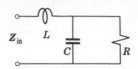

Problems 4-11 through 4-14.

4-13. The circuit determined in Problem 4-11 is operated at frequencies ranging from 800 to 1200 Hz. Compute input impedance through this frequency range; plot real and imaginary components of impedance (two curves). §10*

4-14. If, in Problem 4-11, $R = 600$ ohms and the input impedance is required to be $Z_{in} = 900 + j0$, can the circuit of Problem 4-11 be used for the purpose? If not, devise some other circuit that can be, using only reactive elements. §10*

4-15. In the R-C coupling circuit shown, what is the voltage (magnitude and angle) across the 200,000-ohm resistor? (*a*) Frequency is 100 Hz. (*b*) Frequency is 1000Hz. §8

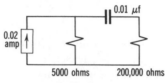

Problem 4-15.

4-16. A circuit diagram gives resistances and reactances in ohms. Find the input impedance. Draw a phasor diagram showing V, I, I_1, and I_2.

   Note: The circuit diagram for this problem has been lost, but fortunately the answers are known. Therefore the problem for you is to reproduce the missing circuit diagram and also draw the phasor diagram. Answers: $Z = 50.0\underline{/20.0°}$, $V = 100\underline{/0°}$, $I = 2.00\underline{/-20.0°}$, $I_1 = 2.00\underline{/-36.9°}$, $I_2 = 0.588\underline{/61.9°}$. §10

4-17. The impedance between terminals of the primary winding of a transformer has the components $R = 3$ ohms, $X = 400$ ohms. Find G and B. §13

4-18. A coil has 100 ohms resistance and 0.050 henry inductance. Find conductance and susceptance, the real and imaginary parts of admittance to this coil, at 500, 1000, and 1500 Hz. Plot G and B and the ratio B/G as functions of frequency. §13*

4-19. Twelve resistors, 1 ohm each, are arranged along the edges of a cube with junctions at the corners. What resistance will be measured between terminals

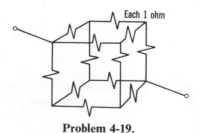

Problem 4-19.

at symmetrically opposite corners of the cube? Use any method or tricks that
occur to you. §10*

4-20. An independent-voltage source operates at a frequency of 400 Hz. The source
has internal resistance R and reactance X as shown. Measurements with a
variable-resistance load give the results shown in the diagram. Find V, R, and
X of the source. §10

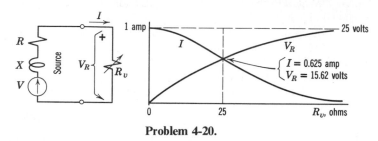

Problem 4-20.

4-21. A resistance network is shown with resistances given in ohms. Find current in
the 8-ohm resistor owing to the 50-volt source. §15*

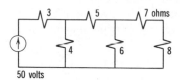

Problem 4-21.

4-22. The figures given in the diagram are ohms of reactance. Resistance is negligible.
What is the reading of the ammeter A? §15*

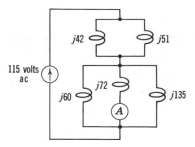

Problem 4-22.

4-23. A coil having an inductance of 0.05 henry and a resistance of 8.0 ohms is in
series with a 1.00-microfarad capacitor. Find the frequency at which
$Z_{in} = R_{in} + j0$ (resonant frequency), and find R_{in} at this frequency. Find also
Y_{in}. §16*

4-24. In the circuit of Problem 4-23, find (a) Z_{in} at a frequency 10% greater than resonant frequency, (b) R_{in} at this higher frequency, (c) Y_{in}, and (d) G_{in} at this frequency.   §18*

4-25. Repeat Problem 4-24 at a frequency 10% less than resonant frequency.   §18

4-26. For the circuit of Problem 4-23, find two frequencies at which $R_{in} = |X_{in}|$. Find Z_{in} and Y_{in} at these frequencies.   §18*

4-27. For the circuit of Problem 4-23, plot $|Y|$ as a function of frequency from 600 to 800 Hz.   §21

4-28. The coil of Problem 4-23, having $L = 0.05$ henry and $R = 8.0$ ohms, is now connected in *parallel* with the 1.00 microfarad capacitor. Find the frequency for which the input impedance to this parallel combination is $Z_{in} = R_{in} + j0$ (resonant frequency), and find R_{in} at this frequency.   §23

4-29. For the circuit of Problem 4-28, plot $|Z|$ as a function of frequency from 600 to 800 Hz.   §23*

4-30. It is required that a series circuit of L, C, and R be resonant at 1 MHz. Its bandwidth (between half-power points) is to be 5000 Hz, and its input impedance at resonance is to be 50 ohms. Find L, C, and R (at 1 MHz).   §21*

4-31. It is required that a series circuit of L, C, and R be resonant at 1.5 MHz. Its bandwidth (between half-power points) is to be 5000 Hz, and C is to be 10 picofarads (10^{-12} farad). Find L and R.   §21*

4-32. Find the input impedance to the two parallel branches shown in the figure, and determine what is necessary to make the input impedance equal $\sqrt{L/C}$.   §23*

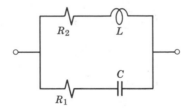

Problem 4-32.

4-33. Between terminals a and b there are two branches in parallel. Branch 1 consists of capacitance C_1 and inductance L_1 in series (with extremely small resistance). Branch 2 consists of capacitance C_2 and inductance L_2 in series (with extremely small resistance), but $L_1 C_1 \neq L_2 C_2$. With both branches connected from a to b, the impedance between a and b is called Z. Find the three resonant frequencies of Z.   §23

4-34. Current to a loud-speaker is 34.0 milliamperes, and terminal voltage is 9.8 volts. An oscilloscope shows the angle between voltage and current to be 42 degrees. Find power input and effective impedance. Draw a phasor diagram.   §24

4-35. A single-phase induction motor running at 245 volts, 60 Hz, 0.80 power factor, receives 1.75 kw active power. A capacitor is to be connected in parallel to provide that the two together will draw 0.50 kva reactive power. Find the necessary capacitance. Draw a phasor diagram.   §25*

4-36. When 2300 volts rms sinusoidal voltage, frequency 60 Hz, is applied to the two terminals of a box, the current is 5 amperes and the power factor is 0.20. It is known that the box contains a capacitor. Power consumption in the box may result because the capacitor has:

(a) A shunt conductance, G being independent of frequency.

(b) A series resistance, R being independent of frequency.

(c) Poor dielectric, having 0.2 power factor at all frequencies.

Find, for each of these three possibilities, what the current and power factor would be if the magnitude of applied voltage stayed the same but the frequency was increased to 180 Hz.   §25*

4-37. Find the power P being delivered from one of these generators to the other. Tell which is producing and which is absorbing electrical power (and how you know). Draw a phasor diagram showing V_1, V_2, and I.   §25*

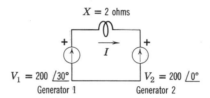

$$X = 2 \text{ ohms}$$

$$V_1 = 200\ \underline{/30°} \qquad V_2 = 200\ \underline{/0°}$$

Generator 1   Generator 2

Problem 4-37.

4-38. In the network shown, find I_1, P_1, Q_1, and S_1 for branch 1; I_2, P_2, Q_2, and S_2 for branch 2; and I, P, Q, and S delivered by the generator.   §26*

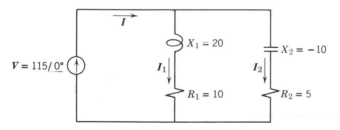

$V = 115\underline{/0°}$ $X_1 = 20$ $X_2 = -10$

I_1 I_2

$R_1 = 10$ $R_2 = 5$

Problem 4-38.

CHAPTER
5

NETWORK FUNCTIONS AND THE s PLANE

1. AN EXPONENTIAL CURRENT

The admittance Y, the impedance Z, and the transfer function or transmittance T were defined in the previous chapter as the ratios of transforms. We write, for instance, $Y = I/V$, or if we wish to emphasize that an admittance is a function of frequency we can write $Y(j\omega) = I(j\omega)/V(j\omega)$. An admittance is a complex quantity (except that of a purely resistive network), for the transforms $I(j\omega)$ and $V(j\omega)$ are themselves complex.†

This definition of Chapter 4 is for use with alternating currents and voltages of sinusoidal form. Another interesting and important form of current and voltage is the exponential. An obvious use of the real exponential is with *natural* components of current or voltage.

A mathematical expression for exponentially varying current can be written

$$i = I_0 e^{st} \qquad (1\text{-}1)$$

---

† Here and hereafter, ordinary italic type (not boldface) will be used for real and complex quantities alike. After all, a real number is merely a special case of a complex number, and it has often been difficult in preceding chapters to know which type to use for a quantity that might sometimes be real and sometimes be complex.

No doubt boldface type is helpful when complex quantities are unfamiliar, but we have now advanced far enough to recognize that transforms of voltage and current are in general complex, and so are admittances, impedances, and other network functions. We shall hereafter write Z rather than $\mathbf{Z}$ for impedance. It will, however, be all the more important to write $|Z|$ for the absolute magnitude when that is the quantity needed.

where I_0 is the current at zero time, and s is a constant. If s is a positive real number, the current increases as in Figure 1a; if s is negative real, the current diminishes as in Figure 1b. There is no discontinuity in this exponential current; it extends through all time. In Figure 1a, for instance, the current began long ago with infinitesimal size, increased gradually until now, and

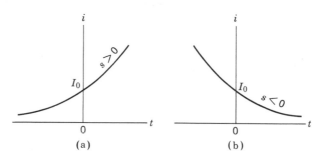

FIGURES 1a,b $i = I_0 e^{st}$ when **(a)** s is positive real and **(b)** s is negative real.

will continue to increase into the future. No doubt at some time to come there will be a catastrophe, such as a fuse blowing, but with this we are not concerned.

2. EXPONENTIAL IMPEDANCE AND ADMITTANCE

Let the exponential current of equation 1-1 flow in a purely *resistive* element; the voltage across the element is then

$$v_R = Ri = RI_0 e^{st} \tag{2-1}$$

If a purely *inductive* element carries the same current its terminal voltage is

$$v_L = L\frac{di}{dt} = sLI_0 e^{st} \tag{2-2}$$

If an exponential voltage $v_C = V_0 e^{st}$ is applied to a *capacitive* element, the entering current is

$$i = C\frac{dv_C}{dt} = sCV_0 e^{st} \tag{2-3}$$

In Section 7 of Chapter 2 a theorem was given regarding alternating current in linear networks. A similar statement for exponential current can now be based on the foregoing equations. *If the current or voltage at any part of any linear network is exponential, the currents and voltages at every part of the network are exponential with the same exponent.* This statement assumes linear elements and steady-state voltages and currents. It is based on the fact that the derivative of an exponential function is itself an exponential function with the same exponent, as in equations 2-2 and 2-3, and that the sum of exponentials with the same exponent is also an exponential.

We find that the admittance and impedance of elements as derived for alternating currents in Chapter 4 appear again in our present work with exponential currents, and almost without change. Thus the impedance of a purely resistive element, defined as the ratio of exponential voltage across the element (equation 2-1) to current through it (equation 1-1) is

$$Z(s) = \frac{v_R}{i} = \frac{RI_0 e^{st}}{I_0 e^{st}} = R \tag{2-4}$$

The impedance of an inductive element, defined in the same way, is, by equation 2-2,

$$Z(s) = \frac{v_L}{i} = \frac{sLI_0 e^{st}}{I_0 e^{st}} = sL \tag{2-5}$$

For a capacitive element, admittance is defined as the ratio of the current of equation 2-3 to the voltage across the element:

$$Y(s) = \frac{i}{v_C} = \frac{sCV_0 e^{st}}{V_0 e^{st}} = sC \tag{2-6}$$

Thus $Z(s)$ and $Y(s)$ for exponential currents are the same functions as $Z(j\omega)$ and $Y(j\omega)$ for alternating currents in the same elements, but with the variable s instead of $j\omega$.

When the elements are combined into networks of branches and loops, the same rules of combination are followed as for impedances and admittances of alternating voltages and currents, and for the same reasons. Thus the impedance of elements in series is the sum of the impedances of the elements, and the admittance of branches in parallel is the sum of the admittances of the branches. It follows that for any network the functions $Y(s)$ and $Z(s)$ are the same as $Y(j\omega)$ and $Z(j\omega)$ with only the variable changed. (See also Section 9.)

In view of the relations of this section, an admittance function, an impedance function, or a transfer function of a network—inclusively, a network function—can be defined as the ratio of *currents and voltages* if they are of exponential form, or alternatively as the ratio of the *transforms* of currents and voltages if they are of sinusoidal form.

3. AN EXAMPLE OF IMPEDANCE AND ADMITTANCE

As an example, Figure 3a shows a circuit of resistance and inductance in series. Its impedance $Z(s)$ is

$$Z(s) = R + sL = L\left(s + \frac{1}{\tau}\right) \tag{3-1}$$

where $\tau = L/R$ by definition. This impedance is plotted in Figure 3b.
   Let the voltage applied to the terminals of the circuit be

$$v = V_0 e^{st} \tag{3-2}$$

(Note that this voltage is applied through all time, from the distant past to the distant future, however hard to visualize this may be, and that the resulting current exists similarly through all time.) The current is then:

$$i = \frac{v}{Z} = \frac{V_0 e^{st}}{R + sL} = \frac{1}{L} \frac{1}{s + 1/\tau} V_0 e^{st} \tag{3-3}$$

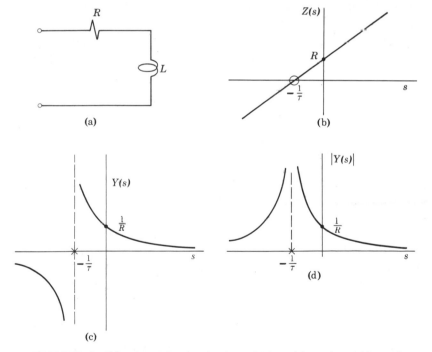

(a)

(b)

(c)

(d)

FIGURE 3 (a) An RL circuit for which $Z(s) = L(s + 1/\tau)$, and (b) $Z(s)$ plotted for real s. (c) $Y(s)$ for the circuit of Figure 3a, and (d) its magnitude $|Y(s)|$.

This is an exponential current with a coefficient that is easy to compute from the circuit parameters and s and V_0 of the voltage.

Figure 3b can be interpreted as showing the magnitude of voltage necessary to produce current of unit size in the circuit, when the exponent s for the applied voltage and also for the resulting current has the value shown on the horizontal axis. If s is zero the voltage and current are unchanging with time, they are constant, and the circuit impedance is just the ordinary dc resistance R; the impedance shown on the vertical axis where $s = 0$ is therefore R.

If s is positive, so that current and voltage have the shape of Figure 1a, the voltage required to produce unit current is relatively greater and the impedance of the circuit is correspondingly greater than R. This is because voltage is needed to keep current rising in the inductance as well as to push current through the resistance of the circuit.

If s is slightly negative the current is slowly decreasing (Figure 1b). This negative rate of change of current produces a negative voltage across the inductance which helps the current to flow through the resistance, and relatively less external voltage is required to keep current going; impedance is less than R.

But there is a critical point at which the voltage produced in inductance by decreasing current is just enough to keep the current flowing through the resistance of the circuit. This critical rate of decay occurs when $s = -1/\tau$. It is called a *zero* of impedance, for the impedance of the circuit to this particular shape of current is zero; no external voltage is required to keep this *natural* current going. This natural value of s is marked, as in Figure 3b, with a small circle. When the current and the voltages have this natural exponential form, energy is delivered from the inductance just as rapidly as it is consumed in the resistance of the circuit. The quantity τ, which in this circuit is equal to L/R, is called the *time constant* of the circuit.

For still more negative values of s the impedance becomes negative. This merely means that when s has so negative a value the current has to decay more rapidly than it naturally wants to, and the applied voltage must be reversed so that it helps the current diminish rapidly.

The *admittance* of a circuit such as that of Figure 3a is perhaps even more commonly used than the impedance. Admittance, the reciprocal of impedance, is

$$Y(s) = \frac{1}{L} \frac{1}{(s + 1/\tau)} \tag{3-4}$$

and $Y(s)$ is plotted in Figure 3c. This curve may be interpreted as showing the value of current I_0 when unit voltage ($V_0 = 1$) is applied at the terminals in Figure 3a; voltage and current are of exponential form, continuous through all time, with exponent s. When s is large and positive the applied voltage increases steeply and the resulting current is rather small. Current is larger if s is small and positive, and yet larger if s is small and negative, and I_0 increases without limit as the critical value of $s = -1/\tau$ is approached. At this critical value there is said to be a *pole* of admittance, and the value of s is marked on the curve with a small cross.

At a pole of admittance, where $s = -1/\tau$, a finite applied voltage would produce an infinite current in response. This is hardly a physical possibility, but it is approached as s approaches a pole.

As s becomes large and negative the admittance becomes small and negative. For reasons that will become evident when complex values of s are considered it is more useful to plot only the magnitude of Y, omitting algebraic sign. This is done in Figure 3d.

4. A TRANSFER FUNCTION

A *transfer function* or *transmittance* is the ratio of a forced response at one place in a network to a driving force (of exponential form) at another place. *Driving-point* impedance or admittance is the ratio of voltage and current at the *same* terminals, whereas a *transfer* function is the ratio of voltages or currents at *different* places in the network.

The transfer function may be the ratio of a voltage to a voltage, or a current to a current, or a voltage to a current or a current to a voltage. As an example, let us find the transfer function that relates voltage across the inductance of Figure 4a to exponential voltage applied at the source. Voltage applied is

$$v = v_0 e^{st} \tag{4-1}$$

Current is known from equation 3-3 to be

$$i = \frac{1}{L} \frac{1}{(s + 1/\tau)} v_0 e^{st} \tag{4-2}$$

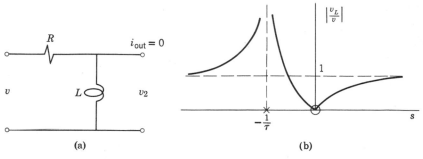

(a)   (b)

FIGURES 4a, b   A circuit and its transfer function.

where $\tau = L/R$. Voltage across the inductance is then, by differentation,

$$v_L = L\frac{di}{dt} = \frac{s}{s + 1/\tau}v_0 e^{st} \tag{4-3}$$

This voltage is now divided by the applied voltage, equation 4-1, to get the desired *transfer function* or *transmittance*:

$$T(s) = \frac{v_L}{v} = \frac{s}{s + 1/\tau} \tag{4-4}$$

The magnitude of this transfer function is plotted in Figure 4b. It has a pole at $s = -1/\tau$. For values of s near this pole the voltage across the inductance is very great because the current is great.

Also the transmittance has a zero at the origin. This means that if the applied voltage is dc, a constant with time, as it is when $s = 0$, there is no voltage across the inductance—an obvious conclusion. Finally, the transmittance approaches 1 for very large s, for then the current is negligibly small and all the applied voltage appears across the inductance.

5. IMPEDANCE AND ADMITTANCE WITH R AND C

In Section 3 we plotted impedance and admittance for a circuit containing resistance and inductance in series. Let us now do the same for a circuit of resistance and capacitance as in Figure 5a. The impedance is easily written

$$Z(s) = R + \frac{1}{sC} = R\frac{(s + 1/RC)}{s} = R\frac{(s + 1/\tau)}{s} \tag{5-1}$$

Here we set $\tau = RC$, this being the time constant of the R and C circuit.

This impedance is plotted as a function of s in Figure 5b. It has a pole at the origin, for the impedance of the capacitor to direct current is infinite. There is a zero of impedance at $s = -1/\tau$, the value of s for which current can continue without voltage. If s is very great, either positive or negative, the capacitance takes negligible voltage and the impedance approaches R, the resistance.

The reciprocal is admittance:

$$Y(s) = \frac{s}{R(s + 1/\tau)} \tag{5-2}$$

Where Z has a zero, Y has a pole, and vice versa. A curve of Y is plotted in Figure 5c, and its absolute magnitude $|Y|$ is Figure 5d.

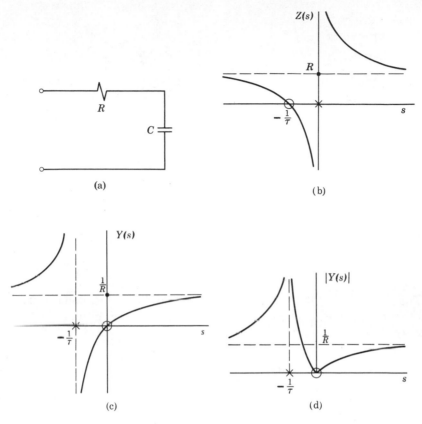

FIGURE 5 **(a)** An RC circuit for which $Z(s) = R(s + 1/\tau)/s$, and **(b)** $Z(s)$ plotted for real s. **(c)** $Y(s)$ for the circuit of Figure 5a, and **(d)** its magnitude $|Y(s)|$.

6. A CIRCUIT WITH R, L, AND C

When a circuit has both inductance and capacitance as well as resistance, as in Figure 6a, the impedance and admittance can have new facets of interest. We write impedance:

$$Z(s) = sL + R + \frac{1}{sC} \tag{6-1}$$

and admittance is the reciprocal:

$$Y(s) = \frac{s}{L[s^2 + (R/L)s + 1/LC]} \tag{6-2}$$

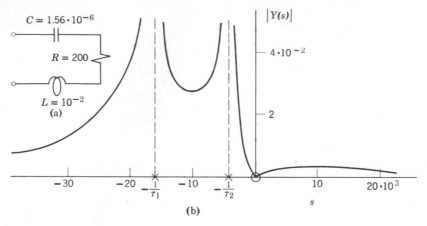

FIGURES 6a, b An RLC circuit for which $Y(s) = \dfrac{s}{L(s + 1/\tau_1)(s + 1/\tau_2)}.$

The admittance of this "double energy" circuit has a denominator that is quadratic in s. This denominator can be expressed as a product of linear factors:

$$Y(s) = \frac{s}{L(s - s_1)(s - s_2)} \tag{6-3}$$

where

$$s_1, s_2 = -\frac{R}{2L} \pm \sqrt{\left(\frac{R}{2L}\right)^2 - \frac{1}{LC}} \tag{6-4}$$

This expression in linear factors is possible for all values of R, L, and C, but s_1 and s_2 are real (and different) only if

$$\left(\frac{R}{2L}\right)^2 > \frac{1}{LC} \quad \text{or} \quad R^2 > 4\frac{L}{C}$$

Let us consider first the situation when s_1 and s_2 are real. This is called the *over-damped* case, for it results from a high value of R compared to L/C. Plotting the magnitude of admittance as a function of s gives, in Figure 6b, a curve with one zero and two poles. The zero at the origin results from the s in the numerator of equation 6-3, and it means that a voltage unchanging with time, for which $s = 0$, will produce no current in the circuit. The poles result from the linear factors in the denominator; the circuit has infinite admittance to current of the form $I_1 e^{s_1 t}$ and to current of the form $I_2 e^{s_2 t}$, or any sum of these. That is, the *natural* current that can flow, once started, without driving voltage is

$$i_n = I_1 e^{s_1 t} + I_2 e^{s_2 t} \tag{6-5}$$

The values of I_1 and I_2 can be any real numbers; they will be found from known conditions in the circuit in Chapter 6.

The time constants of the two exponential components are seen to be

$$\tau_1 = -\frac{1}{s_1} \quad \text{and} \quad \tau_2 = -\frac{1}{s_2} \tag{6-6}$$

Since the time constant of an exponential function is easy to visualize, it is convenient to put equation 6-3 in the form

$$Y(s) = \frac{s}{L(s + 1/\tau_1)(s + 1/\tau_2)} \tag{6-7}$$

Values of the time constants are obtained from equation 6-4. It is evident from this equation that if s_1 and s_2 are real they will be negative quantities $(R, L, \text{and } C \text{ being positive})$; the time constants τ_1 and τ_2 are correspondingly positive.

The physical fact deduced from these poles of admittance is that we may short-circuit together the terminals of the circuit of Figure 6a and if we then start current flowing (perhaps by allowing the capacitance to discharge), the resulting natural current will be composed of either $I_1 e^{-t/\tau_1}$ or $I_2 e^{-t/\tau_2}$, or both. These natural responses will be considered further in Chapter 6; at present our primary interest is to find the admittance of the circuit as a function of s.

Example. The two poles of admittance shown in Figure 6b can be located for specific values of R, L, and C. Given

$$R = 200 \text{ ohms} \qquad L = 10 \text{ millihenrys} \qquad C = 1.56 \text{ microfarads}$$

Then:

$$\frac{R}{2L} = \frac{200}{2 \cdot 10^{-2}} = 10^4, \qquad \frac{1}{LC} = \frac{1}{10^{-2}(1.56)10^{-6}} = 64 \cdot 10^6$$

$$\sqrt{\frac{R^2}{4L^2} - \frac{1}{LC}} = \sqrt{10^8 - 64 \cdot 10^6} = 6 \cdot 10^3 \tag{6-8}$$

$$\frac{1}{\tau_1} = 10^4 + 6 \cdot 10^3 = 16 \cdot 10^3, \qquad \tau_1 = 6.25 \cdot 10^{-5} \tag{6-9}$$

$$\frac{1}{\tau_2} = 10^4 - 6 \cdot 10^3 = 4 \cdot 10^3, \qquad \tau_2 = 25 \cdot 10^{-5} \tag{6-10}$$

Thus we have found the poles and the time constants for the given values of R, L, and C. Next, let us consider where the poles would be located if the damping were different. Suppose R to be much greater, L and C remaining the same. Equation 6-4 shows that one of the poles would then approach the origin while the other would go far to the left, toward $-\infty$; the curve of admittance would be altered correspondingly, with a zero remaining at the origin.

As the other extreme, suppose that R is decreased until $(R/2L)^2 = 1/LC$ and the radical of equation 6-4 becomes zero. This is called the critical value of resistance; it results in *critical damping*, the two factors in the denominator of equation 6-3 become identical, as do the time constants in equation 6-7, and admittance takes the form

$$Y(s) = \frac{s}{L(s + 1/\tau)^2} \tag{6-11}$$

As resistance is decreased toward the critical value, the two poles of admittance (Figure 6b) come together in a single *second-order* pole (Figure 6c). By equation 6-4 this pole is at $s = -R/2L$, and therefore, since R has its critical value,

$$s = -\frac{1}{\sqrt{LC}} \tag{6-12}$$

As an example, let us keep L and C unchanged in Figure 6a but reduce R to its critical value. This gives

$$\left(\frac{R}{2L}\right)^2 = \frac{1}{LC} \quad \text{or} \quad R = 2\sqrt{\frac{L}{C}} = 2\sqrt{\frac{10^{-2}}{1.56 \times 10^{-6}}} = 160 \text{ ohms} \tag{6-13}$$

as the critical value of resistance. The second-order pole is at

$$s = -\frac{1}{\sqrt{LC}} = \frac{1}{[(10^{-2})(1.56 \times 10^{-6})]^{1/2}} = -8000 \tag{6-14}$$

as shown in Figure 6c.

This critically damped case is mathematically interesting, but it is far more important to consider what happens if the circuit resistance is still further reduced to a value below that of critical damping. What then?

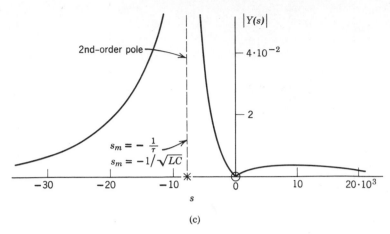

FIGURE 6c $Y(s) = s/L(s + 1/\tau)^2$, **a critically damped circuit.**

7. COMPLEX POLES

If, in equation 6-4, R is so small that $(R/2L)^2$ is less than $1/LC$, the two roots are complex rather than real and can most conveniently be written

$$s_1, s_2 = -\frac{R}{2L} \pm j\sqrt{\frac{1}{LC} - \left(\frac{R}{2L}\right)^2} \tag{7-1}$$

These are the poles, the values of s at which the admittance function is infinite. The two poles are at conjugate complex values of s. The real component of each is $-R/2L$.

In previous diagrams such as Figures 6b and c, poles are shown only at real values of s. To show the poles of equation 7-1 it is necessary for a diagram to include complex values of s. The usual and convenient way is to let the plane of the paper be the s *plane*, measuring real components of s along the horizontal axis, marked σ in Figure 7a, and imaginary components of s along the vertical axis, marked ω, so that all complex values of s find a place in the plane. Poles and zeros of a function—an admittance function, a transfer function, or in general a network function—are then marked with crosses or little circles. Thus, the two real poles and the zero of equation 6-3, which is

$$Y(s) = \frac{s}{L(s - s_1)(s - s_2)} \tag{7-2}$$

are marked on the σ axis of the s plane of Figure 7a.

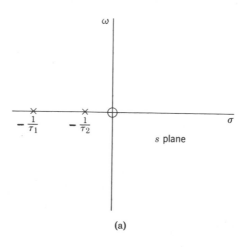

(a)

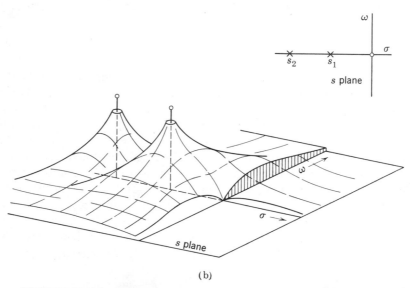

(b)

FIGURE 7 (a) **Real poles; the overdamped case.** (b) **Showing** $|Y(s)|$ **with real poles.**

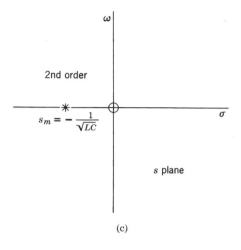

(c)

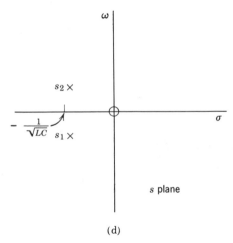

(d)

FIGURE 7 (c) Critical damping. (d) Complex poles.

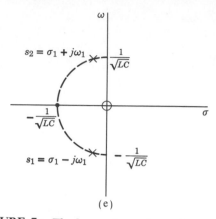

(e)

FIGURE 7e   The locus of complex poles is circular.

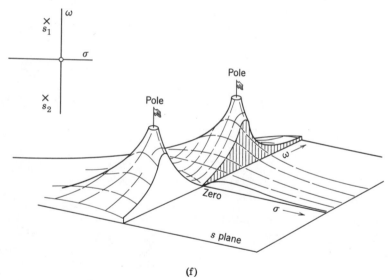

(f)

FIGURE 7f   Showing $|Y(s)|$ with complex poles.

As a matter of notation, s is a complex variable; we call its components σ and ω and write

$$s = \sigma + j\omega \qquad (7\text{-}3)$$

A specific value of the variable s is the constant s_1 with specific components σ_1 and ω_1, so that

$$s_1 = \sigma_1 + j\omega_1 \qquad (7\text{-}4)$$

We shall see that σ, the real component, has to do with the rate of increase or decrease of a function, and that ω is the radian frequency of an oscillation.

Although there is a point in the s plane of Figure 7a for each complex value of s, so that the location of every pole and zero of a function can be marked, the value of the *function* is nowhere shown. $Y(s)$ of equation 7-2 has a value for every s, but these are not shown in Figure 7a. The values of $|Y(s)|$ for *real* s are plotted vertically in Figure 6b, but when the two dimensions of a sheet of paper are used to represent the real and imaginary components of s, as in Figure 7a, we need a third dimension to show values of $Y(s)$.

An obvious extension of Figure 6b would suggest that $|Y(s)|$ for all complex values of s might be plotted as a *surface* above the s plane, in a three-dimensional diagram as sketched in Figure 7b. Perhaps it is more satisfactory to visualize the surface from its poles and zeros than it is to undertake a three-dimensional diagram.

We found that if damping was reduced in the series circuit of R, L, and C the two real poles drew together and met in a second-order pole at $s = -1/\sqrt{LC}$. Figure 7c shows the second-order pole on the real axis. If damping is further decreased the poles move away from the real axis, one upward and one downward, to positions computed from equation 7-1 and shown in Figure 7d. If R is reduced still more (keeping L and C unchanged) the poles of Y approach the imaginary axis as in Figure 7e, following a semicircular locus, and if it were possible for R to become zero the poles would reach the ω axis at values of $s = \pm j1/\sqrt{LC}$.

Figure 7f is a three-dimensional sketch of $|Y(s)|$ when the poles are at complex values of s rather close, as in Figure 7e, to the ω axis.

8. OSCILLATORY NATURAL CURRENT

The previous section shows that a series RLC circuit with low damping has two poles of admittance at complex values of s. The corresponding forms of natural current are

$$i_1 = I_1 e^{s_1 t} = I_1 e^{(\sigma_1 + j\omega_1)t} \tag{8-1}$$

and

$$i_2 = I_2 e^{s_2 t} = I_2 e^{(\sigma_1 - j\omega_1)t} \tag{8-2}$$

These values for i_1 and i_2 are complex, and in the absence of any special convention a current must be real. However, it is shown in Section 10 of Chapter 3 that though neither i_1 nor i_2 has any physical meaning by itself, the sum of the two can be real. For instance, if we let I_1 and I_2 both equal a new constant $\frac{1}{2}I_0$ and add the two natural components, then

$$
\begin{aligned}
i &= \tfrac{1}{2}I_0[e^{(\sigma_1 + j\omega_1)t} + e^{(\sigma_1 - j\omega_1)t}] \\
&= \tfrac{1}{2}I_0 e^{\sigma_1 t}(e^{j\omega_1 t} + e^{-j\omega_1 t}) \\
&= I_0 e^{\sigma_1 t} \cos \omega_1 t
\end{aligned}
\tag{8-3}
$$

Here we have a *real* current; it is *oscillatory*, of damped sinusoidal form, and we shall see in the next chapter that it can be the form of natural current in the *RLC* circuit when R is less than critical. Another possibility is for the coefficients I_1 and I_2 to be opposite and imaginary rather than equal and real, and equation 8-3 becomes a damped sine instead of a damped cosine:

$$
i = I_0 e^{\sigma_1 t} \sin \omega_1 t
\tag{8-4}
$$

In any oscillatory current, the *natural frequency* of oscillation is

$$
\omega_1 = \sqrt{\frac{1}{LC} - \left(\frac{R}{2L}\right)^2}
\tag{8-5}
$$

If R could be reduced to zero, leaving only L and C in the circuit, the extreme of the oscillatory condition would be reached. Current once started in such a lossless circuit would continue to oscillate forever. The oscillatory current of equation 8-3 or 8-4 would be undamped, for σ_1 would be zero (by equation 7-1), and the *undamped* natural frequency ω_0 would be the limit of the damped natural frequency as R approaches zero:

$$
\omega_0 = \sqrt{\frac{1}{LC}}
\tag{8-6}
$$

Equation 6-2 is written specifically for the *RLC* series circuit. However, there are many other networks that give oscillatory responses, and we may now generalize from the foregoing discussion. The *RLC* series circuit can be considered to be a special case of a class, as equation 6-2 is a special case of the more general network function:

$$
F(s) = H \frac{s}{s^2 + 2\zeta\omega_0 s + \omega_0{}^2}
\tag{8-7}
$$

The constants H, ζ, and ω_0 are determined by the parameters of the network (equation 6-2 is an example). The equation is written with these symbols because ζ and ω_0 have interesting physical meanings. These meanings appear when the equation is rewritten to show the poles of $F(s)$:

$$
F(s) = H \frac{s}{(s - s_1)(s - s_2)}
\tag{8-8}
$$

where

$$s_1, s_2 = -\zeta\omega_0 \pm j\omega_0\sqrt{1 - \zeta^2} \qquad (8\text{-}9)$$

Such poles are marked in Figure 8a, the real and imaginary components being:

$$\sigma_1 = -\zeta\omega_0 \qquad \omega_1 = \omega_0\sqrt{1 - \zeta^2} \qquad (8\text{-}10)$$

The currents (or other functions of time) corresponding to this $F(s)$ are exponentials with complex exponents that combine into an oscillation. Equation 8-4 and Figure 8b give the example:

$$i = I_0 e^{\sigma_1 t} \sin \omega_1 t \qquad (8\text{-}11)$$

We are now able to deduce the following general relations; examples of these general relations are seen in the specific case of the RLC series circuit:

1. Given an equation of the form of 8-7, the undamped natural frequency ω_0 is the square root of the last term of the denominator.
2. Having found ω_0, ζ is easily computed from the middle term. Then $\sigma_1 = -\zeta\omega_0$.
3. The natural frequency (including damping) is found from:

$$\omega_1{}^2 = \omega_0{}^2 - \sigma_1{}^2 \qquad (8\text{-}12)$$

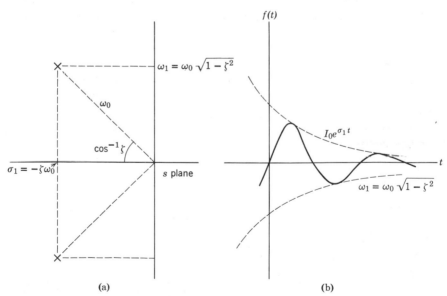

FIGURE 8 (a) Poles of $F(s)$ with the denominator $s^2 + 2\zeta\omega_0 s + \omega_0{}^2$. **(b)** The damped oscillation.

4. On the s plane (Figure 8a), the abscissa of the poles is σ_1 and the ordinate of the upper pole is ω_1.
5. The radial distance of each pole from the origin is ω_0 as seen from equation 8-12.
6. The cosine of the angle to the radius vector (shown in the diagram) is the damping ratio ζ, which determines the number of cycles in which the oscillation dies away. (ζ is the ratio of *actual* damping to *critical* damping in a system; it is widely used in the study of control systems.)

$\llcorner\rightarrow$ <1 underdamped

9. FORCED CURRENT >1 overdamped

As we have seen, special interest attaches to those singular points in the s plane called poles, those values of s at which the network function is infinite and there can be output without input. These poles determine the *natural* response of the network. At every other point in the s plane the ratio of output to input is finite and $F(s)$ has a specific value.

Let us consider, for a moment, that the network function is admittance. This ratio of current to voltage is not often useful for exponential forms such as $I_0 e^{st}$ when s is complex, or even when s is real, but when s is purely imaginary, so that the component of current is expressed by $I_0 e^{j\omega t}$, the current is an alternating, undamped sinusoid. For such values of s, for points that lie along the ω axis of the s plane, the admittances (or other network function values) are those for ordinary steady alternating currents. Thus in Figures 7b and 7f the curves of ordinary ac admittance (as functions of frequency) are emphasized by vertical shading.

Admittance, impedance, or any transfer function can be computed by whatever means is most convenient, and usually the computation is most readily done by the ac circuit rules of Chapter 4. It is perhaps interesting that the vertically hatched curve of Figure 7f, which shows the magnitude of admittance to a lightly damped resonant circuit, has the shape made familiar by the resonance curves of Chapter 4.

10. A NETWORK EXAMPLE

As an example, the admittance function for the network of Figure 10a gives both the natural current and also the forced current, the applied voltage being either direct, or alternating at any frequency.

First the admittance function must be found. The impedance of 2 ohms and $\frac{1}{2}$ henry is $2 + j\omega\frac{1}{2}$, or $2 + s/2$ and this is in series with two branches in parallel. The two branches have impedances of 2 and of $2 + s/2$, so the

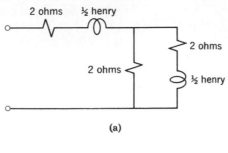

(a)

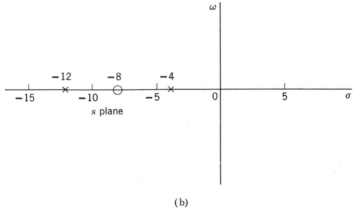

(b)

FIGURE 10 (a) A network. (b) Poles and zeros characterize a network function.

impedance of the two in parallel is $(2)(2 + s/2)/(2 + 2 + s/2)$. Hence the impedance between the input terminals of the total network is

$$Z(s) = \left(2 + \frac{s}{2}\right) + \frac{(2)(2 + s/2)}{(2 + 2 + s/2)} = \frac{(s + 12)(s + 4)}{2(s + 8)} \qquad (10\text{-}1)$$

Input admittance, the reciprocal of impedance, is therefore:

$$Y(s) = \frac{2(s + 8)}{(s + 12)(s + 4)} \qquad (10\text{-}2)$$

The *forced current* in the network depends on the form and magnitude of voltage applied at the network terminals. In this example, let us consider two possibilities. We first find the forced current if the applied voltage is 12 volts dc, unchanging with time, as from a battery.

1. This constant voltage can be considered an exponential function with zero exponent; that is, $v = 12e^{st}$ with $s = 0$. The corresponding dc admittance of the network is found by letting $s = 0$ in equation 10-2, and

$$Y(0) = \frac{2(8)}{(12)(4)} = \frac{1}{3}\,\text{mho} \qquad (10\text{-}3)$$

which can of course be verified by a glance at the network diagram. Steady current produced by the steady application of 12 volts is therefore

$$i = Y(0)\,v = \tfrac{1}{3} \cdot 12 = 4\,\text{amperes} \qquad (10\text{-}4)$$

2. Next consider the steady application of 12 volts ac with frequency 60 cycles per second or $\omega = 377$ radians per second. Admittance is $Y(j\omega)$, and at this frequency

$$Y(j\omega) = \frac{2(8 + j377)}{(12 + j377)(4 + j377)} = \frac{1}{189\underline{/88.8^\circ}}\,\text{mho} \qquad (10\text{-}5)$$

We now find current.

$$i = \mathscr{R}e\, Y(j\omega)V_0\,e^{j\omega t} = \mathscr{R}e\,\frac{1}{189\underline{/88.8^\circ}}\,V_0\,e^{j\omega t} \qquad (10\text{-}6)$$

It is given that voltage is 12 volts ac, and since such a statement invariably gives an rms value we know that $V_0 = \sqrt{2} \cdot 12$ volts.

$$i = \mathscr{R}e\,\sqrt{2}\,(0.0635\underline{/{-}88.8^\circ})e^{j\omega t} \qquad (10\text{-}7)$$

Hence the current at 60 cycles per second, 60 Hz, $\omega = 377$ radians per second, is 63.5 milliamperes, and it lags 88.8 degrees, nearly a quarter of a cycle, behind the applied voltage.

The current at any other frequency can be found by a similar computation using the appropriate value of ω. The current thus computed is the steady or forced current which results from the voltage applied. It has the same s, or the same frequency, as the applied voltage.

11. NATURAL CURRENT

For the natural current in the numerical example of the preceding section, $Y(s)$ is seen in equation 10-2 to have poles at $s = -12$ and $s = -4$. This means that current of the form e^{-12t} is a natural current in the network, and so is a current of the form e^{-4t}. The natural components exist along with the forced components, and if the voltage applied to the network is 12 volts dc, so the forced current is 4 amperes as in equation 10-4, the total current must be

$$i = K_1 e^{-12t} + K_2 e^{-4t} + 4 \qquad (11\text{-}1)$$

If, on the other hand, the voltage applied were 12 volts ac, as in our second example, the total current would have the same natural components as in equation 11-1 but the forced component would be that given by equation 10-7.

In any case the coefficients K_1 and K_2 are to be determined by the requirements of the network. These *initial conditions* will be treated in the next chapter.

12. EXPONENTIALS PICTURED ON THE *s* PLANE

Since a value of *s* defines an exponential function e^{st}, we can say that in this sense the exponential time function corresponds to a point in the *s* plane. This correspondence is suggested in Figure 12a.

The exponential function decays if *s* is negative real, more steeply for larger *s*. If *s* is imaginary, two exponential functions defined by two values of *s* with equal magnitudes but opposite sign add to a sinusoidal function of time. Frequency is proportional to distance above the origin.

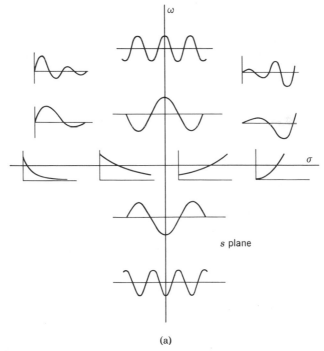

(a)

FIGURE 12a   Wave forms corresponding to various values of *s.* **(Harman and Lytle, courtesy of W. W. Harman.)**

If s is complex, two conjugate values correspond to exponential functions that add to a damped (or expanding) sinusoidal function of time. In the left half of the s plane the waves grow smaller with time because of damping, but in the right half plane the waves grow larger. Frequency is proportional to the distance of the conjugate points above and below the horizontal axis.

13. THE GENERAL NETWORK FUNCTION

All the network functions that we have determined—the admittances, impedances, and transfer functions—have been in the form of rational functions. Equation 10-2 is an example. Indeed, it will be seen in Chapter 9 that network functions of all lumped, linear networks are rational functions and can be put in the form:

$$F(s) = H \frac{(s - s_a)(s - s_b) \cdots (s - s_m)}{(s - s_1)(s - s_2) \cdots (s - s_n)} \tag{13-1}$$

Such a functional form applies to most ordinary networks; *lumped* means that distributed circuits and transmission lines are not included, and *linear* means that R, L, and C do not vary with current or voltage.

Equation 13-1 shows a function of s that has zeros at $s = s_a$, s_b, and so on, and poles at s_1, s_2, and so on. Any one or more of these poles or zeros may be at zero, leaving what is called a free s in the equation. If m is less than n, as it usually is in admittances, one or more of the zeros will be at the *point at infinity* ($s = \infty$; the mathematician makes no distinction among positive and negative, real, imaginary, or complex infinity). The number of poles is always equal to the number of zeros if one counts poles and zeros at infinite s as well as those at finite and zero s. It is possible to have a multiple pole or zero, as for instance if $s_1 = s_2$; in counting, a double pole counts two, and so on.

As an example, the admittance of equation 10-2 has two poles, at $s = -12$ and -4, and a finite zero at $s = -8$. It has also a zero at $s = \infty$, meaning that at very high frequency the admittance approaches zero. Therefore the total number of zeros is 2, and the number of poles is likewise 2. The value of H in this example is 2 and $m = 1$. There is neither a pole nor a zero at the origin; rather, $Y(0) = \frac{1}{3}$, as in equation 10-3.

Since a network function must be of the form of equation 13-1, the function is fully determined, except for the constant H, by the values of s_a, s_b, ..., s_m and s_1, s_2, ..., s_n. But these are the zeros and poles. Thus the network function is fully determined (except for H) when the locations of the poles and zeros are known.

If the network function is computed from the parameters of the network, H appears from the computation. Thus for Y of equation 10-2, H is 2. Alternatively, if poles and zeros are known it is possible to find H without computing from the network parameters if any one finite value of the function is known at any s. For instance, it is known that the admittance to dc of the network of Figure 10a is $\frac{1}{3}$ mho; therefore H must be 2 in order to make $Y(0) = \frac{1}{3}$.

The beauty of a network function is that it provides both the natural response of the network, and its forced response to any disturbance of exponential form. Since we can consider direct current to be an exponential with zero exponent, and sinusoidal alternating current to be the sum of exponentials with imaginary exponents, and any periodic disturbance to be (by Fourier series) the sum of sinusoids and hence of exponentials of finite magnitude, and any nonperiodic signal to be (by Fourier and Laplace integrals) the sum of exponentials of infinitesimal magnitude, there are almost no bounds to the use of network functions within the realm of linear systems.

14. ANALOGY

The three-dimensional sketches of Figures 7b and 7f are intended to help visualization of the $|Y(s)|$ surface above the s plane. Exact heights of the surface can be computed by substituting many numerical values of complex s in the appropriate equations, but a general idea of the shape of the surface can be obtained much more easily by imagining a thin sheet of rubber stretched flat and taut over a large table top. This table top is the s plane, and chalk lines crossing at the middle of the table mark the real and imaginary axes. The rubber sheet is now given the shape (approximately) of the $|Y(s)|$ surface by pushing it up where there are poles and holding it down where there are zeros. A thumbtack into the table top will produce a zero; a pencil stood upright between table and rubber sheet will produce a pole.

This analogy is not exact. It is qualitative, not quantitative. It indicates where the surface slopes upward and where it slopes downward, but it does not accurately indicate how much. An aspect to consider is this: the shape of the entire rubber surface is fully determined by the positions of the thumbtacks and pencils. This strengthens the concept that the entire $Y(s)$ function is fully defined by its poles and zeros. There is, to be sure, the scale factor to be determined arbitrarily; this is the length of the pencils.

Exact analogies to rubber sheets or, better, to electrostatic fields, or to current flow in electrolytic tanks, have been devised with great ingenuity and profit. The analogy, when exact, is not to the function $Y(s)$ but to its logarithm. Pursuit of this fascinating subject must be left for more advanced treatises.†

---

† Tuttle, Pettit, and others.

15. SUMMARY

Exponential currents are important because the *natural* responses of linear systems are exponential in form. They are perhaps more important because steady *alternating* current and steady *direct* current can be expressed in terms of exponential functions.

Network functions (*admittance, impedance, transmittance*) are defined as ratios of *exponential* currents and voltages. These can be written $Y(s)$, $Z(s)$, $T(s)$, and they are the same functions as $Y(j\omega)$, $Z(j\omega)$, $T(j\omega)$ as developed in the previous chapter, with only the variable changed.

A *pole* of Y gives the value of s at which there can be current, called *natural* current, with zero voltage. Natural current has the exponential form $I_0 e^{s_1 t}$, s_1 being the value of s at a pole of $Y(s)$. Once started by some disturbance of the network, natural current continues without forcing voltage.

The admittance of a double-energy circuit has *two poles*. These may both be at *real s*, if damping is heavy, or they may be at *conjugate complex s* if the circuit is lightly damped. An intermediate case with *critical* damping has a single second-order pole at real *s*.

In a *lightly* damped *RLC* circuit with two poles of Y at conjugate complex values of *s*, the natural current is a *damped sinusoidal oscillation*. If R is reduced toward zero, the *damping approaches zero*, the poles approach the imaginary axis of the *s* plane, and the *natural frequency* of oscillation approaches the *undamped natural frequency*.

Natural current and *forced* current comprise the total current of a network.

Forced current is the real component of $Y(j\omega)V_0 e^{j\omega t}$.

A numerical *example* shows both forced and natural components of current computed from the admittance function of a given network. Total current is the sum of these components.

For lumped, linear systems, *network functions* have the form

$$F(s) = H \frac{(s - s_a)(s - s_b) \cdots (s - s_m)}{(s - s_1)(s - s_2) \cdots (s - s_n)} \tag{15-1}$$

s_a and so on are zeros, s_1 and so on are poles. H is determined from the network parameters or from a known value of F.

A crude but helpful *analog* of a network function is a *rubber sheet* over a table. The sheet is thumb-tacked to the table top at zeros of the function and supported on pencils at poles. Better analogs are available, but this rubber-sheet model is easy to construct and even easier to visualize.

PROBLEMS

5-1. For the circuit shown, plot $Y(s)$ as a function of (real) s from $s = -200$ to $s = +100$. Compute enough numerical values to plot the curve.   §3*

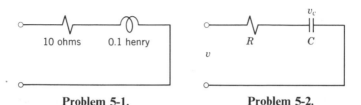

Problem 5-1.   Problem 5-2.

5-2. In the circuit shown, find the transfer function that is by definition $F = V_c/V$, where the applied voltage is $v = Ve^{st}$ and voltage across C is $v_c = V_c e^{st}$.
§4*

5-3. A circuit consists of $L = 0.50$ henry, $R = 100$ ohms, and $C = 2.25 \cdot 10^{-4}$ farad connected in series. Plot $Y(s)$ for the circuit for real values of s. How many poles and how many zeros has $Y(s)$?   §6

5-4. Plot $|Y(s)|$ as a function of real s for a double energy circuit, as in Figure 6b, knowing that $\tau_2 = 3\tau_1$.   §6*

5-5. Figure 6b shows a curve of admittance as a function of s. Plot such a curve for $\tau_2 = 1.25\tau_1$.   §6

5-6. Compute and plot $Y(s)$ for a critically damped circuit of $R, L,$ and C in series in which $L = 0.50$ henry and $C = 0.35$ microfarad.   §6

5-7. Compute and plot $Y(s)$ for a critically damped circuit of $R, L,$ and C in series in which $L = 150$ millihenrys and $C = 0.75$ microfarad.   §6

5-8. A circuit has $L, R,$ and C in series. The amounts are: $L = 10$ millihenrys and $R = 200$ ohms. For the following values of C, compute values of s at poles and zeros of admittance, and indicate the positions of these poles and zeros on a sketch of the complex s plane. Do this for (a) 4 microfarads, (b) 2 microfarads, (c) $\frac{1}{2}$ microfarad, and (d) draw the locus of poles for all positive values of C.
§7*

5-9. A coil with inductance L and resistance R is in parallel with a capacitor of capacitance C (as in Figure 23c of Chapter 4), forming a two-branch circuit with parallel resonance. Write the impedance function $Z(s)$, giving its denominator the form of the denominator in equation 8-7. (a) Find ζ in terms of $L, R,$ and C. (b) Assume that $\zeta = 1/2$, and plot poles and zeros of $Z(s)$ on the complex s plane. (c) How many poles are there, and how many zeros, counting any that may be at s equals infinity?   §8

5-10. The admittance function of a network has poles at $s = -4$ and $s = -12$, and a zero at $s = -8$, as in Section 10. Find a network other than that of Figure 10a, to provide this admittance.   §10

5-11. A transfer admittance is defined as $Y_{12}(s) = I_2/V_1$. The value of this transfer admittance is $Y_{12}(s) = 1/(s + 6)$. (*a*) Devise a network to have this transfer admittance. (*b*) Devise a second and different network to have the same transfer admittance.   §10*

5-12. The admittance function of a network has poles at $s = \pm j10^3$ and at $s = \pm j10^3/\sqrt{2}$, and zeros at $s = 0$, $s = \infty$, and $s = \pm j\sqrt{\frac{2}{3}}10^3$. Find a network that will provide this admittance function.   §13

5-13. A pole of the admittance function of a network corresponds to a certain physical condition in the network, a certain combination of parameters. This is true whether there is a real pole, or an imaginary pair of poles, or a complex pair. It is also true, with significant modification, for a pole at the origin where $s = 0$, or at infinity. (*a*) Discuss this relation between the network and poles of the admittance function, and also (*b*) a somewhat similar relation between the network and zeros of the admittance function. (Section 26 of Chapter 12, on the physical meaning of zeros and poles, may be of interest in this connection.)

                                                                §13

5-14. A network in the diagram is marked Circuit A. Write the input *impedance* function in the form of equation 13-1. Show poles and zeros of this impedance function in a sketch of the *s* plane, and give numerical values of *s* at the poles and zeros.   §13*

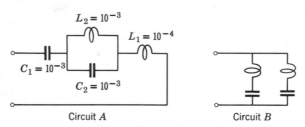

Circuit *A*   Circuit *B*

Problems 5-14 and 5-15.

5-15. Referring to Circuit B in the diagram, if the inductances and capacitances were correctly proportioned, could this network have the same impedance function as the network of Circuit A, and hence the same impedance at all frequencies? (*a*) Explain. (*b*) Compute the *L*'s and *C*'s. (Note: For a shorter problem, part (*b*) can be omitted.)   §14*

CHAPTER
6

THE TIME
DOMAIN

1. THE TRANSIENT INTERVAL

The *natural* component of current discussed in Chapter 5 is added to the *forced* component of Chapter 4 when there is a change in the network, such as the closing of a switch. Since the natural component is exponentially damped, it is *transient* and soon dies away, while the forced component is *steady* and continues as long as the driving source is applied. This chapter will be concerned with the initial interval during which both transient and steady components exist together.

The discrete signals of computers are transient. So are the command signals of automatic control systems when a change of operation is required. So are the pips of radar, and the pulses of sampled-data communication systems, whether used in urban telephone lines or in space among the planets.

Suppose that a switch is closed in a network. There is a source of some kind in the network. Perhaps at the time of closing the switch there are already currents and voltages in some of the inductors or capacitors of the network; these constitute *initial conditions*. Current flows in the network, as determined by the source and the initial conditions.

Natural components and forced components occur together. The forced components satisfy the network equations. The natural components make it possible to meet the initial conditions; they act as a bridge between the initial state and the final state.

There are a number of ways of finding the total current during the transient interval. Since we have already found means of dealing with both forced and natural currents and voltages in the two previous chapters, we can make use of these techniques in the following examples.

Let the switch in the inductive circuit of Figure 1a be suddenly closed at a time that may be called $t = 0$. At all times after closing the switch the circuit equation is

$$Ri + L\frac{di}{dt} = E \tag{1-1}$$

This equation is basic to the methods that we shall use to find the total transient current and we shall refer to it but we shall not use it explicitly. Instead we shall proceed by mere inspection of the circuit.

First, the steady current after a great time is going to be E/R. This is not the total current, for although it would satisfy equation 1-1 to write $i = E/R$, this does not meet the initial condition that $i(0) = 0$.

Both requirements can be met by current that contains a forced component E/R and also a natural component of the form $Ke^{-t/\tau}$, where $\tau = L/R$:

$$i(t) = \frac{E}{R} + Ke^{-t/\tau} \tag{1-2}$$

provided $K = -E/R$, for then

$$i(t) = \frac{E}{R}\left(1 - e^{-(R/L)t}\right) \tag{1-3}$$

In this equation, and in Figure 1b, we can see the current start with its initial value of zero and increase exponentially, as the natural component dies away, to approach its steady value of E/R, and yet equation 1-1 is satisfied at all times.

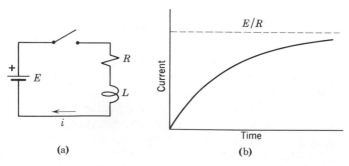

(a)   (b)

FIGURES 1a,b   A circuit with resistance and inductance.

2. FOUR RULES FOR SOLUTION

The reasoning used in this example suggests the following four rules for solution. More formal proof of these rules will come later.

1. *Find the forced or steady response of a circuit.* This is found by any convenient means. It is a solution of the circuit equation, but not the complete solution. It does not (in general) satisfy initial conditions.
2. *Find the natural component of response.* This is found by solving the circuit equation with zero substituted for the driving force. The result, exponential in form,† is the response that can exist without a driving force. Its magnitude is not yet known.
3. *Add the forced component and the natural component.* The sum, with any magnitude of natural component, is a solution of the circuit equation.
4. *Adjust the magnitude of the natural component to fit the initial conditions.* The natural component serves to bridge the gap between initial conditions and steady response.

3. NON-ZERO INITIAL CURRENT

Initial current need not be zero. For instance, closing the switch in Figure 3a starts a transient interval with an initial current that is not zero.

In Figure 3a the steady current is V/R. The initial current is $i(0)$. The coefficient of the natural component is therefore $i(0) - V/R$. The total current is then

$$i = \frac{V}{R} - \left[i(0) - \frac{V}{R}\right]e^{-(R/L)t} \tag{3-1}$$

This result is shown in Figure 3b.

Note that we do not need to know *why* there is an initial current. It is enough to know that it is there.

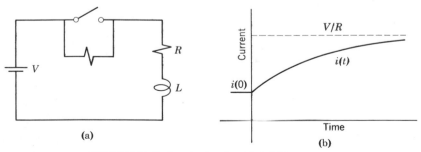

(a)

(b)

FIGURES 3a,b   A circuit with initial current.

---

† Or, in special cases, an exponential times a power of t; see Section 19.

4. NATURAL COMPONENT ALONE

If there is no source in our inductive circuit, but there is initial current, then natural current exists but there is no forced component. In Figure 4a, let the right-hand switch be closed before the left-hand switch is opened. Calling the initial current in the inductive branch $i(0)$, the natural current, which is here the total current, is, as in Figure 4b,

$$i = i(0)e^{-(R/L)t} \tag{4-1}$$

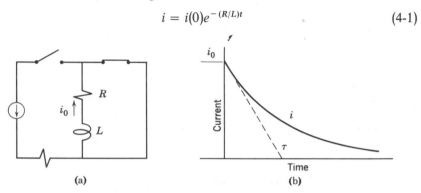

FIGURES 4a,b   Decay of current in an inductive circuit.

5. THE TIME CONSTANT

It is convenient to let $L/R = \tau$ and write equation 4-1 as

$$i = i(0)e^{-t/\tau} \tag{5-1}$$

This form of notation has been used in previous equations and such a coefficient as τ is called the *time constant* of the exponential function.

In general, for any exponential function of the form of equation 5-1, we can say:

1. The initial slope of the function (at $t = 0$) is $-i(0)/\tau$.
2. If this original slope continued, which it does not, the exponential function would reach zero in time τ, as indicated by the straight dashed line in Figure 4b.
3. In fact, the exponential function ebbs to 0.368 of its initial value in time τ.

In the time of *one time constant* the exponential function decreases to $e^{-1} = 0.368$, in two time constants to $e^{-2} = 0.135$, in three time constants to $e^{-3} = 0.0498$, and so on, each being 0.368 times the previous figure.

An extension of this concept tells us that when there is an exponentially decaying component, in a current or voltage or other function, only 36.8% of this component remains after an interval of one time constant. Only about 5% remains after three time constants.†

6. ALTERNATING VOLTAGE

We have considered the current in a circuit of inductance and resistance in series when it is left to come to rest by itself, and also when a constant voltage is suddenly applied. Another possibility, and an important one, is that alternating voltage may be suddenly applied. Figure 6a shows the circuit.

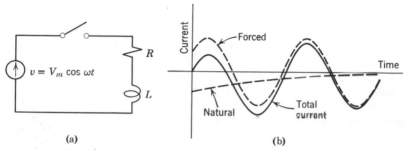

(a)   (b)

FIGURES 6a,b   An inductive circuit with alternating voltage.

Let the switch be closed at $t = 0$, the voltage of the source being

$$v = V_m \cos \omega t \tag{6-1}$$

After the switch is closed, this applied voltage is equated to the sum of the resistance voltage and inductance voltage:

$$Ri + L\frac{di}{dt} = v = V_m \cos \omega t \tag{6-2}$$

This is the same circuit that we have considered before, and its natural component of current is therefore of the same form; although we do not know the constant K, the natural response is exponential:

$$i_n = Ke^{-t/\tau} \quad \text{where} \quad \tau = \frac{L}{R} \tag{6-3}$$

---

† Note the relation to *half-life*, used for the exponential decay of radioactive substances. Half-life is the time in which a quantity decreases to 2^{-1} or 0.50 of its original value.

The forced component is found quite easily by the methods of Section 5 of Chapter 4 (example 1). Impedance is

$$Z = R + j\omega L = |Z|\underline{/\varphi} \tag{6-4}$$

where $|Z| = \sqrt{R^2 + \omega^2 L^2}$ and $\varphi = \tan^{-1} \omega L/R$. The transform of voltage is

$$V_m = V_m\underline{/0°} \tag{6-5}$$

The transform of current is therefore

$$I_m = \frac{V_m\underline{/0°}}{|Z|\underline{/\varphi}} \tag{6-6}$$

Hence the forced component of current as a function of time is

$$i_f = \frac{V_m}{|Z|} \cos(\omega t - \varphi) \tag{6-7}$$

We have now followed Rules 1 and 2, writing the natural and forced components of current. The natural component is to act as a bridge between the initial condition required by the switch

$$i(0) = 0 \tag{6-8}$$

and the forced component which is, at $t = 0$,

$$i_f(0) = \frac{V_m}{|Z|} \cos \varphi \tag{6-9}$$

For the natural component to bridge this gap it is necessary that

$$K = -\frac{V_m}{|Z|} \cos \varphi \tag{6-10}$$

and we add equation 6-7 to 6-3 with this value of K to obtain the total current (by Rules 3 and 4):

$$i = \frac{V_m}{|Z|} [\cos(\omega t - \varphi) - (\cos \varphi)e^{-t/\tau}] \tag{6-11}$$

This total current is shown in Figure 6b. The forced component is sinusoidal. The natural component is exponential. The total current is an offset sinusoidal curve, offset enough to start at zero.

The validity of equation 6-11 can of course be verified by determining that it satisfies the two requirements that must be met by the solution of any differential equation:

1. Equation 6-11 must satisfy the network equation 6-2 at all times, and
2. It must agree with the initial conditions, so that when t is set equal to zero, $i = 0$ in equation 6-11.

7. A CIRCUIT WITH CAPACITANCE AND RESISTANCE

A circuit with capacitance and resistance in series can be treated by the same methods. Let a constant voltage V be applied to the circuit of Figure 7a at $t = 0$. There is no initial charge on the capacitance C.

Admittance of this circuit has one pole, as in Figure 5d of Chapter 5, at $s = -1/\tau$ where $\tau = RC$. The *natural* component of current is therefore

$$i_n = Ke^{-t/\tau} = Ke^{-t/RC} \tag{7-1}$$

as shown in Figure 7b.

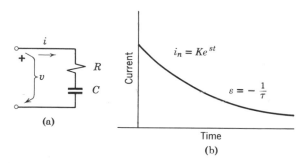

(a)

(b)

FIGURE 7 **(a)** Series R and C. **(b)** Form of the natural response, using s at the pole of the admittance function.

The *forced* component is zero, for the unchanging applied voltage produces no steady-state current to the capacitance. Hence equation 7-1 gives the total current, but K must be found from initial conditions.

Immediately after closing the switch, at the time $t = 0_+$, before any charge has had time to enter the capacitance, all the applied voltage V is across the resistance, and the current is

$$i(0_+) = \frac{V}{R} \tag{7-2}$$

Note that current before closing the switch is zero; immediately after closing the switch the current is V/R. There is a discontinuity in the current function at $t = 0$, and current *just after* closing the switch is called $i(0_+)$.

Since the natural current must bridge the gap between this initial current V/R and the final current zero, K must equal V/R and the total current is

$$i = \frac{V}{R}e^{-t/\tau} \tag{7-3}$$

as in Figure 7b.

This completes our problem, the four rules having been applied. Let us now change the statement of the problem and let there be a known amount of charge initially on the capacitance. We can specify that voltage across the capacitance is initially v_0. Then just after applying a voltage V the current is

$$i(0_+) = \frac{V - v_0}{R} \qquad (7\text{-}4)$$

and the total current in this more general example is

$$i = \frac{V - v_0}{R} e^{-t/\tau} \qquad (7\text{-}5)$$

It is not surprising to note that if the applied voltage V is zero and the initial capacitor voltage v_0 is positive, the capacitance will discharge when the circuit is completed with a negative current. (By definition, positive capacitor voltage results from positive charge placed on the capacitance by positive current.)

8. ALTERNATING APPLIED VOLTAGE

In the same capacitive circuit of Figure 7a, let us now make the initial charge on the capacitor zero, and let the applied voltage be alternating:

$$v = V_m \cos \omega t \qquad (8\text{-}1)$$

In equation 5-10 of Chapter 4 we found the impedance of such a circuit to be

$$Z = R + \frac{1}{j\omega C} = R - j\frac{1}{\omega C} = |Z| \underline{/-\theta} \qquad (8\text{-}2)$$

where

$$|Z| = \sqrt{R^2 + \frac{1}{\omega^2 C^2}} \quad \text{and} \quad \theta = \tan^{-1}\frac{1}{\omega C R}$$

The transform of the voltage is $V_m = V_m \underline{/0°}$; dividing by impedance to get the transform of current we obtain $I_m = (V_m \underline{/0°})/(|Z| \underline{/-\theta})$, and from this transform the expression for forced current as a function of time is seen to be

$$i_f = \frac{V_m}{|Z|} \cos(\omega t + \theta) \qquad (8\text{-}3)$$

The value of this forced component at zero time is

$$i_f(0) = \frac{V_m}{|Z|} \cos \theta \qquad (8\text{-}4)$$

The *total* current at $t = 0_+$ must be the voltage at zero time which, from equation 8-1 is V_m, divided by the circuit resistance R, giving

$$i(0_+) = \frac{V_m}{R} \qquad (8\text{-}5)$$

To provide this necessary total current there must be a natural component of the exponential form of equation 7-1 added to the forced component of the sinusoidal form of equation 8-3. The amount K of this natural component must be the difference between the initial values given by equations 8-5 and 8-4. Total current, then, the sum of the forced and natural components, is

$$i = i_f + i_n = \frac{V_m}{|Z|} \cos(\omega t + \theta) + \left(\frac{V_m}{R} - \frac{V_m}{|Z|} \cos \theta \right) e^{-t/\tau} \qquad (8\text{-}6)$$

This is the solution of our problem, and we can apply the usual two checks. At very large time the exponential term vanishes and we are left with the familiar steady-state solution, with current leading voltage by an appropriate angle. At the other extreme, as time approaches zero, equation 8-6 reduces to V_m/R, as we know it should.

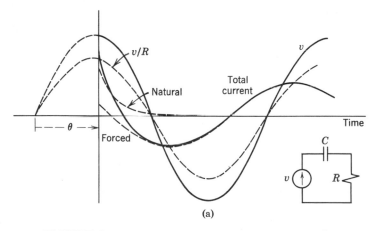

(a)

FIGURE 8a   A capacitive circuit with alternating voltage.

Equation 8-6 describes a current that might be called an offset sinusoidal wave, as was the current in an inductive circuit when alternating voltage was suddenly applied. However, it is perhaps more descriptive to speak of the current to the capacitive circuit as being sinusoidal plus an initial exponential surge. See Figure 8a.

9. A NETWORK EXAMPLE

Figure 9a shows a simple network of two loops. The source is 12 volts constant direct current. The switch is closed at $t = 0$. Immediately upon closing the switch there can be no current in the inductive branch, so $i_2(0_+) = 0$, but in the purely resistive branches the initial current is limited by the resistances and $i_1(0_+) = 3$.

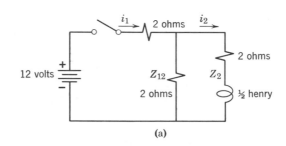

(a)

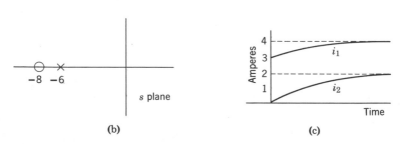

(b)　　　　　　　　　　　　　　　　　　　(c)

FIGURES 9a,b,c　A network, its pole and zero, and its loop currents.

Parallel and series combination gives the impedance function,

$$Z_1(s) = 2 + \frac{2(2 + \frac{1}{2}s)}{2 + 2 + \frac{1}{2}s} = \frac{24 + 4s}{8 + s} \tag{9-1}$$

whence

$$Y_1(s) = \frac{1}{4}\frac{s + 8}{s + 6} \tag{9-2}$$

This input admittance has a zero at $s = -8$, as indicated in Figure 9b, and a pole at $s = -6$.

The *forced component* of current is produced by the constant voltage of the source; voltage being constant, $s = 0$, and the admittance $Y_1(0)$ is $(\frac{1}{4})(\frac{8}{6})$, so

$$i_{1f} = VY_1(0) = 12(\tfrac{1}{4})(\tfrac{8}{6}) = 4 \text{ amperes} \tag{9-3}$$

Since there is only one pole in Figure 9b, at $s = -6$, the *natural component* of current is

$$i_{1n} = K_1 e^{-6t} \tag{9-4}$$

Total current is the sum:

$$i_1 = 4 + K_1 e^{-6t} \tag{9-5}$$

Since *initially* $i_1(0) = 3$, it follows that $K_1 = -1$, and

$$i_1 = 4 - e^{-6t} \tag{9-6}$$

which is shown as a function of time in Figure 9c.

In this example the complete response has been found from the admittance function. Note that no differential equation is used.

Continuing the problem, let us find i_2, the current in the inductive branch of the network. The function that relates current in loop 2 to voltage applied in loop 1 is the transfer admittance y_{21},

$$y_{21}(s) = \frac{I_2(s)}{V_1(s)} = \frac{I_2(s)}{I_1(s)} \cdot \frac{I_1(s)}{V_1(s)} \tag{9-7}$$

By equation 9-2, $I_1/V_1 = (s + 8)/4(s + 6)$. From the general relation of current division $I_2/I_1 = Z_{12}/(Z_{12} + Z_2)$, which in this network becomes

$$\frac{I_2}{I_1} = \frac{2}{2 + 2 + \frac{1}{2}s} = \frac{4}{s + 8} \tag{9-8}$$

It therefore follows that

$$y_{21}(s) = \frac{I_1}{V_1} \cdot \frac{I_2}{I_1} = \frac{s + 8}{4(s + 6)} \frac{4}{s + 8} = \frac{1}{s + 6} \tag{9-9}$$

From this transfer admittance, two facts are obtained. First, that the natural component of current in the inductive branch is

$$i_{2n} = K_2 e^{-6t} \tag{9-10}$$

and, second, that the forced component of current is

$$i_{2f} = E y_{21}(0) = \frac{12}{6} = 2$$

(This forced component, to be sure, could have been determined by a glance at the network diagram.)

Finally, then, the total current is $i_2(t) = 2 + K_2 e^{-6t}$, and since we are given as an initial condition that $i_2(0) = 0$, K_2 must be equal to -2, and

$$i_2(t) = 2 - 2e^{-6t} \tag{9-11}$$

This result is shown in Figure 9c also.

10. A NETWORK WITH TWO INDUCTANCES

When both loops of a network have inductance there can be two poles in the admittance function and two natural exponential terms. A new complication is introduced into the solution, because there are then two initial conditions to be satisfied, and two natural exponential components to be used for the purpose. The two initial conditions can be the initial currents in the two inductances; these correspond to the initial *distribution of energy* in the network, the initial *state* of the network. There are a number of ways to formulate and to solve the network equations (loop equations, node equations, state-variable equations), all of which will be discussed in later chapters, but perhaps the easiest if not the most systematic is to continue as in previous examples and use the methods that we already know.

Let a constant direct voltage of 24 volts be applied by closing a switch at $t = 0$ in the network of Figure 10a. Find the currents $i_1(t)$ and $i_2(t)$, and the

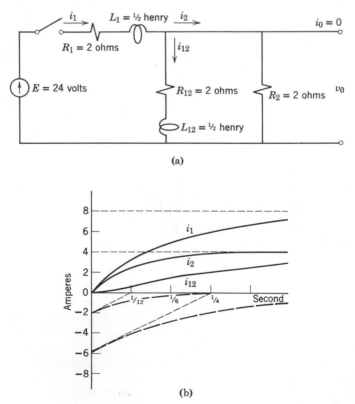

(a)

(b)

FIGURE 10   (a)   A network of two loops, and   (b)   Its currents.

output voltage $v_0(t)$ after the switch is closed. It is given that there is no current in either inductance before the switch is closed.

Solution: We begin by writing the input impedance:

$$Z_1(s) = 2 + \frac{1}{2}s + \frac{2(2 + \frac{1}{2}s)}{2 + 2 + \frac{1}{2}s} \tag{10-1}$$

from which admittance, the reciprocal, is

$$Y_1(s) = \frac{2(s + 8)}{(s + 4)(s + 12)} \tag{10-2}$$

This admittance function has poles at -4 and -12, and a zero at -8, as we found for the same network in Section 10 of Chapter 5.

From this admittance function we can find the *forced* component of entering current. Since the battery voltage is 24 and the dc impedance $Z_1(0) = 3$, the forced current is 8 amperes. Also the forms of the natural components are known from the poles of admittance; adding these to the forced component, the total entering current is

$$i_1 = 8 + K_1 c^{-4t} + K_2 e^{-12t} \tag{10-3}$$

in which two coefficients remain to be evaluated from initial conditions.

It can next be seen that the total entering current i_1 must divide between i_2 and i_{12}, and the fraction of i_1 to become i_{12} is (by equation 9-5 of Chapter 4):

$$\frac{i_{12}(s)}{i_1(s)} = \frac{Z_2}{Z_2 + Z_{12}} = \frac{2}{\frac{1}{2}s + 4} = \frac{4}{s + 8} \tag{10-4}$$

The fraction of i_1 that becomes i_{12} depends on the value of s, as follows:

Forced: $s = 0$ Natural: $s = -4$ Natural: $s = -12$

$$\frac{i_{12}(s)}{i_1(s)} = \frac{4}{8} = \frac{1}{2} \qquad \frac{4}{4} = 1 \qquad \frac{4}{-4} = -1$$

Hence, by applying these three factors to the appropriate terms of i_1 in equation 10-3 we obtain for the branch current:

$$i_{12} = 4 + K_1 e^{-4t} - K_2 e^{-12t} \tag{10-5}$$

To this point we have used network relations based on Kirchhoff's laws. We have found the forced components of i_1 and i_{12}, but we cannot find the coefficients of the natural components until initial conditions are introduced. The initial state is that the network is at rest; that is, $i_1(0) = 0$ and

$i_{12}(0) = 0$. Letting $t = 0$ in equations 10-3 and 10-5, and setting the results equal to these initial zero values,

$$8 + K_1 + K_2 = 0 \tag{10-6}$$

$$4 + K_1 - K_2 = 0 \tag{10-7}$$

from which

$$K_1 = -6 \quad \text{and} \quad K_2 = -2 \tag{10-8}$$

Finally, then,

$$i_1 = 8 - 6e^{-4t} - 2e^{-12t} \text{ amperes} \tag{10-9}$$

$$i_{12} = 4 - 6e^{-4t} + 2e^{-12t} \text{ amperes} \dagger \tag{10-10}$$

However, the required current is not i_{12} but i_2. Since $i_2 = i_1 - i_{12}$, we merely subtract to obtain

$$i_2 = 4 - 4e^{-12t} \text{ amperes} \tag{10-11}$$

These results are plotted in Figure 10b. A number of aspects are well worth considering. Notice the manner in which the components add to three quite

---

† As an alternative technique, less interesting because it does not make full use of the network function, use equation 10-3:

$$i_1 = 8 + K_1 e^{-4t} + K_2 e^{-12t} \tag{10-13}$$

Differentiate this equation:

$$\frac{di_1}{dt} = -4K_1 e^{-4t} - 12K_2 e^{-12t} \tag{10-14}$$

The initial conditions are:

$$i_1(0) = 0 \quad \text{and} \quad \left.\frac{di_1}{dt}\right|_{t=0} = \frac{24}{1/2} = 48 \tag{10-15}$$

Letting $t = 0$ in equations 10-13 and 14 and equating to these initial values:

$$0 = 8 + K_1 + K_2 \tag{10-16}$$

$$48 = -4K_1 - 12K_2 \tag{10-17}$$

whence

$$K_1 = -6 \quad \text{and} \quad K_2 = -2 \tag{10-18}$$

giving

$$i_1 = 8 - 6e^{-4t} - 2e^{-12t} \tag{10-19}$$

and this is equation 10-9. This technique can be used to find equation 10-10 and any of the currents of a network, using initial first derivatives of the various currents, and if necessary the initial higher derivatives also, for which numerical values may be obtained from the differential equations if they are not apparent from the diagrams.

different currents. See that the two branch currents add to the total entering current. Note that the initial slope of i_{12} is zero, and consider the explanation. It is interesting that one natural component has completely disappeared from i_2. Why?

The output voltage v_0 is also required. In this example the output voltage is extremely simple, being merely the 2 ohms of R_2 times i_2 :

$$v_0 = 8 - 8e^{-12t} \text{ volts} \tag{10-12}$$

11. A NODE-CURRENT EXAMPLE

Figure 11a shows a circuit in which a resistor with conductance G and a capacitor C are in parallel. If a known voltage is applied to the terminals, the problem is rather trivial. Let us rather assume that it is the input current that is known and input voltage is to be found.

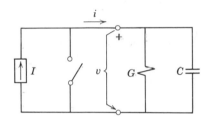

FIGURE 11a **An example in which current is given, and voltage is to be found.**

Input current enters the network for which the admittance is

$$Y = G + sC \tag{11-1}$$

Letting $C/G = \tau$, input impedance is

$$Z = \frac{1}{G + sC} = \frac{1}{C(s + 1/\tau)} \tag{11-2}$$

Since input impedance has a pole at $s = -1/\tau$, there can be a natural component of input voltage of the form

$$v_n = Ke^{-t/\tau} \tag{11-3}$$

Note that we are now finding the duals of quantities required in previous solutions.

The forced component of input voltage can be found from the input current, which must of course be given. Let us first assume that the source current is a constant direct current of value I, and the switch in Figure 11a is opened at $t = 0$. Then, either by inspection of the circuit, or by letting $s = 0$ in equation 11-2, we recognize that the forced component of input voltage is

$$v_f = IZ(0) = \frac{I}{G} \qquad (11\text{-}4)$$

Adding, the total input voltage is

$$v = \frac{I}{G} + Ke^{-t/\tau} \qquad (11\text{-}5)$$

with the K to be found from initial conditions.

Because of the switch in the network, initial input voltage must be zero; $v(0_+) = 0$. Using this initial condition in equation 11-5, we find that when $t = 0$,

$$\frac{I}{G} + K = 0 \qquad K = -\frac{I}{G} \qquad (11\text{-}6)$$

so from equation 11-5, $v(t)$ is

$$v = \frac{I}{G} - \frac{I}{G} e^{-t/\tau} \qquad (11\text{-}7)$$

The analogy—the duality—to equation 1-2 is obvious.

12. A DOUBLE-ENERGY CIRCUIT

The circuit of Figure 12a contains two energy-storage elements, inductance and capacitance. This is an important circuit and seems often to be considered typical of general circuit problems.

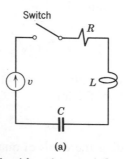

(a)

FIGURE 12a   A circuit with resistance, inductance, and capacitance.

A circuit equation based on Kirchhoff's voltage law, to apply after the switch is closed, gives

$$L\frac{di}{dt} + Ri + v_0 + \frac{1}{C}\int_0^t i\, dt = v \tag{12-1}$$

Each term is a voltage; the third term is voltage across the capacitance owing to initial charge on the capacitance, and the fourth term is the change of capacitor voltage due to current in the circuit after the switch is closed (at $t = 0$).

Let us suppose, as a first example, that the applied voltage is an unchanging battery voltage, so that $v = V$, a constant. Since the problem cannot be solved without initial conditions of energy storage (or the equivalent), let it be given that there is no initial charge on the capacitor, so that $v_0 = 0$, and there is no initial current in the circuit, so that when $t = 0$, $i = 0$.

The four rules for solution may be followed. First, *steady current* is to be found. But the capacitor blocks any steady current from a battery and for this example $i_f = 0$.

Second, the *natural current*, which will also be the total current, is to be found. Here we use Section 6 of Chapter 5 (page 109) which tells us that the admittance of the *RLC* circuit has two poles (Figure 6b of Chapter 5), that the natural current therefore has two exponential components, and that it is of the form

$$i = K_1 e^{-t/\tau_1} + K_2 e^{-t/\tau_2} \tag{12-2}$$

where

$$\frac{1}{\tau_1} = \frac{R}{2L} + \sqrt{\left(\frac{R}{2L}\right)^2 - \frac{1}{LC}} \quad \text{and} \quad \frac{1}{\tau_2} = \frac{R}{2L} - \sqrt{\left(\frac{R}{2L}\right)^2 - \frac{1}{LC}} \tag{12-3}$$

When, in the previous chapter, we were considering only the location of the poles of Y and the corresponding form of natural current, the constants K_1 and K_2 could have any values; now, for the solution of a specific problem, we recognize that they are to be determined from the given initial state of the circuit.

To proceed with this evaluation, two equations containing K_1 and K_2 are needed. One of these can be equation 12-2 and the other can be obtained by differentiating equation 12-2 to give

$$\frac{di}{dt} = -\frac{K_1}{\tau_1} e^{-t/\tau_1} - \frac{K_2}{\tau_2} e^{-t/\tau_2} \tag{12-4}$$

Theoretically, the values of K_1 and K_2 can be obtained from this pair of equations if the current and its derivative could be known at any time; actually, we use the given initial conditions to find the current and its derivative at *zero* time. Putting $t = 0$ in equations 12-2 and 12-4:

$$K_1 + K_2 = i(0_+) \qquad\qquad -\frac{K_1}{\tau_1} - \frac{K_2}{\tau_2} = \frac{di}{dt}\bigg|_{t=0_+} \tag{12-5}$$

But we are given that $i(0_+) = 0$, so

$$K_1 + K_2 = 0 \tag{12-6}$$

The other initial condition, that the capacitor is initially uncharged, is introduced into the mathematics as follows. Let $t = 0$ in equation 12-1. We are given that $i(0) = 0$ and $v_0 = 0$; the integral is necessarily zero. This leaves $L(di/dt) = V$ at $t = 0_+$, or

$$\frac{di}{dt}\bigg|_{t=0_+} = \frac{V}{L} \tag{12-7}$$

Using this initial derivative in equation 12-5,

$$-\frac{K_1}{\tau_1} - \frac{K_2}{\tau_2} = \frac{V}{L} \tag{12-8}$$

Finally, equations 12-6 and 12-8 are solved simultaneously to give[†]

$$K_2 = -K_1 = \frac{V}{\sqrt{R^2 - 4L/C}} \tag{12-9}$$

We have now completed the solution for current in the *RLC* circuit that is initially at rest, with a constant voltage suddenly applied. The answer is equation 12-2 where K_1 and K_2, equal in magnitude but opposite in sign, are given in equations 12-9, and the time constants τ_1 and τ_2 are from equations 12-3. An example shows the truly remarkable way in which these values fit together to provide a surge of current that (1) satisfies the initial conditions, (2) satisfies the differential equation, and (3) bridges the gap between the initial and final conditions.

[†] When, in more complicated networks, it is necessary to evaluate a number of the coefficients *K* from an equal number of simultaneous equations, these can be obtained by repeated differentiation of the expression for current (equation 12-2). Set the results equal to the initial current, the initial first derivative of current, the initial second derivative of current, and so on, as determined from the given initial conditions substituted into the network equation (or equations) such as equation 12-1. See Skilling (4) in bibliography.

13. A DOUBLE-ENERGY EXAMPLE

Let us put numbers into the circuit of Figure 12a. Let

$$L = 10 \text{ millihenrys} = 10^{-2} \text{ henry}$$
$$C = 1.56 \text{ microfarads} = 1.56 \cdot 10^{-6} \text{ farad}$$
$$R = 200 \text{ ohms}$$

These same numerical values were used in Section 6 of Chapter 5 to locate the poles of admittance of this circuit, and the values then computed can be used here. There are two poles at the following values of s:

$$s_1 = -\frac{1}{\tau_1} = -16 \cdot 10^3$$

$$s_2 = -\frac{1}{\tau_2} = -4 \cdot 10^3 \tag{13-1}$$

These numerical values of parameters are in fact reasonable for an audio-frequency circuit, particularly if the inductor being modeled has a ferromagnetic core, and the time constants are seen to be of the order of fractions of milliseconds. It may be considered that the poles were computed from equations 12-3.

Let there be no initial charge on the capacitor ($v_0 = 0$), and let the switch be open until $t = 0$, at which instant it is closed to apply 12 volts from an independent source of constant voltage (modeling a battery). Find the current.

1. Because of capacitance in the circuit, there is no forced current from the constant applied voltage.

2 and 3. Hence the natural component is the total current, and

$$i = K_1 e^{-16t \cdot 10^3} + K_2 e^{-4t \cdot 10^3} \tag{13-2}$$

4. The coefficients are now evaluated from equations 12-9:

$$K_2 = \frac{12}{\sqrt{4 \cdot 10^4 - 4(10^{-2})/(1.56)10^{-6}}} = 0.10$$

$$K_1 = -0.10$$

and

$$i = 0.10(e^{-4t \cdot 10^3} - e^{-16t \cdot 10^3}) \tag{13-3}$$

This current is plotted in Figure 13a. Note the components and their time constants. The curve shows the general shape of an overdamped surge of

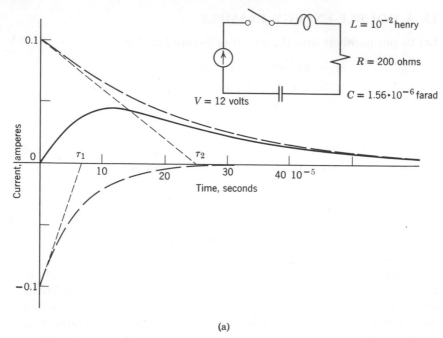

(a)

**FIGURE 13a   Sudden application of a constant voltage to a circuit with
L, R, and C in series.**

current. With different parameters, of course, the surge would be somewhat
different. In a circuit with very large resistance, the current approximates a
step, with a steep rise and a long tail. If, at the other extreme, the circuit
resistance is reduced toward the critical value, the surge of current is
rounded and approaches the shape shown in Figure 15a.

14. CRITICAL DAMPING

If a circuit contains resistance, inductance, and capacitance, as in the two
foregoing sections, but the values of these parameters are so adjusted that
the circuit is critically damped, and $R^2 = 4L/C$, we find that the mathematical forms developed in the previous sections are not helpful. Several things
happen.

First, the two poles of the admittance function, shown in Figure 6b of
Chapter 5 (page 110) come together in a single second-order pole (Figure 6c
of Chapter 5). It follows that τ_1 and τ_2 approach equality in equation 12-2
(page 145).

But equation 12-9 shows that K_1 and K_2 become infinitely large, though opposite in sign. The current of equation 12-2 is then indeterminate, being the difference between two infinitely large quantities that approach equality.

There are various ways out of this difficulty. The most direct is to apply l'Hôpital's rule to evaluate the limit of a ratio as both numerator and denominator approach zero. If the RLC circuit is initially without stored energy, equations 12-2 and 12-9 apply and can be written (for the circuit not critically damped) as

$$\frac{i}{V} = \frac{e^{-t/\tau_1} - e^{-t/\tau_2}}{\sqrt{R^2 - 4L/C}} \qquad (14\text{-}1)$$

If R^2 is now made to approach $4L/C$, so that the circuit becomes critically damped, the numerator and denominator may be differentiated with respect to R and l'Hôpital's rule gives

$$i = \frac{V}{L} te^{-t/\tau} \qquad (14\text{-}2)$$

where $\tau = \sqrt{LC}$, the limit of both τ_1 and τ_2.

This is a special case, for the circuit initially at rest and with no forced component of current. The general form for the critically damped circuit (which may be deduced from Section 16) is:

$$i = i_f + k_1 e^{-t/\tau} + k_2 te^{-t/\tau} \qquad (14\text{-}3)$$

where k_1 and k_2 are to be evaluated from initial conditions.

An example will show that the critically damped surge is not greatly different in fact from the surge of current in another RLC circuit that has *nearly* critical damping, though the mathematical form of the critically damped surge is so very different.

A practical circuit can hardly, in physical fact, have a mathematically precise critical value of damping. If the actual value is reasonably close to critical it is sometimes most convenient to model the real circuit as exactly critical, and sometimes as slightly overdamped, depending on which mathematical formulation is the easiest to work with. For computing current it is easiest to let the model have critical damping, but for finding roots (as in the root-locus method of determining stability) a model with two slightly different real roots is likely to be more tractable.

15. A CRITICALLY DAMPED EXAMPLE

The circuit is similar to that of Section 13 and Figure 13a; $L = 10^{-2}$ henry and $C = 1.56 \cdot 10^{-6}$ farad, as before, but R is to have a value that will give critical damping. Applied voltage is a constant 12 volts. There is no initial current and no initial charge on the capacitor.

The resistance is critical:

$$R^2 = 4\frac{L}{C} = 4\frac{10^{-2}}{1.56 \cdot 10^{-6}} \qquad R = 160 \text{ ohms} \qquad (15\text{-}1)$$

Hence

$$\tau = \sqrt{LC} = \sqrt{(10^{-2})(1.56)10^{-6}} = 1.25 \cdot 10^{-4} \qquad (15\text{-}2)$$

With the given initial conditions,

$$k_2 = \frac{V}{L} = \frac{12}{10^{-2}} = 1200 \qquad (15\text{-}3)$$

Hence, forced current being zero,

$$i = 1200te^{-8000t} \qquad (15\text{-}4)$$

This result is plotted in Figure 15a.

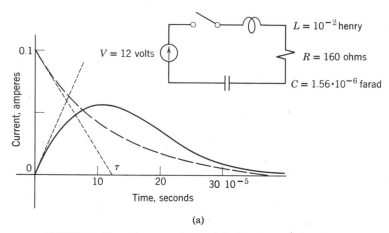

(a)

FIGURE 15a   Response in a critically damped circuit.

16. OSCILLATORY RESPONSE

If resistance of the *RLC* circuit is *less* than critical, the poles of admittance are complex, as considered in Section 7 of Chapter 5, and the natural current is oscillatory. The general oscillatory form can be derived from the exponential form of equation 6-5 of Chapter 5 (page 111) as follows. We have

$$i_n = K_1 e^{s_1 t} + K_2 e^{s_2 t} \qquad (16\text{-}1)$$

and we know from equation 7-1 of the previous chapter that if s_1 and s_2 are complex, they are conjugates; hence

$$i_n = K_1 e^{(\sigma_1 + j\omega_1)t} + K_2 e^{(\sigma_1 - j\omega_1)t} \tag{16-2}$$

where

$$\sigma_1 = -\frac{R}{2L} \quad \text{and} \quad \omega_1 = \sqrt{\frac{1}{LC} - \left(\frac{R}{2L}\right)^2}$$

Using Euler's expansion of $e^{j\omega_1 t}$,

$$\begin{aligned}
i_n &= K_1 e^{\sigma_1 t}(\cos \omega_1 t + j \sin \omega_1 t) + K_2 e^{\sigma_1 t}(\cos \omega_1 t - j \sin \omega_1 t) \\
&= e^{\sigma_1 t}[(K_1 + K_2)\cos \omega_1 t + j(K_1 - K_2)\sin \omega_1 t] \\
&= A e^{\sigma_1 t} \sin(\omega_1 t + \varphi)
\end{aligned} \tag{16-3}$$

where

$$A = 2\sqrt{K_1 K_2}$$

$$\varphi = \tan^{-1} \frac{K_1 + K_2}{j(K_1 - K_2)} \tag{16-4}$$

Equations 16-1 and 16-3 are equivalent, but the former is convenient when s_1 and s_2 are real, and the latter when they are complex.

In the latter case, when the current is oscillatory and equation 16-3 is used, it is more convenient to compute the values of A and φ from the initial conditions of the problem than it is to compute the K's and then find A and φ from them. This is illustrated in the next section. However, it may be noted that for A and φ to be real, so that current is real, it is necessary for K_1 and K_2 to be conjugates if complex.

17. AN OSCILLATORY EXAMPLE

Let us again use the RLC circuit that has appeared in several foregoing examples, but this time let the resistance of the circuit be only 100 ohms. With this low damping the current is oscillatory and from equation 16-2:

$$\sigma_1 = -\frac{R}{2L} = -\frac{100}{2(10^{-2})} = -5 \cdot 10^3 \tag{17-1}$$

$$\omega_1^2 = \frac{1}{LC} - \left(\frac{R}{2L}\right)^2 = 64 \cdot 10^6 - 25 \cdot 10^6 = 39 \cdot 10^6$$

$$\omega_1 = (6.25)10^3 \tag{17-2}$$

Hence

$$i_n = Ae^{-5000t}\sin(6250t + \varphi) \tag{17-3}$$

As in previous examples we shall close a switch at $t = 0$ to connect a source of 12 volts (constant) to the given $\dot{R}LC$ circuit, the circuit being initially without stored energy. The circuit capacitor blocks any forced current from the constant voltage source, so the natural current of equation 17-3 is the total current. A and φ are found from initial conditions.

The initial current is used to find φ. Given that $i(0_+) = 0$, let $t = 0$ in equation 17-3:

$$0 = A \sin \varphi \tag{17-4}$$

and this is satisfied by letting

$$\varphi = 0 \tag{17-5}$$

The initial derivative of current is used to find A. Immediately after closing the switch, the applied voltage V is entirely across the inductance L (since

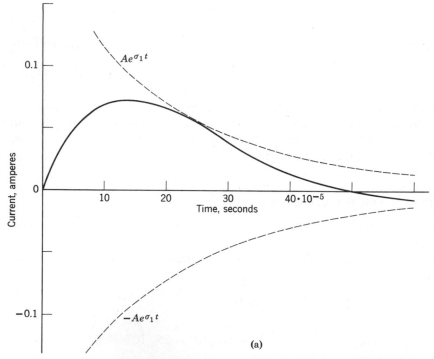

(a)

FIGURE 17a   Oscillatory response of an RLC circuit; $R = 100$ ohms, L and C are unchanged.

the initial capacitor voltage $v_0 = 0$), so at $t - 0_+$, $di/dt - V/L$. (See equation 12-7.) This we set equal to the derivative of current from equation 17-3 (or 16-3); let $\varphi = 0$ as in equation 17-5,

$$\frac{di}{dt} = A(\sigma_1 e^{\sigma_1 t} \sin \omega_1 t + e^{\sigma_1 t} \omega_1 \cos \omega_1 t) \qquad (17\text{-}6)$$

and at $t = 0$,

$$A\omega_1 = \frac{V}{L} \qquad (17\text{-}7)$$

or

$$A = \frac{V}{\omega_1 L} = \frac{12}{(6.25)10^3(10^{-2})} = 0.192 \qquad (17\text{-}8)$$

The resulting expression for current,

$$i = 0.192 e^{-5000t} \sin 6250t \text{ ampere} \qquad (17\text{-}9)$$

is plotted in Figure 17a. It is interesting to compare this oscillatory response with the overdamped and critical responses shown in Figures 13a and 15a. Note that the initial slopes, di/dt, are the same in the three cases. The maxima become later and higher as resistance is decreased.

18. OTHER PROBLEMS

It is tempting to consider others of the innumerable problems that offer points of interest. In the *RLC* circuit alone we might look at the response when damping is very light and the oscillatory current resembles Figure 18a. Responses with an initial current, or an initial charge on the capacitor might be determined. Total current when the applied voltage is alternating rather than direct could be computed. Voltage could be found across one or more of the elements. Calculation of terminal voltage when the current is a discontinuous step function might be considered.

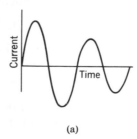

(a)

FIGURE 18a   Oscillatory response with light damping.

When networks of many circuits are used there are many points of interest, but the algebraic complications increase very rapidly, and a disorganized approach to problems of this kind is discouraging. Electronic computers are sometimes useful, but as complications increase it may become a time-consuming business even to manage the programming.

Fortunately there are several means of so organizing network problems that routine labor is reduced to a minimum, and the intellectual work becomes straightforward and clear. In the next two chapters we shall consider three such means: the state-variable equations, the loop equations, and the node equations. The state-variable method is particularly adapted to solution by an electronic computer, but computer solution may be used, or not, as desired, with any of the the the three formulations. Computer programs are often available ("canned" programs) that have been previously prepared to produce numerical answers.

19. LINEAR EQUATIONS, GENERAL

A number of statements have been made in this chapter on the strength of examples, or on arguments of plausibility, rather than formal proof. A precise statement will now be given that can be proved by mathematical reasoning and from which the statements of this chapter can be derived. The four rules for solution that are given in Section 2, for instance, are justified by the following statement.

A differential equation of nth order (a linear equation with constant coefficients) can be written:

$$A_0 x + A_1 \frac{dx}{dt} + A_2 \frac{d^2 x}{dt^2} + \cdots + A_n \frac{d^n x}{dt^n} = f(t) \tag{19-1}$$

The particular integral is

$$x_f = \frac{1}{A_n} e^{s_n t} \int e^{(s_{n-1} - s_n)t} \cdots \int e^{(s_1 - s_2)t} \int e^{-s_1 t} f(t) (dt)^n \tag{19-2}$$

where $s_1, s_2, \ldots, s_n$ are the n roots of the *characteristic* equation

$$A_0 + A_1 s + A_2 s^2 + \cdots + A_n s^n = 0 \tag{19-3}$$

If the roots of the characteristic equation are distinct, the complementary function is

$$x_n = K_1 e^{s_1 t} + K_2 e^{s_2 t} + \cdots + K_n e^{s_n t} \tag{19-4}$$

but if two or more roots are equal so that s_m appears q times, we omit from this expression the corresponding q terms and substitute the following:

$$k_1 e^{s_m t} + k_2 t e^{s_m t} + \cdots + k_q t^{q-1} e^{s_m t} \tag{19-5}$$

Then the solution is the sum

$$x = x_f + x_n \tag{19-6}$$

Values of K and k are not defined in this statement of the solution of the differential equation, and are to be found from given conditions, usually initial conditions.

Several of the differential equations of this chapter are of the first order, and by way of illustration the foregoing general solution can be written for the first-order equation as follows. If

$$A \frac{dx}{dt} + Bx = f(t) \tag{19-7}$$

where A and B are constant, $f(t)$ is given, and x is the response, a function of t, to be found. Then

$$x = \frac{1}{A} e^{-t/\tau} \int e^{t/\tau} f(t) \, dt + Ke^{-t/\tau} \tag{19-8}$$

where $\tau = A/B$.

That equation 19-8 is indeed a solution of the differential equation 19-7, valid for any value of K, can be proved by substitution. It is a formal statement of the response of any of the single-energy circuits of this chapter to any form of disturbance.

20. SUMMARY

After a change in a circuit or network there is a *transient* interval during which the total current comprises *both* the *forced* component and the *natural* component. The *forced* component satisfies the *differential equation* of the network and the *natural* component makes it possible to satisfy the given *initial* conditions.

In Section 2 there are four *rules* for finding total current.

The duration of an exponential voltage or current can be estimated from its *time constant*.

In a number of *examples* the four rules for solution are applied to *single-energy* circuits containing either inductance or capacitance but not both. Examples are shown with and without *initial current*, some having *constant* and some having *sinusoidal* driving force. The driving force may be either an input *voltage* or an input *current*.

The examples include simple inductive *networks* of two loops.

The *double-energy* circuit containing both inductance and capacitance as well as resistance is considered. The natural response can be either a *surge* if the circuit is overdamped, or an *oscillation* in a lightly damped circuit; the intermediate condition of *critical* damping is mathematically distinct but its practical usefulness is greatest for modeling a near-critical physical circuit for which the alternative mathematical forms are awkward.

The mathematical solution of the *general* linear differential equation with constant coefficients is given.

Solution of more complicated problems is postponed until neater formulation of the equations has smoothed the way, and until Laplace transformation is available.

PROBLEMS

6-1. A circuit of R and L in series is supplied by a constant source, as in Figure 1a. What is (*a*) the forced (steady state) component of current, (*b*) the natural (transient) component, and (*c*) the total current, all at $t = 0.18$ second? $E = 10$ volts, $R = 7.5$ ohms, $L = 1.35$ henrys; the switch is closed at $t = 0$.   §2*

6-2. A circuit consisting of R and L in series is carrying 1.00 ampere at time $t = 0$. If $R = 15$ ohms and $L = 450$ millihenrys ($L = 0.450$ henry), what is the current at (*a*) 0.03 second, (*b*) 0.10 second, (*c*) 1.00 second?   §4

6-3. If, in the circuit of Problem 6-2, current is 0.10 ampere at time $t = 0.10$ second, what was the current at $t = 0$?   §4*

6-4. In the circuit shown, the switch, previously open, is closed at time $t = 0$. Compute and plot the forced component, natural component, and total current as functions of time, given $R = 100$ ohms, $L = 1$ henry,

$$E_m = 86 \text{ volts, and } \omega = 1000 \text{ rn/sec.} \qquad \S6$$

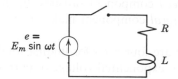

Problems 6-4 and 6-5.

6-5. In the circuit of Problem 6-4, the switch can be closed at any time; let us therefore say that the switch is closed at $t = 0$ and $e = E_m \sin(\omega t + \theta)$. There can be any ratio of R to L. Find the conditions for (*a*) maximum transient component of current (compared to the steady state), and (*b*) zero transient component.   §6*

6-6. The switch s in the circuit shown has been open for a long time. (*a*) Can the switch be closed without producing any transient (natural component) of current? (*b*) If so, when?   §6*

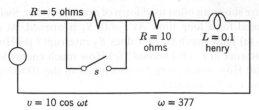

$v = 10 \cos \omega t$ $\qquad$ $\omega = 377$

Problem 6-6.

6-7. A generator of constant current $I = 1$ ampere is switched into the parallel RC combination shown at $t = 0$. The initial voltage on the capacitor is zero. $R = 20$ ohms, $C = 0.1$ microfarad. Find and sketch the voltage across the RC combination as a function of time. $\qquad$ §11*

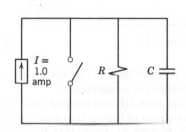

Problem 6-7. $\qquad\qquad\qquad\qquad$ **Problem 6-8.**

6-8. I is a constant-current source of 2 amperes. The switch is closed until $t = 0$ and is then opened. Find current in R thereafter; compute and sketch. $\qquad$ §11

6-9. Section 11 gives a solution for voltage from a source of constant current when the initial capacitor voltage is zero, that is, $v(0) = 0$. Generalize the solution, finding $v(t)$ when the initial capacitor voltage is not zero but $v(0) = v_0$.

§11*

6-10. Given

$$\frac{dv}{dt} + v = t \quad \text{and} \quad v(0) = 1$$

Find v. Of what electric circuit is this the equation? $\qquad$ §11*

6-11. To a circuit of R and L in series we apply the voltage shown: for $t < 0$, $v = 0$, and $i = 0$; for $0 < t < 1$, $v = t$. Find i from $t = 0$ to $t = 1$ second; compute and sketch, showing the time constant. $\qquad$ §8

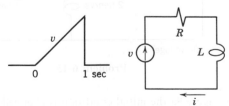

Problem 6-11.

6-12. The figure shows an often used form of arc suppressor. Switches S_1 and S_2 have been closed for a long time. At $t = 0$, S_1 is opened. At $t = 0.1$ second, S_2 is opened. (*a*) How much current does S_2 interrupt? (*b*) How much energy is in the inductance at $t = 0.1$ second? (*c*) How much energy is in the inductance at $t = 0$? (*d*) How much energy is dissipated in the 100-ohm resistance between $t = 0$ and $t = 0.1$ second?   §3*

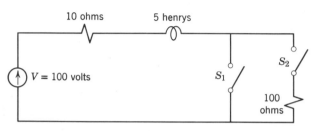

Problem 6-12.

6-13. Show by substitution that the current of equation 6-11 satisfies the differential equation 6-2 as well as the initial condition of equation 6-8.   §6

6-14. Write the circuit equation (similar to equation 1-1) that is based on Kirchhoff's voltage law as applied to the capacitive circuit of Figure 7a to which a source of alternating voltage is applied (as in Section 8). Show that the current of equation 8-6 satisfies this circuit equation, and that it also satisfies the initial condition of equation 8-5.   §8

6-15. (*a*) For the network shown, find the input $Y(s)$. (*b*) From the source voltage and $Y(0)$, find i_f. (*c*) Plot poles and zeros of $Y(s)$ in the s plane. (*d*) From the poles, write i_n. (*e*) Write the total current i, and evaluate its coefficients from the initial conditions: $i_1(0) = 0$ and $i_2(0) = 0$. After $t = 0$, the applied voltage is the constant value V.   §10*

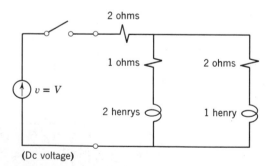

Problem 6-15.

6-16. In Section 9, Figure 9a the initial condition is changed as follows: just before closing the switch at $t = 0$ there is a circulating current of 4 amperes that is down in Z_2 and up in Z_{12}, so that $i_2(0_+) = +4.0$ amperes. Compute and plot $i_1(t)$ and $i_2(t)$. §10

6-17. In the circuit shown, with R, L, and C in series, $R = 500$ ohms, $L = 5$ millihenrys, $C = 0.10$ microfarad. (a) Plot $Y(s)$ for real s, including all finite poles and zeros. Compute enough points for reasonable accuracy. (b) Plot $i(t)$, given $v = 10$ volts and the initial values $v_0 = 0$, $i(0) = 0.01$ ampere. (c) Write $|Z(\omega)|$. §13*

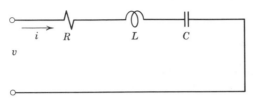

Poblems 6-17, 6-18, and 6-19.

6-18. In the same circuit, but with $R = 100$ ohms, $L = 0.50$ henry, and $C = 2.25 \cdot 10^{-4}$ farad, $v = 0$. When $t = 0$, $i = 1.0$ ampere and $v_0 = 0$. Plot $i(t)$. §13

6-19. In the same circuit, with $R = 100$ ohms, $L = 0.50$ henry, and $C = 2.25 \cdot 10^{-4}$ farad, $v = 0$. When $t = 0$, $i = 0$, and the initial capacitor voltage $v_0 = 50$ volts. Plot $i(t)$. §13*

6-20. In the circuit shown, the switch, after being closed for a long time, is opened at $t = 0$. (a) What is the form of current thereafter? (b) What is i at $t = 0$? What is di/dt at $t = 0$? (c) Show in the s plane the poles and zeros of $Y(s)$ for the circuit into which the source feeds after $t = 0$. §15*

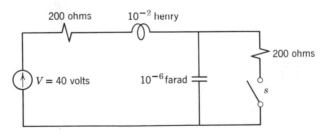

Problems 6-20 and 6-21.

6-21. Continuing Problem 6-20, compute and plot i as a function of time. §15*

6-22. A source and a switch are in series with $L = 10$ millihenrys, $C = 3.5$ microfarads, and $R = 75$ ohms. Until $t = 0$ the switch is open and the capacitor is uncharged. After $t = 0$ the switch is closed and the source has a constant voltage of 5.0 volts (dc). (a) Find current in the circuit as a function of time, and plot. (b) Plot the poles and zeros of $Y(s)$ in the s plane, and also plot the poles and zeros if R is changed to 25 ohms; to 100 ohms; to 150 ohms. §17*

6-23. A source and a switch are in series with $L = 0.30$ millihenry, $C = 0.0012$ microfarad, and $R = 75$ ohms. The switch is closed at $t = 0$. There is no initial charge on the capacitor. The voltage of the source is constant at 1.0 volt (dc). (a) Find current in the circuit as a function of time, and plot to show 3 or 4 cycles. (b) Plot the poles and zeros of $Y(s)$ in the complex s plane, and also plot the poles and zeros if R is changed to 500 ohms; to 1000 ohms; to 2000 ohms.

§17

6-24. In a critically damped circuit, initially at rest, natural current has the characteristic form shown in Figure 15a. If the resistance in the circuit is decreased by about 1%, (a) in what way is this curve changed? (b) How are the poles in the s plane moved? (c) Is exact computation of current made harder or easier by this change?   §17*

6-25. Solve equation 1-1 for current in an inductive circuit by means of the general equations of Section 19.   §19

CHAPTER
7

STATE-VARIABLE EQUATIONS

1. EQUATIONS OF STATE

A network can be characterized by a set of first-order differential equations. This statement can be broadened to include other physical systems, but we are particularly interested in electric networks. In electric networks the variables are usually the currents and voltages that give the *state* of the network, and by this we mean that they give the amount of energy stored in each inductance and capacitance at every instant. Inductor currents and capacitor voltages, though not the only possible state variables, are the most common. (Other possibilities will be mentioned later.)

Let us indicate the state variables by the symbols $x_1, x_2, \ldots$; the letter x is used because the variable may be either a current or a voltage (or perhaps even an electric charge or a magnetic flux) and the state variables are of course functions of time. Input currents and voltages are written $u_1, u_2, \ldots$, also functions of time. The essence of the method is that the derivative of each state variable can be written in terms of all the state variables and all the inputs; that is,

$$\frac{dx_1}{dt} = a_{11}x_1 + a_{12}x_2 + \cdots + b_{11}u_1 + b_{12}u_2 + \cdots$$

$$\frac{dx_2}{dt} = a_{21}x_1 + a_{22}x_2 + \cdots + b_{21}u_1 + b_{22}u_2 + \cdots \qquad (1\text{-}1)$$

$$\vdots$$

$$\frac{dx_n}{dt} = a_{n1}x_1 + a_{n2}x_2 + \cdots + b_{n1}u_1 + b_{n2}u_2 + \cdots$$

This set of equations is often seen written in matrix notation, a kind of shorthand that will be explained in Chapter 9. In the notation of that chapter,

$$[\dot{x}] = [a][x] + [b][u] \tag{1-2}$$

In a linear network the a and b coefficients are constants, not functions of current or voltage. We shall limit the present discussion to non-time-varying networks so that the a's and b's are also not functions of time. These limitations are not necessary, for the method can be extremely general, but they are surely advisable in an introductory study.

2. FORMULATION AND SOLUTION

The first step in writing equations of state must be selection of the state variables. The number of state variables (and hence the value of n in equations 1-1) must be equal to the number of independent energy storage elements in the network, and these can be counted. A possible (and usual) choice is to use each independent inductor current and each independent capacitor voltage as a state variable. (Inductors in series are not independent; the same current flows in both. Capacitors in parallel are not independent; the same voltage is across both. A closed loop consisting of nothing but k capacitors has only $k - 1$ independent voltages, for the sum of all voltages must be zero. A star of k inductors has only $k - 1$ independent currents.) Write, therefore, the appropriate number of equations such as 1-1, and identify each variable x with the proper current or voltage, and each u with an input.

The coefficients, the a's and b's, are now to be found, and the method shown in the next section is a technique that can be used. It is a method that emphasizes the physical meaning. Then the set of equations must be solved simultaneously for the state variables. This can be done by any convenient means. Ordinary classical solution of simultaneous differential equations is possible; the Laplace transformation methods of Chapter 14 are convenient for linear systems; electronic computers are rapid and easy if available.

As in the solution of any differential equations, there must be a set of known conditions, a known *state*, from which to begin. This is often a set of *initial conditions* that exist at the reference time $t = 0$ when a switch is operated or some other change is made in the network. The set of equations 1-1 shows how the state variables change from this known state, and simultaneous solution gives us the several state variables as functions of time. From these state variables, any desired voltages or currents in the network can be found.

The last step of the solution is then to find any or all of the other currents and voltages of the network—those that were not selected to be state variables. This is not difficult; each current or voltage is related to the state variables and the inputs by linear equations (in a linear system):

$$y_1 = c_{11}x_1 + c_{12}x_2 + \cdots + d_{11}u_1 + d_{12}u_2 + \cdots$$
$$y_2 = c_{21}x_1 + c_{22}x_2 + \cdots + d_{21}u_1 + d_{22}u_2 + \cdots \qquad (2\text{-}1)$$
$$\ldots, \text{etc.}$$

where each x is a state variable, each y is another voltage or current of the network, and each u an input, all functions of time. The c and d coefficients are constants which (as will be shown) are easily found. Equations 2-1 can be formalized in the matrix equation

$$[y] = [c][x] + [d][u] \qquad (2\text{-}2)$$

in which $[y]$ is called the *output vector*.

When state equations are applied to a network it is the state of energy storage that is of interest. The behavior of a network at each instant is determined by the form (or topology) of the network, by its inputs, and by its state of energy storage at that instant. It may be part of the solution to draw the form of the network, including only resistors (not inductors or capacitors) in a sort of skeleton. This skeleton network has no memory (as inductors and capacitors have memory). Every current and voltage in a purely resistive network can be found at any instant by knowing only the inputs at that instant. Past history makes no difference; there is no state of energy storage. To such a skeleton (if the system is linear) the sources and the energy storage elements can be connected one at a time and the results can then be superimposed to give the total response of the network. (Superposition is further discussed in Chapter 10.)

A great many different techniques and methods are used by different authors in formulating and solving state-variable equations. We shall now consider a few examples, and if examples in other books look quite different it is not because of a difference in the concept; it is only a matter of different ways and means.

3. AN EXAMPLE OF FORMULATION

The best procedure for solving a problem depends on what information is known. For this example the network is shown in Figure 3a, the constant value of each parameter, each R, L, and C, is known, and the only source has a known voltage, v. Voltage across the capacitor C_2 is called v_{C2} and voltage across the inductor L_3 is v_{L3} ; reference directions of these voltages are shown in the diagram, the higher polarity being at the tail of each arrow marked $+$.

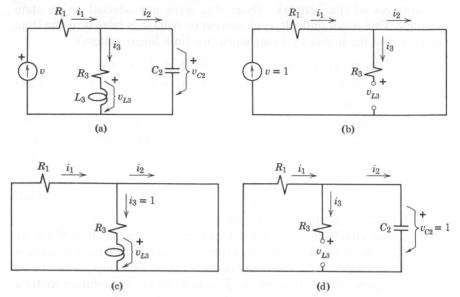

FIGURE 3 (a) A network for state equations. (b) Using v only.
(c) Using L_3 only. (d) Using C_2 only.

There are two state variables, and we shall take the inductor current i_3 as one, and v_{C2} as the other. Using the form of equations 1-1 we write the state equations for this network with coefficients that are as yet unknown:

$$\frac{di_3}{dt} = a_{11}i_3 + a_{12}v_{C2} + b_{11}v \tag{3-1}$$

$$\frac{dv_{C2}}{dt} = a_{21}i_3 + a_{22}v_{C2} + b_{21}v \tag{3-2}$$

To make these state equations useful, the coefficients must be found. We know they are constants, for the network is linear and does not vary with time. Hence if we can find b_{11}, for example, for *any* values of the voltages and currents, it will be the same for *all* values of the voltages and currents. Let us, then, assume some purely arbitrary values for which we can easily tell the value of di_3/dt and find what b_{11} must be.

The element equations tell us that under any circumstances:

$$\frac{di_3}{dt} = \frac{v_{L3}}{L_3} \tag{3-3}$$

$$\frac{dv_{C2}}{dt} = \frac{i_2}{C_2} \tag{3-4}$$

A convenient assumption for finding b_{11} is that $v = 1$, while both i_3 and v_{C2} are zero. Let us (just to find b_{11}) assume these values and proceed as follows:

To find b_{11} and b_{21}:

Let

$$v = 1 \qquad i_3 = 0 \qquad v_{C2} = 0 \tag{3-5}$$

To give the form of the network (the topology) draw Figure 3b, showing resistances and putting a short circuit where zero voltage is assumed and an open circuit where we assume zero current. Include the source. Evidently, under these assumed conditions,

$$v_{L3} = -i_3 R_3 = 0 \quad \text{and} \quad i_2 = \frac{v}{R_1} = \frac{1}{R_1} \tag{3-6}$$

Use the assumptions of equations 3-5 in the state equation 3-1; set the derivative equal to equation 3-3 and, with equation 3-6,

$$b_{11} = \frac{di_3}{dt} = \frac{v_{L3}}{L_3} = 0 \tag{3-7}$$

Thus b_{11} is found to be zero.

Use the assumptions of equations 3-5 in the state equation 3-2, and set the derivative equal to equation 3-4; then with equation 3-6:

$$b_{21} = \frac{dv_{C2}}{dt} = \frac{i_2}{C_2} = \frac{1}{R_1 C_2} \tag{3-8}$$

Thus we have evaluated b_{21}.

To find a_{11} and a_{21}:

Let

$$v = 0 \qquad i_3 = 1 \qquad v_{C2} = 0 \tag{3-9}$$

Draw the resistance network, without the source but with L_3, Figure 3c, from which:

$$v_{L3} = -i_3 R_3 = -R_3 \qquad i_2 = -i_3 = -1 \tag{3-10}$$

Using the assumptions in the state equations, as before, we thus obtain:

$$a_{11} = \frac{di_3}{dt} = \frac{v_{L3}}{L_3} = -\frac{R_3}{L_3} \tag{3-11}$$

$$a_{21} = \frac{dv_{C2}}{dt} = \frac{i_2}{C_2} = -\frac{1}{C_2} \tag{3-12}$$

To find a_{12} and a_{22}:

Let

$$v = 0 \qquad i_3 = 0 \qquad v_{C2} = 1 \tag{3-13}$$

Draw the resistance network of Figure 3d, this time with C_2, from which:

$$v_{L3} = 1 \qquad i_2 = -\frac{1}{R_1} \tag{3-14}$$

Then

$$a_{12} = \frac{di_3}{dt} = \frac{v_{L3}}{L_3} = \frac{1}{L_3} \tag{3-15}$$

$$a_{22} = \frac{dv_{C2}}{dt} = \frac{i_2}{C_2} = -\frac{1}{R_1 C_2} \tag{3-16}$$

All of the coefficients for the state equations of this network have now been evaluated, and we combine equations 3-7, 3-8, 3-11, 3-12, 3-15, and 3-16 with equations 3-1 and 3-2 to write:

$$\frac{di_3}{dt} = -\frac{R_3}{L_3} i_3 + \frac{1}{L_3} v_{C2} \tag{3-17}$$

$$\frac{dv_{C2}}{dt} = -\frac{1}{C_2} i_3 - \frac{1}{R_1 C_2} v_{C2} + \frac{1}{R_1 C_2} v \tag{3-18}$$

Thus we have completed the required formulation of the state equations. Solution, which means finding i_3 and v_{C2}, remains to be done.

There are other ways to formulate these state equations instead of following the method shown. For instance, network topology can be expressed algebraically (by the connection equations of Section 7 of Chapter 1) instead of graphically, but this fails to take advantage of what we already know about resistance networks and tends to be correspondingly longer and more obscure. Also, loop or node equations (Chapter 8) can be solved simultaneously to yield the first-order state equations. Or a single network equation of the nth order can be found, and by repeated differentiations of a solution of this equation a set of state equations can be obtained. These are among the alternatives that must be mentioned because they are often used or at least described, but for this introduction to state variables a single simple method is perhaps enough.

4. ANOTHER EXAMPLE

The network of Figure 4a introduces certain new ideas in the state-variable solution, although it has only one energy storage element and hence one state variable. All currents in the network are zero initially; physically this may mean that a switch is closed to connect a 12-volt battery to a network that was previously without current, but mathematically all that we need know is that (for whatever reason) $i_2 = 0$ when $t = 0$. The following formulation is arranged to emphasize the method; little explanation is given, for the steps are the same as in the previous example.

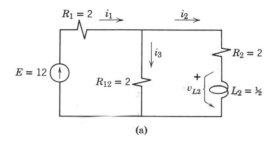

(a)

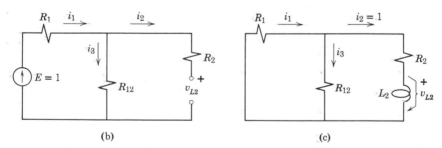

(b)   (c)

FIGURE 4 (a) A network; values are volts, ohms, and henrys.
(b) Letting $E = 1$ and $i_2 = 0$. (c) Letting $E = 0$ and $i_2 = 1$.

Given the network in Figure 4a, select i_2 as the state variable and write the state equation:

$$\frac{di_2}{dt} = a_{11}i_2 + b_{11}E \qquad (4\text{-}1)$$

Write the element equation:

$$\frac{di_2}{dt} = \frac{v_{L2}}{L_2} \qquad (4\text{-}2)$$

To find b_{11}, let $E = 1$ and $i_2 = 0$. Draw the resistance network of Figure 4b with the source $E = 1$. From that figure, with these conditions, write:

$$v_{L2} = \frac{R_{12}}{R_1 + R_{12}} \qquad (4\text{-}3)$$

Hence

$$b_{11} = \frac{di_2}{dt} = \frac{v_{L2}}{L_2} = \frac{R_{12}}{(R_1 + R_{12})L_2} = \frac{2}{2} = 1 \qquad (4\text{-}4)$$

To find a_{11}, let $E = 0$ and $i_2 = 1$. Draw the resistance network, including L_2 as in Figure 4c, and write that with these conditions:

$$v_{L2} = -1\left(R_2 + \frac{R_1 R_{12}}{R_1 + R_{12}}\right) = -3 \qquad (4\text{-}5)$$

Hence

$$a_{11} = \frac{di_2}{dt} = \frac{v_{L2}}{L_2} = -\frac{3}{1/2} = -6 \qquad (4\text{-}6)$$

The state equation is therefore

$$\frac{di_2}{dt} = -6i_2 + 12 \qquad (4\text{-}7)$$

This completes the formulation of the state equation. In this example the single first-order equation can be solved by integration.

Solution:

$$dt = \frac{di_2}{-6i_2 + 12} \qquad (4\text{-}8)$$

Integrating, and introducing the initial condition that when $t = 0$, $i_2 = 0$ to establish the lower limit:

$$t = -\left[\frac{1}{6}\ln(-6i_2 + 12)\right]_0^{i_2} = -\frac{1}{6}\ln\frac{-6i_2 + 12}{12} \qquad (4\text{-}9)$$

$$i_2 = 2 - 2e^{-6t} \qquad (4\text{-}10)$$

Thus we have found the state variable from which (together with the input) it is possible to compute any current or any voltage by a linear relation. The same current was computed as equation 9-11 (page 139) of Chapter 6 by quite another method, and it is plotted in Figure 9c of that chapter. The method used in Chapter 6 was to express admittance as a function of the complex frequency s and to work with transforms (or phasors) of voltage and current; such a solution is in the *frequency domain*, as contrasted with the state-variable solution in the present chapter which makes no reference to admittance, impedance, transformation, frequency, or the complex variable s, but procedes by solving differential equations in the *time domain*.

To illustrate the determination of another current in the network from the state variable i_2, let us find the source current i_1. Since the network is linear,

$$i_1 = ci_2 + dE \qquad (4\text{-}11)$$

where c and d are constants as yet unknown. (This is a specific application of equation 1-3.) To evaluate d, let $i_2 = 0$ and $E = 1$; this is the situation represented by Figure 4b. In this situation, equation 4-11 says that $i_1 = d$ and, from Figure 4b, $i_1 = E/(R_1 + R_{12})$. Hence we find d:

$$d = i_1 = \frac{1}{R_1 + R_{12}} = \frac{1}{4} \qquad (4\text{-}12)$$

To evaluate c, let $i_2 = 1$ and $E = 0$; this is represented in Figure 4c. Now $i_1 = c$, and from Figure 4c, $i_1 = R_{12}/(R_1 + R_{12})$. Hence

$$c = i_1 = \frac{R_{12}}{R_1 + R_{12}} = \frac{1}{2} \qquad (4\text{-}13)$$

Since c and d are constant, these values which are found for assumed values of variables must be valid in general and equation 4-11 becomes

$$i_1 = \tfrac{1}{2}i_2 + \tfrac{1}{4}E = 1 - e^{-6t} + \tfrac{12}{4} = 4 - e^{-6t} \qquad (4\text{-}14)$$

This input current i_1, like the inductor current i_2, was found in Chapter 6 by a computation in the frequency domain, and the same result was expressed as equation 9-6 (page 139) and plotted in Figure 9c.

5. AN EXAMPLE WITH TWO INDUCTANCES

A slightly more complicated network leads to state equations that cannot be solved by mere integration. This network is used to illustrate an iterative or step-by-step solution. In this example the solution will be carried out by means simple enough for manual operation, but anyone wishing to do so can readily program the problem on either a digital or analog computer. Adaptability to computer solution is one of the greatest advantages of the state-variable formulation of network equations.

Figure 5a shows a network. L_1 is a model of an inductor with negligible resistance. The switch is closed at $t = 0$ at which time i_1 and i_2 are zero. Values of parameters are shown. Voltage of the source is independent of current. It is required to find i_2, from which we can compute the voltage across R_2, this being the output of the network.

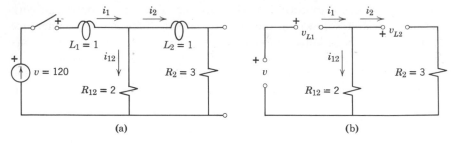

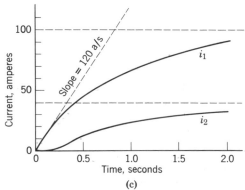

FIGURE 5 **(a)** Voltage, resistance, and inductance are in volts, ohms, and henrys. **(b)** The skeleton resistive network. **(c)** The resulting currents.

There are two state variables, and the obvious choices are i_1 and i_2. Formulation of the state equations proceeds exactly as in the previous examples, and will be indicated but not discussed.

State equations with undetermined coefficients are:

$$\frac{di_1}{dt} = a_{11} i_1 + a_{12} i_2 + b_{11} v \qquad (5\text{-}1)$$

$$\frac{di_2}{dt} = a_{21} i_1 + a_{22} i_2 + b_{21} v \qquad (5\text{-}2)$$

Element equations are:

$$\frac{di_1}{dt} = \frac{v_{L1}}{L_1} \qquad \frac{di_2}{dt} = \frac{v_{L2}}{L_2} \qquad (5\text{-}3)$$

Figure 5b shows the skeleton network of resistance only.

Let $v = 1$, $i_1 = 0$, $i_2 = 0$. Then

$$b_{11} = \frac{di_1}{dt} = \frac{v_{L1}}{L_1} = \frac{v}{L_1} = 1 \tag{5-4}$$

$$b_{21} = \frac{di_2}{dt} = \frac{v_{L2}}{L_2} = 0 \tag{5-5}$$

Let $v = 0$, $i_1 = 1$, $i_2 = 0$. Then $v_{L1} = -R_{12}i_{12}$ and

$$a_{11} = \frac{di_1}{dt} = \frac{v_{L1}}{L_1} = -\frac{R_{12}}{L_1} = -2 \tag{5-6}$$

By Kirchhoff's voltage law, $v_{L2} = R_{12}i_{12}$ so

$$a_{21} = \frac{di_2}{dt} = \frac{v_{L2}}{L_2} = \frac{R_{12}}{L_2} = 2 \tag{5-7}$$

Let $v = 0$, $i_1 = 0$, $i_2 = 1$. Then $v_{L1} = R_{12}i_2 = R_{12}$

$$a_{12} = \frac{di_1}{dt} = \frac{v_{L1}}{L_1} = \frac{R_{12}}{L_1} = 2 \tag{5-8}$$

Also, $v_{L2} = -(R_2 + R_{12})i_2$, so

$$a_{22} = \frac{di_2}{dt} = \frac{v_{L2}}{L_2} = -\frac{R_2 + R_{12}}{L_2} = -5 \tag{5-9}$$

Introducing the six coefficients thus determined, the state equations are:

$$\frac{di_1}{dt} = -2i_1 + 2i_2 + 120 \tag{5-10}$$

$$\frac{di_2}{dt} = 2i_1 - 5i_2 \tag{5-11}$$

The initial conditions are $i_1(0) = i_2(0) = 0$.

It would now be possible to solve these two equations simultaneously by conventional means, and if this were done (it is shown in Chapter 15) the result would be

$$i_1 = 100 - 96e^{-t} - 4e^{-6t} \quad \text{amperes} \tag{5-12}$$

$$i_2 = 40 - 48e^{-t} + 8e^{-6t} \quad \text{amperes} \tag{5-13}$$

These currents are plotted in Figure 5c, and it can be seen that they meet the following requirements that initial conditions and inspection of the network show to be necessary:

1. Both currents have initial values of zero.
2. The final steady-state value of i_1 is 100 amperes, and that of i_2 is 40.
3. Initial slope of i_2 is zero.
4. Initial slope of i_1 is 120 amperes/second, which is correctly equal to v/L_1.

However, this formal solution of the state equations is not the object of our present discussion. Rather, we wish to consider a numerical solution of a kind that might be obtained from a computer.

6. NUMERICAL SOLUTION

State equations such as 5-10 and 11 are particularly well adapted to numerical solution of an iterative or step-by-step kind. This solution could conveniently be programmed on an electronic computer, but if a computer is not available paper and pencil and patience will do the job. The following method, attributed to Euler, is clear and simple though not very precise.

Step 1. Start column 1 of Table 7-6 at $t = 0$ with the given initial currents $i_1 = 0$ and $i_2 = 0$. Use these values in the state equations to compute the derivatives di_1/dt and di_2/dt at $t = 0$. Assume that these initial derivatives continue through an arbitrary increment of time Δt, taken to be 0.1 second, and compute the corresponding increments of the currents Δi_1 and Δi_2. Add these increments to i_1 and i_2 at the top of column 1 to obtain i_1 and i_2 at the top of column 2.

Step 2. Repeat step 1. Begin with the values of i_1 and i_2 which have been computed (approximately) for $t = 0.1$. Compute new derivatives from the state equations. Assume these new derivatives continue through a second arbitrary step of 0.1 second, and compute the increments of current in this second step. Add these increments of current to i_1 and i_2 at the beginning of

TABLE 7-6   An Iterative Computation From State Equations 5-10 and 5-11

			0.1	0.2	0.3	0.4	0.5	0.6	0.7
①		$t = 0$	0.1	0.2	0.3	0.4	0.5	0.6	0.7
②		$i_1 = 0$	12.0	21.6	29.8	36.9	43.3	49.0	54.1
③		$i_2 = 0$	0	2.40	5.52	8.72	11.74	14.53	17.07
④	$\dfrac{di_1}{dt} = -2i_1 + 2i_2 + 120 = 120$		96.0	81.6	71.4	63.6	56.9	51.1	45.9
⑤	$\dfrac{di_2}{dt} = 2i_1 - 5i_2$	$= 0$	24.0	31.2	32.0	30.2	27.9	25.35	22.85
⑥		$\Delta t = 0.1$	0.1	0.1	0.1	0.1	0.1	0.1	0.1
⑦		$\Delta i_1 = 12.0$	9.6	8.16	7.14	6.36	5.69	5.11	4.59
⑧		$\Delta i_2 = 0$	2.40	3.12	3.20	3.02	2.79	2.54	2.28

step 2 to obtain i_1 and i_2 at $t = 0.2$ which is the end of step 2 and also the beginning of step 3; enter these new currents at the top of column 3.

Step 3. Repeat step 2, and continue as long as desired. Table 7-6 gives the essential numbers for 20 steps, 2.0 elapsed seconds, and the results are shown in Figure 5c. The results obtained by this numerical computation are indistinguishable in the diagram from values computed from equations 5-12 and 5-13, as they are within the width of the printed line.

This step-by-step method is approximate. The initial slope of i_1 is shown in Figure 5c, and it is evident that the true slope diminishes appreciably during the first step, before $t = 0.1$ second. Is this too long a step? How long a step may be used?

The length of step is a matter of judgment, weighing accuracy against labor (or machine cost). If the output has exponential components, as in the present example, a rapidly vanishing component can be lost in a long step, and the increment of time should be less than the shortest time constant. If an output is oscillatory, each increment of time should be a small part of a cycle.

In the present example the time constants of exponential components are shown by equations 5-12 and 5-13 to be 1 second and $\frac{1}{6}$ second. A time increment of 0.1 second per step was arbitrarily chosen, and is rather long.

It is entirely possible to change the length of step in the course of a computation. We might have started with steps of 0.05 second and changed to 0.10 second at a later time. Clearly the longer the step the less the computational work, but the greater the error.

It does not follow, however, that accuracy can be increased indefinitely by shortening the steps. If steps are very short the round-off error can become greater than the errors of assumptions. How much error results if such a number as 6.543 is rounded off to 6.54? This is not the place to consider

TABLE 7-6 (*continued*)

0.8	0.9	1.0	1.1	1.2	1.3	1.4	1.5	1.6	1.7	1.8	1.9	2.0
58.7	62.8	66.5	69.9	73.0	75.7	78.1	80.3	82.3	84.1	85.7	87.1	88.4
19.35	21.4	23.3	24.9	26.4	27.8	29.0	30.1	31.1	32.0	32.8	33.5	34.2
41.3	37.2	33.6	31.0	26.8	24.2	21.8	19.6	17.6	15.8	14.2	12.8	
20.65	18.6	16.5	15.3	14.0	12.4	11.2	10.1	9.1	8.2	7.4	6.7	
0.1	0.1	0.1	0.1	0.1	0.1	0.1	0.1	0.1	0.1	0.1	0.1	
4.13	3.72	3.36	3.10	2.68	2.42	2.18	1.96	1.76	1.58	1.42	1.28	
2.06	1.86	1.65	1.53	1.40	1.24	1.12	1.01	0.91	0.82	0.74	0.67	

round-off error, but even the results obtained in Table 7-6 are slightly influenced by the number of figures carried through the calculations.

Instead of always using shorter increments of time to gain accuracy, it is sometimes better to use a more sophisticated method of approximation. There are closer approximations than the Euler method used here for illustration. An electronic computer that does not mind a longer calculation could be programmed to use a "second-order Euler approximation" in which slopes at the beginning and end of a step are averaged, or the still more recondite Runge-Kutta method.

7. ANALOG SOLUTION

If it were possible to have a box that multiplied, a box that added, and a box that integrated, we could use them to solve state equations. We could feed x_1 into the box that multiplied, which would be so adjusted that $a_{11} x_1$ came out. Then this could be fed into the adding box along with $a_{12} x_2, b_{11} u_1$ and any other functions required by the state equation:

$$\frac{dx_1}{dt} = a_{11}x_1 + a_{12}x_2 + \cdots + b_{11}u_1 + \cdots \tag{7-1}$$

The output of the adding box would be dx_1/dt, which would then go into the integrating box, from which the output would be x_1. This x_1 we feed back to supply the input to the multiplying box, as suggested in Figure 7a. We may also display x_1 on a screen, for it is part of the answer to our state-variable problem.

We should, of course, need to solve the other state equations simultaneously, and in doing so we would obtain $a_{12} x_2$ and other quantities that are needed to compute x_1. When all interconnections are completed there remain only the inputs such as u_1 that have to be supplied to the computing system. This is suggested by Figure 7b for a set of two state equations with one input.

It is of course necessary for the computing system to start with the given initial conditions of the state variables. If these are not zero, the proper values must be fed into the system.

The boxes that we need for our hypothetical computing device are in fact closely approximated by the units of an actual *analog computer*. Inside each computer unit is an amplifier with heavy feedback so that the output of the unit is an excellent approximation of addition, multiplication, or integration. An amplifier with high gain can be made to integrate by using capacitive feedback. Resistive feedback makes an amplifier multiply by a desired constant. There are various practical matters to be taken care of, such as providing high input impedance, constancy of operation, possible change of

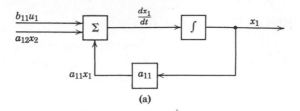

(a)

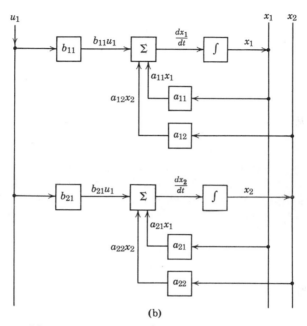

(b)

**FIGURE 7   (a)   A loop for integrating one of a set of state equations.
(b)   Two state equations; others could be added.**

scale, and possible reversal of sign, but these requirements can all be arranged in the *operational amplifiers* of an analog computer.

Output is conveniently displayed on an oscilloscope screen. Inputs and
initial conditions are repeatedly cycled into the computer so that the output
is seen as a wave standing on the screen. It is then easy to study the effect of
changing network parameters, or inputs, or initial conditions, merely by
adjusting the computer settings and watching the resulting change of output.

Analog computers have other kinds of elements besides the three that
have been mentioned. It is possible to solve state equations that are nonlinear or time-varying, but these are definitely less convenient than the simple
equations with constant coefficients that have been considered here.

8. COMPARISON OF SOLUTIONS

It has been suggested that the high-speed electronic digital computer offers something more than speedy calculation by old methods; perhaps the computer promotes new ways of thinking. The present chapter brings this suggestion into focus, for here we have clearly before us a type of solution that is not merely facilitated by the computer but that is exceptionally adapted to the computer's way of working.

A real physical system, an actual electrical network, does what it does at a particular moment because that is the only behavior possible under the conditions at that moment. As time passes, the conditions change, and the behavior changes correspondingly. The actual network knows only the state and the input at the moment of action, and on these it acts.

The iterative solution of a digital computer is quite similar except that it proceeds in a series of finite steps rather than by continuous action. It is assumed that conditions remain unchanged for a short increment of time, and then change suddenly. This procedure by steps introduces a certain imprecision, but this can be made quite small. It was doubtless one of the great triumphs of human thought when a formal solution of the differential equations of physical problems was found to be possible, but a step-by-step solution is remarkably attractive because of its similarity to real physical behavior.

Since a network acts in compliance with the value that each parameter has now, it is immaterial what the parameter was in the past, or what it will be in the future. To an actual network, constancy of parameters, linearity of equations, is not relevant. Since a step-by-step solution can follow any variations that are known to the programmer, a digital computer is not bothered by changing parameters, and nonlinearity is not a great difficulty. It is surely safe to say that for many problems, particularly nonlinear problems for which a formal solution is impracticable, the knowledge that a high-speed electronic digital computer is at hand may well influence the scientist's way of thinking.

The digital computer offers great numerical precision, and its printouts may contain as many significant figures as are justified by the data. An analog computer, however, is attractive in other ways. It does not employ an approximate method of finite differences for its action at each instant is determined by its state at that instant, as in the actual network. An analog computer is quick, easy, and (compared with a digital computer) cheap. It may give a convenient display of results on a screen. It can perhaps work with variable parameters and nonlinear relations, although its capacity for doing so depends on its construction.

It might be mentioned that neither analog computation nor digital computation by finite differences is exactly new. Mechanical analog machines have been in use for decades, and digital solution by finite differences for centuries. It is the greater speed obtainable from electronic devices that is new. Now computing machines are so fast, so readily available, and so large in problem-solving capacity that their numerical results compete with the elegance of formal solutions.

9. CHOICE OF STATE VARIABLES

The selection of state variables is an interesting subject. There are as many independent state variables as there are independent energy-storage elements in a network, and this suggests that the currents through inductances and the voltages across capacitances can be the state variables. Such currents and voltages do, indeed, provide a convenient set of state variables, but other sets are equally possible and are sometimes preferred.

It is evident that instead of using voltage across a capacitance as a state variable, the electric charge q entering the branch that contains the capacitance may be used; one is merely a constant times the other. If this same branch contains inductance, current in the branch can be another state variable, and this current is dq/dt. Thus a charge and its time derivative can be the two state variables of a branch containing L and C in series.

Following this pattern, it is sometimes convenient to take an electrical quantity and its first, second, and higher time derivatives as the state variables of a network. This possibility can be considered for use if the transfer function, or other network function, is known as an nth-order differential equation, and for some reason (possibly for machine solution) a set of n first-order state equations is desired.

We have seen that the state variable *capacitor-voltage* can be changed to another state variable *capacitor-charge* by a simple proportionality. More generally, each of a set of state variables is a linear combination of the state variables of another set. This suggests, correctly, that a very wide choice of state-variable sets is permissible.

The intriguing mathematics of state vectors will not even be outlined here. If the n state variables of a system are considered to be the components of a vector in n dimensional space, this *state space* can be the subject of very extensive mathematical investigation. Many books have been written. Perhaps it will be well to close this chapter with the suggestion that the linear transformation that converts from one set of state variables to another is related to a change of orthagonal axes in state space.

10. SUMMARY

One way (the first of several to be mentioned) to bring order to network relations is to write the *state equations*.

The state of a system (for us, the energy distribution in a network) is given by a set of *state variables*.

The derivatives of the state variables can be equated to the sum of state variables and inputs, each multiplied by the appropriate coefficient giving a set of *first-order* differential equations. These state equations are solved for the state variables.

The *number* of state variables is equal to the number of independent energy-storage elements in a network. It is often convenient, though not necessary, to select *inductor currents* and *capacitor voltages* as the state variables.

Any network voltage or current can be found from the state variables.

Formulation of the state equations requires evaluation of the coefficients. These can be found from the topology and parameter values of the network, and the input quantities. A systematic method of evaluation is shown, making use of the skeleton *network of resistances*.

An example shows *solution* of state equations *by integration*, and the computation of output values from state variables.

Another example shows the *solution* of state equations by a method of *finite differences*. An approximate numerical solution is obtained, by a technique that might be used on an electronic computer. Adaptability to a programmed *digital computer* is one of the advantages of the state-variable method.

The use of an *analog computer* for solution of state-variable equations is considered.

A network can have many *different sets* of state variables.

PROBLEMS

7-1. A circuit consists of inductance L and resistance R in series with a source of constant voltage V. It is desired to use the methods of this chapter to write a state equation for this circuit. Let us follow the example of Section 3. (*a*) Note that there is one state variable and write an equation in the form of equation 3-1. Also write an element equation somewhat similar to 3-3. (*b*) Writing appropriate equations similar to 3-5, 3-6, and 3-7, obtain b_1. (*c*) Writing appropriate equations similar to 3-9, 3-10, and 3-11, obtain a_1. (*d*) Write a state-variable equation similar to equation 3-17, and show that this can be made the same as the familiar $Ri + L(di/dt) = v$.   §3

7-2. Continuing Problem 7-1, solve the state-variable equation obtained in that problem to give the current i, the initial current being zero. Write $v_L = ci + dv$, similar to equation 4-11 (or 1-3). Then (a) find the coefficients c and d by methods similar to those in equations 4-13 and 4-12. (b) Finally, write an expression for v_L, the voltage across the inductance. §4*

7-3. Generalizing Problem 7-2, solve the same state equation from Problem 1 for current i, with the assumption that the initial current is $i(0)$, not zero. Find the voltage across the inductance from $v_L = ci + dv$ by evaluating the coefficients c and d. §4

7-4. Voltage v produces current i through resistance R, inductance L, and capacitance C in series. Write the state equations for the circuit. Noting the two energy-storage elements, the two equations can be:

$$\frac{di}{dt} = a_{11}i + a_{12}v_C + b_1 v$$

$$\frac{dv_C}{dt} = a_{21}i + a_{22}v_C + b_2 v$$

where v_C is capacitor voltage. Using the element equations $v_L = L(di/dt)$ and $i = C(dv_C/dt)$, evaluate the coefficients a_{11}, a_{12}, b_1, a_{21}, a_{22}, and b_2. §4*

7-5. Continuing Problem 7-4, solve the two state equations for current, given that the applied voltage is a constant value V after $t = 0$, and that initial current and initial capacitor voltage are both zero. Use either (a) classical solution of the differential equations (as in Chapter 6), or (b) solution by transformation (as in Problem 15-1), or (c) a numerical solution, as in this chapter, or (d) a numerical solution on an electronic digital computer, or (e) an analog computer solution. In options c, d, or e, select arbitrary values for L, C, and R. If a computer is used, give R a range of values to include both oscillatory and overdamped currents. Compare with Sections 12 and 16 of Chapter 6. §6

7-6. Write state equations for the network in the diagram. Voltage applied, v_1, is the constant value $V = 24$ volts. Using equations $di_2/dt = a_{11}i_2 + a_{12}i_1 + b_1 V$ and $di_1/dt = a_{21}i_2 + a_{22}i_1 + b_2 V$, find the six coefficients. §6*

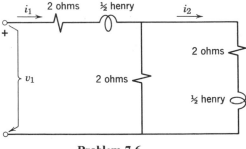

Problem 7-6.

7-7. Continuing Problem 7-6, solve the state equations to obtain the entering current i_1, knowing initial currents to be zero. This can be done (as in Problem 7-5) either by classical means, or by transformation (as in Problem 15-2), or numerically, with or without computer. Compare the result with equation 10-9 and Figure 10b of Chapter 6.   §6

7-8. The switch is opened at $t = 0$ after having been closed for a long time. Impressed current is constant at 10^{-2} ampere. Compute $v(t)$. Sketch $v(t)$, showing a time constant on the time scale.   §6

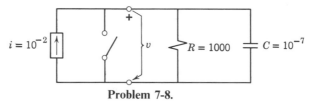

Problem 7-8.

7-9. A 10-volt battery voltage is suddenly applied to the terminals of the circuit shown. There is no initial current in the circuit. Use the two currents in inductances as state variables and write the state equations; find the six coefficients of these equations.   §6

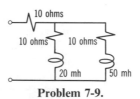

Problem 7-9.

7-10. Continuing Problem 7-9, find the entering current after the switch is closed. Equations can be solved (as in Problem 7-5) by any desired method. (See Problem 15-3.)   §6*

7-11. In the diagram shown, the switch is closed at time $t = 0$, after having been open for a long time. The source is constant at 12 volts. Find the switch current as a function of time after closing the switch. (a) First write the state equation using the capacitance voltage as variable: $dv_C/dt = av_C + bV$, and evaluate the coefficients. Then (b) solve the state equation and obtain i_{12}.   §6*

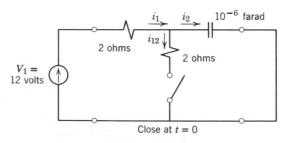

Problem 7-11.

7-12. A network diagram is shown with Problem 6-15, page 158. The switch is closed with the network initially at rest. Write the state-variable equations for this network.   §6

7-13. Continuing Problem 7-12, solve the state equations to obtain the entering current after closing the switch. As in Problem 5, any means of solution may be used, either classical, by transformation (as in Problem 15-4), numerical, or by computer.   §6*

7-14. The network shown is initially without current or charge. A 10-volt battery is suddenly connected to this network. Write a pair of state equations for this network, evaluating all coefficients.   §6

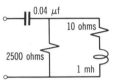

Problem 7-14.

7-15. Continuing Problem 7-14, solve for current entering the network after the battery is connected. As in Problem 5, any means of solution may be used, either classical, by transformation (as in Problem 15-5), numerical, or by computer.   §6*

7-16. In the network shown, find the output voltage $v_o(t)$. There is no output current. There is no initial charge on the capacitor. The input voltage $v_i(t)$ is a step function, being 50 volts for positive time and zero for negative time. Use the state-variable method to find the output voltage.   §6

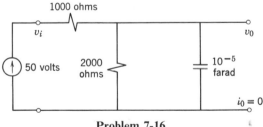

Problem 7-16.

CHAPTER
8

LOOP AND
NODE EQUATIONS

1. TIME AND FREQUENCY DOMAINS

The analysis of networks can also be organized in *loop* equations, or in the dual *node* equations. These, also, like state-variable equations, begin with the voltage and current relations of elements, and with Kirchhoff's two laws.

Elements in series are first added to give *branch* equations. It would be possible to write and solve these branch equations together with the *connection* equations from Kirchhoff's laws, but in most cases it is simpler to use the branch and connection equations only to derive the loop and node equations.

Loop and node equations can be written in either the time domain, involving voltages, currents, and their time derivatives in differential equations, or in the frequency domain with purely algebraic equations. State equations were discussed in the previous chapter in the time domain, partly because it is customary to do so, and partly because this method leads to sets of first-order differential equations which are readily handled on digital or analog computers. In the present chapter, however, loop and node equations are treated in the frequency domain, using admittance and impedance with transforms of voltage and current.

Our use of admittance and impedance does not limit the application of loop and node equations to sinusoidal voltages and currents. Y and Z may be considered to be $Y(s)$ and $Z(s)$; then voltage and current can be considered exponential functions of time: real exponentials if s is real,

182

alternating if s is imaginary, dc if s is zero. Indeed, with the Laplace concepts of Chapter 13 or 14, any voltage or any current, steady or transient, can be used in the loop and node equations. Our only limit is to a linear system, and this is necessary in any case.

Loop and node equations, then, will be written in terms of the frequency variables s or $j\omega$. It must be remembered that these are variables, and the solutions of the equations are in general functions of frequency, although in present examples we may choose to give a specific numerical value to the frequency ω. Indeed we shall often use direct current with $\omega = 0$ for illustration.

2. BRANCH EQUATIONS

Consider a network. For example we may speak of the network of Figure 2a, with six branches. All twelve currents and voltages of this network can be found if all sources (only one is shown) and all parameters are known. Twelve equations are necessary to define this network.

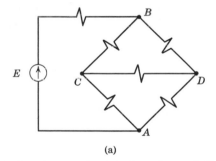

(a)

FIGURE 2a A network of six branches; a bridge circuit.

There are six *branch* equations of the form

$$V_{BC} = Z_{BC}I_{BC}$$
$$\cdots$$
$$V_{BA} = Z_{BA}I_{BA} + E$$

 (2-1)

There are also six *connection* equations.

Three of these, from Kirchhoff's current law, are *junction* equations of the form

$$I_{BC} + I_{BD} + I_{BA} = 0$$

 (2-2)

It might be expected that there would be four junction equations as there are four junctions, but one of these is not independent and can be derived from the other three.

Finally, there are three *loop* equations, derived from Kirchhoff's voltage law, of the form

$$V_{BC} + V_{CD} + V_{DB} = 0 \tag{2-3}$$

(More than three loops can be found, but only three are independent.)

These twelve equations uniquely define the network, and they can be solved simultaneously for the twelve unknown quantities, namely the six branch currents and six branch voltages that appear in the foregoing equations. But this simultaneous solution is rather tiresome, and a good deal of dull work can be saved by using techniques more ingenious than the branch method.

3. LOOP EQUATIONS

The ingenuity of the loop method lies in the selection of currents to be determined. By calling the current around a loop I_1, and that around another loop I_2, and so on, we need only as many currents as there are independent loops.

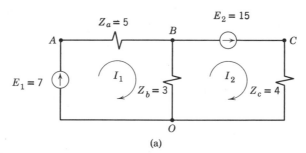

(a)

FIGURE 3a   A network of two circuits. Numerical values are in volts and ohms.

Thus in the network of Figure 3a there are two loops, and we assume two loop currents as shown. Current in Z_a is I_1. Current in Z_c is I_2. Current in Z_b is $(I_1 - I_2)$. Using Kirchhoff's voltage law to equate the applied electromotive force to the sum of the voltages around a loop,

$$Z_a I_1 + Z_b(I_1 - I_2) = E_1 \tag{3-1}$$

This equation applies to loop 1. An alternative way of writing the same equation, a way that will be found more expeditious, is to look at loop 1 and ask, what voltage is produced in loop 1 by current in loop 1? The answer is, $(Z_a + Z_b)I_1$. Then, what voltage is produced in loop 1 by current in loop 2? The answer is $-Z_bI_2$, the sign being negative because in Z_b the arrow direction of I_2 is opposite to the arrow direction of I_1. This leads to the equation

$$(Z_a + Z_b)I_1 - Z_bI_2 = E_1 \tag{3-2}$$

which is obviously the same as equation 3-1.

Now, what voltage is produced in loop 2 by I_2? If this were the only loop, and loop 1 were open, we should write $(Z_b + Z_c)I_2 = E_2$. When we take into account also the voltage produced in loop 2 by I_1, however, the equation becomes

$$-Z_bI_1 + (Z_b + Z_c)I_2 = E_2 \tag{3-3}$$

Equations 3-2 and 3-3 are loop equations for this network. They can be solved simultaneously for the two loop currents, from which all desired currents and voltages are easily found.

Note that there are two equations to be solved simultaneously by the loop method. By the branch method of the preceding section, there would be six simultaneous equations: three branch equations and three connection equations. If state variables were used, as in Chapter 7, the number of simultaneous equations would depend on the kinds of elements in the branches; there might be as many as six. Compared to the branch method, the loop method is economical because the junction equations are implicit in the choice of loop currents.

4. A LOOP EXAMPLE

Let the impedances shown in Figure 3a be pure resistances with the values shown, in ohms. The electromotive forces are in volts; they may be either dc values or rms alternating voltages, for the solution is the same.

$$8I_1 - 3I_2 = 7 \tag{4-1}$$

$$-3I_1 + 7I_2 = 15 \tag{4-2}$$

These are linear equations that can be solved by any convenient means. The method of determinants is so much easier and neater than any other that it is highly recommended. (Anyone not entirely familiar with the use of determinants for solving simultaneous equations will find enough information for this particular purpose in Appendix 1.) To proceed:

$$I_1 = \frac{\begin{vmatrix} 7 & -3 \\ 15 & 7 \end{vmatrix}}{\begin{vmatrix} 8 & -3 \\ -3 & 7 \end{vmatrix}} = \frac{49 + 45}{56 - 9} = \frac{94}{47} = 2 \tag{4-3}$$

$$I_2 = \frac{\begin{vmatrix} 8 & 7 \\ -3 & 15 \end{vmatrix}}{D} = \frac{120 + 21}{D} = \frac{141}{47} = 3 \tag{4-4}$$

(D, the denominator of I_2, is the same as the denominator of I_1.) The results are the two loop currents in amperes.

All unknown quantities in the network are now easily found. Current in the central branch is $I_{BO} = I_1 - I_2 = -1$ ampere; the negative sign indicates that 1 ampere is flowing upward. If the bottom node (node O) is taken to be the reference node, at an assumed zero potential, then the potential at node B is -3 volts; potential at node A is $-3 + 5 \cdot 2 = 7$ volts, which is also the electromotive force of the source E_1; at node C the potential is $-3 + 15 = 12$ volts, which can also be found (across Z_c) as $4 \cdot 3 = 12$ volts. Note that the most convenient way to specify all the voltages of a network is to give the potentials at the various nodes with reference to some one node that is arbitrarily assumed to be at zero potential.

5. STANDARD NOTATION FOR LOOP EQUATIONS

It is customary, and quite helpful, to use a standard system of symbols for writing the loop equations:

$$Z_{11}I_1 + Z_{12}I_2 + Z_{13}I_3 + \cdots + Z_{1L}I_L = V_1$$
$$Z_{21}I_1 + Z_{22}I_2 + Z_{23}I_3 + \cdots + Z_{2L}I_L = V_2$$
$$Z_{31}I_1 + Z_{32}I_2 + Z_{33}I_3 + \cdots + Z_{3L}I_L = V_3 \tag{5-1}$$
$$\cdot \quad \cdot \quad \cdot \quad \cdot \quad \cdot \quad \cdot \quad \cdot \quad \cdot \quad \cdot \quad \cdot \quad \cdot \quad \cdot$$
$$Z_{L1}I_1 + Z_{L2}I_2 + Z_{L3}I_3 + \cdots + Z_{LL}I_L = V_L$$

This is a set of L simultaneous equations applying to the L loops of a network; the network can be any linear network of L independent loops. The following conventions are used.

1. Each loop current is numbered, as I_1 and I_2.
2. The total impedance about loop 1 is Z_{11}. This is called the *self-impedance* of the loop. It is the voltage produced in loop 1 by unit current in loop 1, all other loops being open. In Figure 3a, as an example, Z_{11} is $Z_a + Z_b$.

3. Certain branches are common to two (or more) loops. If current in one loop produces voltage in another loop there is said to be *mutual impedance*, and Z_{12} is the voltage produced in loop 1 by unit current in loop 2 (all other loops being open). In Figure 3a, Z_{12} is $-Z_b$. Z_{21} is also $-Z_b$. The negative sign arises from the fact that arrows drawn (arbitrarily) in Figure 3a show unit I_2 and unit I_1 to be in opposite directions in the mutual impedance, and they hence produce opposite voltages. (It would not be wrong to reverse one of the arbitrary reference arrows in the figure, and final results would be the same, but reference arrows for loop currents are often drawn clockwise.) In the example, $Z_{12} = Z_{21} = -3$. If $Z_{mn} = Z_{nm}$ the network is called *reciprocal* (or *symmetrically bilateral*), and this is usual (even with transformers). Certain transistor and vacuum-tube connections can be modeled as *unilateral*, or perhaps *unsymmetrically bilateral*, even though linear.

4. Each impedance (as Z_{23}) has a double subscript. The first is the number of the loop in which voltage is produced; the second is the number of the loop carrying current. Hence in equation 1 the first subscript of each Z is 1, while the second subscript of each Z agrees with the number of the associated loop current.

5. Each term in each equation is a voltage; each equation is a statement of Kirchhoff's voltage law setting the impedance drops equal to the applied voltages.†

6. V_1 is a voltage in loop 1 that is not taken into account by the terms of the left-hand side of the equation. It may be an applied electromotive force or source, as in Figure 3a and equations 3-2 and 3-3. It may, however, be a voltage arising from any other cause, or the sum of several such voltages. It *must* include all voltages that appear in loop 1, other than the impedance drops of the left-hand side of the equation. A source in a mutual branch will appear in two equations.

7. Signs in the equations assume that the reference direction of V_1 is the same as the reference direction of current in the loop. Thus, other loops being open, positive voltage produces positive current.

---

† Equations 5-1 are written in terms of impedances and transforms of currents and voltages. More basically, Kirchhoff's voltage law can be written as differential equations:

$$\left(R_{11}i_1 + L_{11}\frac{di_1}{dt} + \frac{1}{C_{11}}\int i_1\,dt\right) - \left(R_{12}i_2 + L_{12}\frac{di_2}{dt} + \frac{1}{C_{12}}\int i_2\,dt\right) - \cdots = v_1$$

$$(5\text{-}2)$$

$$-\left(R_{21}i_1 + L_{21}\frac{di_1}{dt} + \frac{1}{C_{21}}\int i_1\,dt\right) + \left(R_{22}i_2 + L_{22}\frac{di_2}{dt} + \frac{1}{C_{22}}\int i_2\,dt\right) - \cdots = v_2$$

These equations 5-2 transform to equations 5-1 as in Section 27 of Chapter 4. These differential equations are used again, and are transformed to functions of s, in Chapter 15, Section 4.

6. AN EXAMPLE WITH THREE LOOPS

Figure 6a shows a resistance bridge circuit of familiar form. It is supplied from a source of constant voltage E. R_x is the resistance of the galvanometer or other instrument used to indicate balance. The problem is to find the galvanometer current I_x through R_x, the bridge being unbalanced. Bridge resistances are given with the diagram.

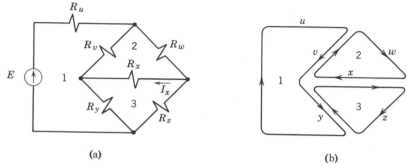

(a)   (b)

FIGURES 6a,b An unbalanced bridge: $R_u = 0.10$; $R_v = 1.0$; $R_w = 0.5$; $R_x = 1.0$; $R_y = 1.0$; $R_z = 1.0$; $E = 10$ volts.

Solution: This seems to be a network with three obvious loops as indicated in Figure 6b. Of course, loops do not have to be selected in the obvious way and there might be advantages in taking other loops, but we shall start work with the three indicated. The three loop equations are to be written; let us write them in the form of the general loop equations 5-1.

$$(R_u + R_v + R_y)I_1 - R_v I_2 - R_y I_3 = E$$
$$-R_v I_1 + (R_v + R_w + R_x)I_2 - R_x I_3 = 0 \qquad (6\text{-}1)$$
$$-R_y I_1 - R_x I_2 + (R_x + R_y + R_z)I_3 = 0$$

The first term is the voltage in loop 1 owing to I_1, the second term the voltage in loop 1 owing to I_2, the third term is the voltage in loop 1 owing to I_3; altogether they equal E. The other equations are similar. Numerical values are now introduced:

$$2.10I_1 - 1.00I_2 - 1.00I_3 = E$$
$$-1.00I_1 + 2.50I_2 - 1.00I_3 = 0 \qquad (6\text{-}2)$$
$$-1.00I_1 - 1.00I_2 + 3.00I_3 = 0$$

Current through R_x is the difference of two loop currents; if the reference direction of I_x is taken from right to left as in the diagram, then

$$I_x = I_2 - I_3 \qquad (6\text{-}3)$$

The loop equations must be solved for I_2 and I_3. This is most easily done by determinants:

$$I_2 = \frac{\begin{vmatrix} 2.1 & E & -1.0 \\ -1.0 & 0 & -1.0 \\ -1.0 & 0 & 3.0 \end{vmatrix}}{\begin{vmatrix} 2.1 & -1.0 & -1.0 \\ -1.0 & 2.5 & -1.0 \\ -1.0 & -1.0 & 3.0 \end{vmatrix}} \tag{6-4}$$

Designating the denominator and numerator of I_2 as D and N_2 respectively,

$$D = (2.1)(2.5)(3.0) - (1.0) - (1.0) - (2.1 + 2.5 + 3.0)$$
$$= 6.15 \tag{6-5}$$

$$N_2 = -E \begin{vmatrix} -1.0 & -1.0 \\ -1.0 & 3.0 \end{vmatrix} = 4.0E \tag{6-6}$$

Similarly,

$$I_3 = \frac{\begin{vmatrix} 2.1 & -1.0 & E \\ -1.0 & 2.5 & 0 \\ -1.0 & -1.0 & 0 \end{vmatrix}}{D} = \frac{N_3}{D} \tag{6-7}$$

$$N_3 = E \begin{vmatrix} -1.0 & 2.5 \\ -1.0 & -1.0 \end{vmatrix} = 3.5E \tag{6-8}$$

$$I_x = I_2 - I_3 = \frac{N_2}{D} - \frac{N_3}{D}$$

$$= E \frac{4.0 - 3.5}{6.15} = \frac{E}{12.3} \tag{6-9}$$

Since E is given as 10 volts, $I_x = 10/12.3 = 0.813$ ampere. This is the current through the galvanometer of the bridge circuit.

7. CHOICE OF LOOPS

In Figure 6b the most obvious loops were chosen. Possibly a little cleverness would save some drudgery. It would clearly be easier if I_x were just one loop current; that is, if only one loop passed through R_x. Then we would need to solve for only one loop current instead of two. At the same time it is helpful in expanding the determinants if E is in only one loop. Let us therefore select loops† as in Figure 7a.

Loop equations are now written:

$$(R_u + R_v + R_y)I_1 - R_vI_2 - (R_v + R_y)I_3 = E$$
$$-R_vI_1 + (R_v + R_w + R_x)I_2 + (R_v + R_w)I_3 = 0 \quad (7\text{-}1)$$
$$-(R_v + R_y)I_1 + (R_v + R_w)I_2 + (R_v + R_w + R_y + R_z)I_3 = 0$$

The principle is exactly the same, of course. First the voltages in loop 1 resulting from I_1 are written. Then we write the voltage in loop 1 owing to I_2: this is $-R_vI_2$, and the minus sign is used because the reference direction of I_2 through R_v is opposite to the reference direction of I_1 through R_v. Next the voltage in loop 1 caused by I_3 is written; both R_v and R_y are in both loops 1 and 3, and the reference directions are opposite, so the term is $-(R_v + R_y)I_3$. The sum of these voltage drops is equated to the electromotive force E.

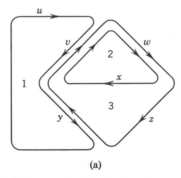

(a)

FIGURE 7a   A different set of loops for the network of Figure 6a.

There is nothing unusual about the first two terms of the second equation. The third term is the voltage in loop 2 that is due to I_3; loops 2 and 3 have R_v and R_w in common, and in both these elements the reference direction of I_3 is the *same* as that of I_2. Hence I_3 will produce a voltage of the same polarity as will I_2, and the sign of this mutual-resistance term is consequently the *same* (positive) as the sign of the self-resistance term.

The third equation requires no further explanation. The three equations are solved for I_2, preferably using determinants, and the result comes out the same as I_x in the first part of this example. The work required by this method is found to be a little less than that for the first method.

---

† Limitations on the selection of loops are discussed in Section 13.

Two different choices of loops have now been made in the same network, and many others are possible.[†]

8. NODE EQUATIONS

In writing loop equations we start with the concept of loop currents. This makes it unnecessary to give any attention to Kirchhoff's current law, for loop currents necessarily add to zero at every node, and Kirchhoff's current law is automatically satisfied. The loop-current concept therefore reduces the number of equations that must be solved simultaneously from the $2B$ equations of the branch method to L, the number of independent loops, which is usually about one-fourth as many.

The ingenuity of the node method lies in considering that there is a potential at each node, and the voltage from any node to any other node is the potential difference. Branch voltages that are obtained by means of this consideration will necessarily satisfy Kirchhoff's voltage law, for around any loop such voltages add to zero. It is now only necessary to satisfy Kirchhoff's current law at each node, for the voltage law is automatically satisfied. Thus the number of simultaneous equations is reduced to the number of independent nodes N, a number much smaller than $2B$ and comparable with L.

Whether the node method or the loop method is the more convenient depends on the network. Some networks have fewer loops than nodes, some fewer nodes than loops, and other factors also affect the relative convenience as will be seen.

9. A NODE EXAMPLE

Figure 9a shows a network with two independent nodes. We choose A and B as the two independent nodes with potentials V_A and V_B respectively, and assign zero potential to the reference node O.

[†] The bridge network (see J. Clerk Maxwell, Section 347) is the three-loop network of perfect symmetry. Figure 7b shows it drawn in the familiar manner, and as a lattice, and in a triangular form to emphasize symmetry. Finally it is indicated as a three-dimensional figure, each branch being the edge of a regular tetrahedron, to show that the branches are geometrically indistinguishable.

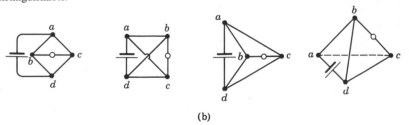

(b)

FIGURE 7b   Four ways of drawing the same bridge network.

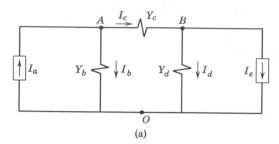

(a)

FIGURE 9a   A network with two independent nodes.

Kirchhoff's current law is the basis of the node equations, and for node A, considering the directions of the arrows,

$$I_b + I_c = I_a \tag{9-1}$$

Since $I_b = Y_b V_A$ and $I_c = Y_c(V_A - V_B)$, we can write

$$Y_b V_A + Y_c(V_A - V_B) = I_a \tag{9-2}$$

This is one of the node equations; note that each term is a current.

There is an alternative way of writing this equation and, as with the loop equations, the alternative will be found quicker. First, assume that potential is zero at every node except node A and write the admittance currents *from* node A. (This can be visualized by supposing every node except node A to be short-circuited to node O.) We write

$$(Y_b + Y_c)V_A = I_a \tag{9-3}$$

In fact, however, this is not the whole story at node A, for there are really potentials at other nodes that will have their effects, too. In the circuit of Figure 9a, the effect of potential at node B is the additional current $- Y_c V_B$; this is current *from* node A that results when there is potential V_B at node B and zero potential at node A and all other nodes. (This can be visualized by supposing the short circuit to be removed from node B, while node A, node O, and all other nodes are short-circuited together.)

Now we simply add the current from node A when there is potential at node A alone, the current from node A owing to potential at node B alone, and (in networks that have other nodes) additional terms for currents from node A owing to potentials at node C alone, node D alone, and so forth. These must equal source current *to* node A. Since the simple network of Figure 9a has only two independent nodes, we write

$$(Y_b + Y_c)V_A - Y_c V_B = I_a \tag{9-4}$$

which is obviously the same as equation 9-2.

Similarly, we write currents *from* node B. First we write currents owing to potential at node B alone that appear when there is potential only at node B (all other node potentials being zero). To this we add the current that flows *from* node B when there is potential at node A only, and so on. The sum is equated to source current supplied *to* node B. In our network the result is

$$(Y_c + Y_d)V_B - Y_c V_A = -I_e \tag{9-5}$$

Equations 9-4 and 9-5 are the two node equations.

These two node equations are to be solved for the two unknown voltages V_A and V_B, and the unknown currents of the network are then easily found. In this example the two given sources are independent current sources; it is simpler in these first examples to use independent *current* sources with *node* equations and independent *voltage* sources with *loop* equations but, as we shall see, this is not at all necessary.

10. STANDARD NOTATION FOR NODE EQUATIONS

There is a standard form for writing node equations similar to the standard form for loop equations. For a network of N independent nodes,

$$
\begin{aligned}
Y_{AA} V_A + Y_{AB} V_B + Y_{AC} V_C + \cdots + Y_{AN} V_N &= I_A \\
Y_{BA} V_A + Y_{BB} V_B + Y_{BC} V_C + \cdots + Y_{BN} V_N &= I_B \\
Y_{CA} V_A + Y_{CB} V_B + Y_{CC} V_C + \cdots + Y_{CN} V_N &= I_C \\
&\quad\quad\quad\quad\quad\quad\quad\quad\quad\quad\quad\quad\quad\quad (10\text{-}1) \\
\cdot \quad \cdot \quad \cdot \quad \cdot \quad \cdot \quad \cdot \quad & \cdot \\
Y_{NA} V_A + Y_{NB} V_B + Y_{NC} V_C + \cdots + Y_{NN} V_N &= I_N
\end{aligned}
$$

Y_{AA} is called the *self-admittance* at node A, and in the network of Figure 9a it is equal to $(Y_b + Y_c)$. Note that Y_{AA} is the *sum of all admittances attached to node A*. Y_{BB}, Y_{CC}, . . . are self-admittances at the other nodes.

Y_{AB} is the *mutual admittance* between nodes A and B. In our example $Y_{AB} = -Y_c$. Y_{BA} also equals $-Y_c$. Both Y_{AB} and Y_{BA} are the *sum of all admittances connected directly between nodes A and B*, but written with a negative sign.

I_A is current flowing *toward* node A; it is often, as in our example, a known current from a source. Under other circumstances, however, it may be current that is not from a source, and all that can be said with confidence is that I_A is current to node A that is not taken into account by the left-hand member of the equation.

In equations 10-1 the first subscript attached to a Y shows the node at which current is being considered; the second subscript shows the node potential producing that current. Thus $Y_{PQ} V_Q$ is a current at node P resulting from a potential at node Q. For ordinary reciprocal elements,

$Y_{PQ} = Y_{QP}$. This is evident when the two admittances are physically those of the same element (as both Y_{AB} and Y_{BA} are the admittance Y_c in Figure 9a), and it is still true when the coupling between nodes is not directly conductive but is by way of transformers or coupled coils. It is ordinarily not true when the coupling is by way of transistors or electron tubes.

The generalities of this section will have more meaning when applied to specific networks. Let us look at an example or two.

11. AN EXAMPLE WITH THREE NODES

The unbalanced bridge network provides an example of node solution, as in Section 6 it illustrated the loop method. The network is shown in Figure 11a. The conductances of each branch are given (the reciprocals of the resistances of Figure 6a), and the bridge is operated by an independent current source of 100 amperes shunted by a 10-mho conductance (equivalent to the independent voltage source of Figure 6a; see Chapter 10).

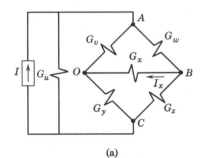

(a)

FIGURE 11a   Solution of bridge network by node method: $G_u = 10.0$; $G_v = 1.0$; $G_w = 2.0$; $G_x = 1.0$; $G_y = 1.0$; $G_z = 1.0$; $I = 100$ amperes.

One of the nodes is selected as reference node and is marked O. Since the object of the solution is to find current through G_x, it will be most convenient to select one end or the other of this branch as node O, for then only one other node voltage will have to be found. The other nodes are lettered A, B, and C.

Three admittances are seen to be attached to node A, and one source of current. Current outward through the admittances, owing to potential at node A only, is $(G_u + G_v + G_w)V_A$, so this is the first term of equations 11-1. Current to node A through G_w from node B, potential at node A being zero, is $G_w V_B$; the *outward* current from node A is written in equations 11-1 as $- G_w V_B$. Current to node A from node C, node A being at zero potential, is $G_u V_C$, and the outward current from node A is $- G_u V_C$. The total of these currents outward from node A through the three admittances must equal the current I into node A from the source; this gives the first of equations 11-1.

The second equation is written for node B. The third equation is written for node C. These require no further explanation except perhaps to point out that the source current, being *from* node C, appears in the right-hand side of the third equation with a negative sign.

$$(G_u + G_v + G_w)V_A - G_w V_B - G_u V_C = I$$
$$(G_w + G_x + G_z)V_B - G_w V_A - G_z V_C = 0 \tag{11-1}$$
$$(G_u + G_y + G_z)V_C - G_z V_B - G_u V_A = -I$$

Numerical values are next introduced into the equations, and at the same time the terms are arranged in order, to agree with the form of equations 10-1.

$$13V_A - 2V_B - 10V_C = I$$
$$-2V_A + 4V_B - 1V_C = 0 \tag{11-2}$$
$$-10V_A - 1V_B + 12V_C = -I$$

V_B can be computed as the ratio of two determinants, N_B/D, where

$$D = \begin{vmatrix} 13 & -2 & -10 \\ -2 & 4 & -1 \\ -10 & -1 & 12 \end{vmatrix} = 624 - 20 - 20 - (400 + 48 + 13) = 123$$

and

$$N_B = \begin{vmatrix} 13 & I & -10 \\ -2 & 0 & -1 \\ -10 & -I & 12 \end{vmatrix} = 10I - 20I - (13I - 24I) = I$$

Thus

$$V_B = \frac{N_B}{D} = \frac{I}{123}$$

Finally

$$I_x = G_x V_B = 1 \cdot \frac{I}{123} = \frac{I}{123} \tag{11-3}$$

Since I is given as 100 amperes, $I_x = 100/123 = 0.813$ ampere, and this is the same result that was obtained by the loop method in Section 6.

12. FORCED CURRENTS AND FORCED VOLTAGES

A network containing sources of known voltage can be solved as easily by the node method as by the loop method, although the latter has been used for illustration. An example will show that the node method may be even easier than the loop method in some cases.

The two-loop network of Figure 3a, for example, can be solved by the node method. It turns out to be a three-independent-node problem, but an easy one. Let us try this problem, redrawing the network as Figure 12a.

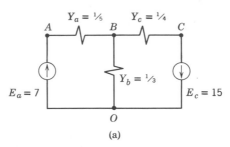

(a)

FIGURE 12a   A network in which two node voltages are forced.

Any node can be selected for reference; we choose the one marked O. The usual equation for currents at node B can be written:

$$(Y_a + Y_b + Y_c)V_B - Y_aV_A - Y_cV_C = 0 \qquad (12\text{-}1)$$

Because the voltage of the sources is known, two of the nodes have *forced* potentials:

$$V_A = E_a = 7 \quad \text{and} \quad V_C = -E_c = -15 \qquad (12\text{-}2)$$

V_B is the only remaining unknown. It can be found, using numerical values, as follows:

$$(\tfrac{1}{5} + \tfrac{1}{3} + \tfrac{1}{4})V_B - \tfrac{1}{5}7 + \tfrac{1}{4}15 = 0 \qquad (12\text{-}3)$$

$$V_B = -3 \qquad (12\text{-}4)$$

The remarkable ease of this solution results from the fact that two of the node voltages are so simply specified. To some extent this is due to the way in which the diagram is drawn. It was not by accident that the network was arranged in the particular configuration shown, and a few trials with other electrically equivalent arrangements will show the value of care in planning the layout before starting the solution.

In a similar manner, the loop method of solution can be applied to networks containing sources of known current, although the illustrative examples of loop solutions in this chapter have had independent voltage sources only. When there is a source in which current is known, current in the branch containing that source is *forced* to have the designated value. In the network of Figure 9a, for example, although there are three loops, the currents in two of the loops are forced, and only one remains for solution. The loop-equation solution is therefore easy.

13. TOPOLOGY

Some networks have more nodes than loops, others more loops than nodes, and it is of course advantageous to use the method of solution that requires the smaller number of simultaneous equations. To judge the better method, or even to know how many equations are adequate for the solution of a given problem, it is necessary to have some dependable means of counting independent nodes and independent loops.

In determining the number of nodes and loops, a field of mathematics called *topology* gives useful information. If we let

E = number of elements

N = number of independent nodes

L = number of independent loops

then in any fully connected network

$$N + L = E \qquad (13\text{-}1)$$

It is easy to count the elements E on the network diagram. It is also easy to count the nodes, between elements; count the number of nodes and subtract 1 to obtain N (or subtract S if the network is not fully connected).†

There are three ways to find the number of independent loops L. First and easiest, in a *planar* network, the number of independent loops is equal to the number of *windows*. A planar network is one that can be drawn on a plane, on a flat sheet of paper, with no crossing of conductors. (Thus the lattice network of Figure 7b is planar because it can be redrawn as a bridge, but if another element connected the midpoint of *a-c* to the midpoint of *b-d* it would be nonplanar.) A window is a loop that has no conductors within it. Most simple networks are planar and it is easy to count windows; the bridge network of Figure 7b has three.

Another way to find L is to imagine one branch of the network to be cut, as if by snipping with wire cutters. This will open at least one of the possible loops. Then cut another branch, opening another loop. Continue cutting branches until not a single loop remains, but do not cut in such a way as to isolate any node from the network. Enough branches have been cut when there is no possible closed path for current left in the network, but every node remains conductively coupled to every other node. What remains is

---

† If there are S subnetworks that influence each other only by mutual magnetic or electric fields (transformer coupling, for instance) and are not conductively connected (for example, no common ground), the net is not "fully connected" and the number of independent nodes N is equal to the total number of nodes minus S. (There must be a reference node in each subnetwork.) The rule given in the text is the special case for the fully connected net with $S = 1$.

sometimes called a *tree*. The number of cuts is the number of independent loops L.

Third, L can be computed from equation 13-1. This requires counting elements and nodes, but these are usually obvious.

In the network of Figure 12a, for instance, $E = 5$ and $N = 3$ so $L = 2$. In Figure 11a, $E = 7$ while $N = 3$ and $L = 4$. In Figure 13a, counting nodes at A and F, $N = 5$, $E = 9$, and $L = 4$; note that 4 cuts will open all loops without isolating any nodes.

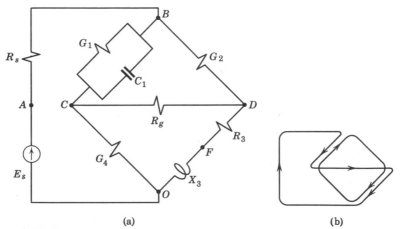

(a)   (b)

FIGURE 13 (a) A Maxwell bridge network. (b) Two loops of a three-loop network.

It may seem unnecessary to go to this trouble to determine the number of independent loops, but in writing loop equations a definite guide is sometimes needed. The use of too many loop equations is only a waste of time, but the use of too few may lead to the wrong answer.†

14. DUALITY

Duality has been several times mentioned. Two networks are called duals if the loop equations of one have the same form as the node equations of the other. Two equations have the same form if a mere change of the letter symbols can make the equations identical.

---

† For instance, the two loops of Figure 13b for the bridge network look hardly less probable than the three loops of Figure 7a, and students have not unreasonably attempted to solve the bridge network with only these two loops. It was not obvious that the answer thus obtained was wrong. A correct answer results if any third loop is considered together with the two shown in Figure 13b.

Thus the networks of Figures 14a and 14b are duals. The loop equations of the former are:

$$(R_1 + R_{12} + j\omega L_{11})I_1 - R_{12}I_2 = E$$

$$-R_{12}I_1 + \left(R_{12} - j\frac{1}{\omega C_{22}}\right)I_2 = 0 \tag{14-1}$$

The node equations of Figure 14b are

$$(G_1 + G_{12} + j\omega C_{11})V_1 - G_{12}V_2 = I$$

$$-G_{12}V_1 + \left(G_{12} - j\frac{1}{\omega L_{22}}\right)V_2 = 0 \tag{14-2}$$

The similarity of form of the equations is evident. In detail, we recognize a duality between elements. In the following list, elements on the same line are the duals of each other:

resistance	conductance
inductance	capacitance
voltage	current

A rule for finding the dual of a network is illustrated in Figure 14c. On the network diagram, mark a node for the dual network within each window of the original network, and one more node (to be the reference node) outside all loops of the original network. Through each element of the original network, draw a line; each of these lines is to terminate on the nodes that have just been indicated for the dual network. Each of these lines represents an element of the dual network that is itself the dual of the original element through which the line was drawn. That is, if the line was drawn through capacitance, the element of the dual network is inductance; if through a source of independent voltage it is a source of independent current, and so on according to the above list.

Duality is a form of analogy. Two physical systems of different nature, such as an electric circuit and a pendulum, for instance, are called analogs (but not duals) if their action is described by equations of the same form. If two physical systems of the same nature, such as two electrical networks, have equations of the same form but yet are physically different (as the networks of Figures 14a and b, for example) they are called duals. Dual mechanical systems exist, as well as dual electrical systems. [For further discussion, see Gardner and Barnes or Guillemin (1).]

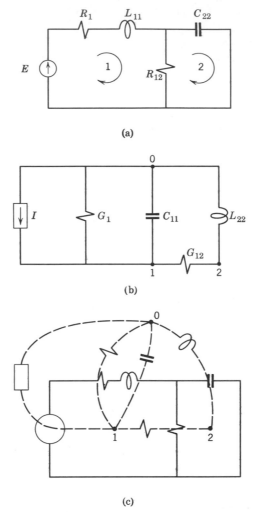

FIGURE 14 (a) A two-loop network. (b) A two-node-pair network, the dual of (a). (c) The derivation of (b) from (a).

15. SUMMARY

Loop or *node* equations, as well as state-variable equations, provide practical means of solving network problems. Both loop and node equations are derived from the less elegant *branch* and *connection* equations.

Loop, node, branch, and state equations can all be expressed in the *time domain* or transformed into the *frequency domain*. There are advantages in considering state equations in the time domain, and loop or node equations in the frequency domain. Transforms of voltage and current are used in the frequency domain, with admittance and impedance which are *functions of the frequency variables* s *or* jω.

Each *loop* equation expresses the voltage about a closed loop, using loop currents that necessarily satisfy Kirchhoff's current law at every node. The number of simultaneous equations is equal to the number of independent loops, L.

Each *node* equation expresses the current to or from a node, using node potentials that necessarily satisfy Kirchhoff's voltage law about every loop. The number of simultaneous equations is equal to the number of independent nodes, N.

Whether L or N is smaller depends on the network, but either method is more *economical* of equations than is the branch method.

Standard formulations are given for loop and node equations.

Examples of numerical solution are given, but the great advantage of these formal expressions is in *generalization* rather than numerical solution.

Network *topology* gives rules for finding the *number* of independent *loops* and independent *nodes* in a network, and gives the general relation that $N + L = E$.

Two networks are *dual* if the loop equations of one are similar in form to the node equations of the other. *Duality* is a special case of *analogy*.

PROBLEMS

8-1. Find the current in each resistor of the network shown. Use the loop method.
§4*

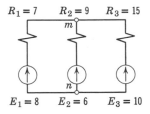

$$R_1 = 7 \qquad R_2 = 9 \qquad R_3 = 15$$

m

n

$$E_1 = 8 \qquad E_2 = 6 \qquad E_3 = 10$$

Problem 8-1.

8-2. Solve for current in each branch of Figure 3a, using the loop method, after making the following change in the network: Z_b is now $j3$ ohms of reactance instead of 3 ohms of resistance. E_1 and E_2 are in phase with each other and have rms values. §4*

8-3. In the example of Section 6, the bridge would be balanced if R_z were 0.5 ohm. To find the effect of unbalance, let $R_z = 0.5 + r$ and solve for I_x in terms of r. Is galvanometer current proportional to unbalance?   §6*

8-4. Repeat the example of Section 6, using the loop method but selecting loops differently from either Figure 6b or 7a.   §7

8-5. The loop equations of a network are:

$$8I_1 - 5I_2 - I_3 = 110\underline{/0°}$$
$$-5I_1 + 12I_2 \qquad = 0$$
$$-I_1 \qquad + 7I_3 = 115\underline{/0°}$$

(a) Draw the network, carefully labeling each element. (b) Compute current in the 115-volt source.   §7*

8-6. Write loop equations for the circuit shown, voltage V_1 being applied by a source attached to the terminals.   §7

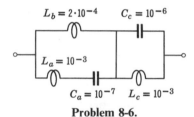

Problem 8-6.

8-7. Solve the equations of Problem 8-6 for input current I, in terms of voltage V_1 at frequency $\omega = 10^5$. Find input impedance (V_1/I_1).   §7*

8-8. Write node equations for the network shown. Solve the node equations for V_1 in terms of I_1, and compute input impedance.   §9*

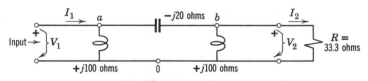

Problems 8-8 and 8-9.

8-9. (a) Write loop equations for the network shown. Solve the loop equations for I_1 in terms of V_1, and compute input impedance. (b) Use the ladder method to compute input impedance to the network. Draw a phasor diagram approximately to scale, showing V_1, V_2, V_{ab}; I_1, I_2, I_{ab}, I_{bo}, and I_{ao}.   §11*

8-10. In the network of Figure 9a, $Y_b = 6$, $Y_c = 3$, and $Y_d = 2$ mhos. $I_a = 36$ amperes, and $I_e = 8$ amperes. Write and solve the node equations; find V_{AB}.   §9*

8-11. Write loop equations for the network of Figure 9a (in terms of source currents), using numerical values from Problem 8-10, and solve; find V_{AB}.   §12*

8-12. Write the node equations for the network of Figure 9a with node B instead of node O as the reference node. Use numerical values from Problem 8-10, and solve the node equations for V_{AB}.   §9

8-13. A circuit consists of R, L, C, and a source, all in *series*. $R = 1$ ohm, $L = 2$ millihenrys, $C = 1000$ microfarads, and current from the source is $I = 2\underline{/0°}$ with $\omega = 1000$ radians/s. Node a is between the source and R, node b between R and L, node c between L and C, and node O between C and the source. Write the standard form of the three *node* equations for the independent nodes a, b, and c, using numerical values for admittances. Use determinants to find V_b.

§12*

8-14. How many elements, independent nodes, and independent loops are there in Figure 7c of Chapter 1 and 5a of Chapter 9? Count loops by counting windows, by cutting, and check by equation 13-1.   §13

8-15. In Problem 4-19, page 98, twelve resistors are arranged along the edges of a cube. Counting only the circuit shown (not any external connection), find (*a*) the number of elements, nodes, junctions, and branches, and (*b*) the number of independent nodes and independent loops. Use equation 13-1, and check by cutting loops and by counting windows.   §13*

8-16. (*a*) Draw the dual of the bridge network of Figure 2a. (*b*) Draw the dual of the network of Figure 9a; then find the dual of the dual. Do you regain Figure 9a? (*c*) Draw the dual of the Maxwell bridge, Figure 13a. Is the result a Maxwell bridge?   §14*

8-17. Write loop equations for the network of Problem 7-6, page 179, applying an independent voltage source whose transform is $V_1(s)$. Write the equations in terms of the complex frequency variable s (rather than $j\omega$). Solve for $I_1(s)$, the transform of entering current. (See also Problem 15-2.)   §6

8-18. Write node equations for the network of Problem 7-8, page 180, letting the independent source current be $I(s)$. Write in terms of the variable s. Solve for $V(s)$, the transform of the node voltage.   §11

8-19. Write loop equations for the network of Problem 7-9, page 180, assuming an independent voltage source with transform $V(s)$. Write the equations in terms of the complex frequency variable s, and solve for $I(s)$, the transform of the entering current. (See also Problem 15-3.)   §6

8-20. Write loop equations for the network of Problem 7-11, page 180, assuming an independent voltage source with transform $V_1(s)$. Write the equations in terms of the complex frequency variable s and solve for $I_1(s)$, the transform of the entering current, and also for $I_2(s)$.   §6

8-21. Write loop equations for the network of Problem 6-15, page 158, assuming an independent voltage source with transform $V(s)$. Write the equations in terms of s, and solve for the transform of entering current, and also for that of current through the 1-henry inductance. (See also Problem 15-4.)   §6

8-22. Write loop equations for the network of Problem 7-14, page 181, assuming an independent voltage source with transform $V(s)$. Write the equations in terms of s. Solve for $I(s)$, the transform of entering current. (See also Problem 15-5.)

§6

8-23. Write loop equations for the network of Problem 7-16, page 181, assuming an independent voltage source with transform $V_i(s)$. Write the equations in terms of s. Solve for $V_o(s)$, the transform of the output voltage.   §6

CHAPTER
9

MATRIX SOLUTIONS

1. SOLVING THE LOOP EQUATIONS

When the loop method is used, there are L simultaneous loop equations, L being the number of independent loops. These are given as equations 5-1 in Chapter 8 and are repeated here:

$$
\begin{aligned}
Z_{11}I_1 + Z_{12}I_2 + Z_{13}I_3 + \cdots + Z_{1L}I_L &= V_1 \\
Z_{21}I_1 + Z_{22}I_2 + Z_{23}I_3 + \cdots + Z_{2L}I_L &= V_2 \\
Z_{31}I_1 + Z_{32}I_2 + Z_{33}I_3 + \cdots + Z_{3L}I_L &= V_3 \\
\quad \cdot \quad \cdot \quad \cdot \quad \cdot \quad \cdot \quad \cdot \quad \cdot \quad \cdot \quad \cdot \quad \cdot \quad \cdot \quad \cdot \\
Z_{L1}I_1 + Z_{L2}I_2 + Z_{L3}I_3 + \cdots + Z_{LL}I_L &= V_L
\end{aligned}
\tag{1-1}
$$

The Z's and V's are known; the currents are to be found. There are L unknown currents.

By the use of Cramer's rule (if this is not familiar, see Appendix 1), each current is the ratio of determinants. I_1, for example, is

$$
I_1 = \frac{N_1}{D}
\tag{1-2}
$$

where the denominator D is the determinant

$$
D =
\begin{vmatrix}
Z_{11} & Z_{12} & Z_{13} & \cdots & Z_{1L} \\
Z_{21} & Z_{22} & Z_{23} & \cdots & Z_{2L} \\
Z_{31} & Z_{32} & Z_{33} & \cdots & Z_{3L} \\
\cdot & \cdot & \cdot & & \cdot \\
Z_{L1} & Z_{L2} & Z_{L3} & \cdots & Z_{LL}
\end{vmatrix}
\tag{1-3}
$$

and the numerator N_1 is the determinant

$$N_1 = \begin{vmatrix} V_1 & Z_{12} & Z_{13} & \cdots & Z_{1L} \\ V_2 & Z_{22} & Z_{23} & \cdots & Z_{2L} \\ V_3 & Z_{32} & Z_{33} & \cdots & Z_{3L} \\ \cdot & \cdot & \cdot & \cdot & \cdot \\ V_L & Z_{L2} & Z_{L3} & \cdots & Z_{LL} \end{vmatrix} \qquad (1\text{-}4)$$

N_1 can be partially expanded, and the expression for I_1 becomes

$$I_1 = \frac{N_1}{D} = \frac{V_1\Delta_{11}}{D} + \frac{V_2\Delta_{21}}{D} + \frac{V_3\Delta_{31}}{D} + \cdots \qquad (1\text{-}5)$$

Δ_{11} is the cofactor (the minor with appropriate sign) of V_1 in equation 1-4, or of Z_{11} in equation 1-3; Δ_{21} is the cofactor of V_2 or of Z_{21}, and so forth. (If this is not clear, see Appendix 1.) Thus, for example,

$$\Delta_{21} = - \begin{vmatrix} Z_{12} & Z_{13} & Z_{14} & \cdots & Z_{1L} \\ Z_{32} & Z_{33} & Z_{34} & \cdots & Z_{3L} \\ Z_{42} & Z_{43} & Z_{44} & \cdots & Z_{4L} \\ \cdot & \cdot & \cdot & \cdot & \cdot \\ Z_{L2} & Z_{L3} & Z_{L4} & \cdots & Z_{LL} \end{vmatrix} \qquad (1\text{-}6)$$

In a similar manner, Cramer's rule gives for I_2:

$$I_2 = \frac{N_2}{D} = \frac{V_1\Delta_{12}}{D} + \frac{V_2\Delta_{22}}{D} + \frac{V_3\Delta_{32}}{D} + \cdots \qquad (1\text{-}7)$$

Δ_{12} being the cofactor of Z_{12}, and so on. Δ_{12}, for instance, is

$$\Delta_{12} = - \begin{vmatrix} Z_{21} & Z_{23} & Z_{24} & \cdots & Z_{2L} \\ Z_{31} & Z_{33} & Z_{34} & \cdots & Z_{3L} \\ Z_{41} & Z_{43} & Z_{44} & \cdots & Z_{4L} \\ \cdot & \cdot & \cdot & \cdot & \cdot \\ Z_{L1} & Z_{L3} & Z_{L4} & \cdots & Z_{LL} \end{vmatrix} \qquad (1\text{-}8)$$

I_3 and other currents can be expressed in similar formal arrays. The solution for the currents of the network is now, in a formal sense, complete. Let us try to see what it means.

2. DRIVING-POINT AND TRANSFER ADMITTANCES

The appearance of equations 1-5 and 1-7 can be simplified by introducing new symbols. Let us introduce y's that are defined to have the following meanings:

$$y_{11} = \frac{\Delta_{11}}{D} \quad y_{12} = \frac{\Delta_{21}}{D} \quad y_{13} = \frac{\Delta_{31}}{D} \quad \cdots$$

$$y_{21} = \frac{\Delta_{12}}{D} \quad y_{22} = \frac{\Delta_{22}}{D} \quad y_{23} = \frac{\Delta_{32}}{D} \quad \cdots \qquad (2\text{-}1)$$

The equations for I_1, I_2, and the other currents then become simply

$$I_1 = y_{11} V_1 + y_{12} V_2 + \cdots + y_{1L} V_L$$
$$I_2 = y_{21} V_1 + y_{22} V_2 + \cdots + y_{2L} V_L \qquad (2\text{-}2)$$

$\cdot \quad \cdot \quad \cdot \quad \cdot \quad \cdot \quad \cdot \quad \cdot \quad \cdot \quad \cdot$

These lower case y's are so significant that they are given special names. If the two subscripts are alike, as y_{11} and y_{22}, they are called driving-point admittances. If the subscripts are unlike, they are called transfer admittances. The reason for these names will be seen as we consider the physical meanings of the quantities.

To have something specific to consider, let us refer to the two-loop network of Figure 2a. In a network of two loops there are only two loop

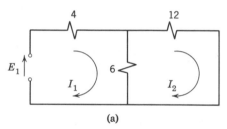

(a)

FIGURE 2a   **A network of two loops. Resistance values are shown in ohms.**

equations, and consequently the solutions for current in equations 2-2 have only two terms each:

$$I_1 = y_{11} V_1 + y_{12} V_2 \qquad (2\text{-}3)$$

$$I_2 = y_{21} V_1 + y_{22} V_2 \qquad (2\text{-}4)$$

If, as indicated in the diagram, there is an electromotive force E_1 in loop 1, making $V_1 = E_1$, but no electromotive force in loop 2 (making $V_2 = 0$), equations 2-3 and 4 for current are merely

$$I_1 = y_{11} E_1 \qquad (2\text{-}5)$$

$$I_2 = y_{21} E_1 \qquad (2\text{-}6)$$

Suppose, now, that admittance was measured at the E_1 terminals of the network: the input admittance looking into the network would be I_1/E_1. But equation 2-5 tells us that I_1/E_1 is equal to y_{11}. That is, y_{11}, the driving-point admittance as defined in equations 2-1, is the admittance seen by a source of electromotive force located in the loop in question. It is the ratio of the transform of current in a given loop to the transform of voltage applied to that loop when there are no voltages applied in any other loops of the network. That is the physical meaning of *driving-point admittance.*

Transfer admittance is the ratio of current in some other loop to the driving voltage. Thus in Figure 2a and equation 2-6, y_{21} is the ratio of the current I_2 produced in loop 2 to the electromotive force E_1 that is in loop 1. The admittance y_{21} would be measured in the network of Figure 2a by measuring I_2 and E_1 (with the appropriate phase angle if the current were alternating) and dividing one by the other.

If the network had a third loop, there might also be a transfer admittance y_{31}; this would be the ratio of the current in loop 3 produced by the voltage in loop 1 to the voltage in loop 1. Also, there might be a y_{32}.

It must be particularly noted that these interpretations of y_{11} or y_{21} as ratios of a particular current to a driving voltage have meaning only in a network with a single driving voltage that is located in only one loop. It was the elimination of all sources but E_1 that permitted equations 2-5 and 2-6 to be written. The driving electromotive force can be considered to be in any loop of a net, but it can be in only one loop at a time while the above measurements of driving-point and transfer admittance are being made.† Of course y_{11} and y_{21} are defined by equations 2-1 whether there are one or many sources.

3. EXAMPLE OF LOOP SOLUTION

It is entirely possible to use the methods of the foregoing section to solve numerical problems, although their outstanding value is in theoretical work. Let us take the net of two loops shown in Figure 2a as an example and compute the currents by means of driving-point and transfer admittances.

Let it be given that the three impedances of the net are pure resistance with the values shown in the figure. The voltage E_1 is 24 volts.

Self-impedance of loop 1 is the sum of the two resistances in that loop: $Z_{11} = 10$. Self-impedance of loop 2 is 18. The loop equations are

$$10I_1 - 6I_2 = 24$$
$$-6I_1 + 18I_2 = 0 \tag{3-1}$$

---

† The object of this paragraph is to limit discussion to "one-terminal-pair networks" which are passive networks with only one pair of terminals at which a source can be applied. Later, in our discussion of "two-terminal-pair networks," similar phraseology will be used. However, it will then be necessary to stipulate conditions at the nondriving terminal pair, and we shall need such terms as "short-circuit driving-point admittance" or "short-circuit transfer admittance."

Equations 2-5 and 2-6 give the solution for currents. To use these equations we must find y_{11} and y_{21} as defined by equations 2-1. Both driving-point admittance and transfer admittance have the following determinant in their denominators:

$$D = \begin{vmatrix} 10 & -6 \\ -6 & 18 \end{vmatrix} = 180 - 36 = 144 \tag{3-2}$$

The cofactor Δ_{11} of this determinant (that which is left after striking out the first row and the first column) is 18. The cofactor Δ_{12} (obtained by striking out the first row and the second column and changing sign) is 6. Hence

$$y_{11} = \frac{\Delta_{11}}{D} = \frac{18}{144} = \frac{1}{8}$$

$$y_{21} = \frac{\Delta_{12}}{D} = \frac{6}{144} = \frac{1}{24} \tag{3-3}$$

The loop currents, then, by equations 2-5 and 6, are

$$I_1 = y_{11} E = \tfrac{1}{8} \cdot 24 = 3 \text{ amperes} \tag{3-4}$$
$$I_2 = y_{21} E = \tfrac{1}{24} \cdot 24 = 1 \text{ ampere}$$

Current in the 6-ohm resistor is $I_1 - I_2 = 3 - 1 = 2$ amperes. All currents are now known.

It will be recognized that there are other and possibly simpler ways of finding the currents in this net. That is not important; the significant fact is that we can see here exactly what is meant by driving-point admittance, $\tfrac{1}{8}$ mho in this example, the ratio of I_1 to E_1, and by transfer admittance, here $\tfrac{1}{24}$ mho, the ratio of I_2 to E_1.

Also, it will later be found quite valuable that this general method can be adapted to use on electronic computers that will solve extremely complicated network problems very rapidly. The heart of this method of solution is clearly the computation of a determinant and certain minors, and this is a job that can be done automatically.

We have now seen that the y's of equations 2-2 have a simple physical meaning. It is no longer necessary to think of them as abstract ratios of determinants, as we did when we wrote their definitions in equations 2-1. Now we can, if we have the physical network in front of us, merely measure the y's themselves.[†]

[†] It must be clearly understood that $y_{11}, y_{21}, \ldots$ (driving-point and transfer admittances) are not at all the same thing as $Y_{AA}, Y_{AB}, \ldots$ (self-admittances and mutual admittances) as in equations 4-1. Y_{AB} and the others designated by capital letters are the admittances of certain individual elements of a net. The lower case y's, however, refer to admittances at network terminals. To be sure, the y's can be computed from the Y's, but they are equal only in the extreme and rather absurd limit of a "network" consisting of just one element.

4. SOLVING THE NODE EQUATIONS

The past several pages have been occupied with the solution of loop equations in their most general form. Next we must consider the solution of the general node equations, for these are just as important and just as useful as the loop equations. The general node equations were written as equations 10-1 in Chapter 8 and are

$$
\begin{aligned}
Y_{AA}V_A + Y_{AB}V_B + Y_{AC}V_C + \cdots + Y_{AN}V_N &= I_A \\
Y_{BA}V_A + Y_{BB}V_B + Y_{BC}V_C + \cdots + Y_{BN}V_N &= I_B \\
Y_{CA}V_A + Y_{CB}V_B + Y_{CC}V_C + \cdots + Y_{CN}V_N &= I_C \qquad (4\text{-}1) \\
\cdot \qquad \cdot \qquad \cdot \qquad \cdot \qquad \cdot \qquad \cdot \qquad & \cdot \\
Y_{NA}V_A + Y_{NB}V_B + Y_{NC}V_C + \cdots + Y_{NN}V_N &= I_N
\end{aligned}
$$

Since we have just been through the solution of the loop equations in detail, it is apparent that a solution of the node equations for the node voltages is

$$
\begin{aligned}
V_A &= z_{AA}I_A + z_{AB}I_B + \cdots + z_{AN}I_N \\
V_B &= z_{BA}I_A + z_{BB}I_B + \cdots + z_{BN}I_N \qquad (4\text{-}2)
\end{aligned}
$$

where

$$
z_{AA} = \frac{\Delta_{AA}}{\Delta} \quad z_{AB} = \frac{\Delta_{BA}}{\Delta} \quad z_{AC} = \frac{\Delta_{CA}}{\Delta} \quad \cdots
$$

$$
z_{BA} = \frac{\Delta_{AB}}{\Delta} \quad z_{BB} = \frac{\Delta_{BB}}{\Delta} \quad z_{BC} = \frac{\Delta_{CB}}{\Delta} \quad \cdots \qquad (4\text{-}3)
$$

In these expressions the denominator is the determinant derived from equations 4-1:

$$
\Delta = \begin{vmatrix}
Y_{AA} & Y_{AB} & \cdots & Y_{AN} \\
Y_{BA} & Y_{BB} & \cdots & Y_{BN} \\
\cdot & \cdot & \cdots & \cdot \\
Y_{NA} & Y_{NB} & \cdots & Y_{NN}
\end{vmatrix} \qquad (4\text{-}4)
$$

and the numerators are the appropriate cofactors of this determinant.

Equations 4-2 give the potentials V_A, V_B, and so forth, at the nodes of a network in terms of the admittance elements of the network and the currents that are imposed on that network. The similarity of this solution to the solution for loop currents is an illustration of *duality*.

As we solved for loop currents in Section 2 in terms of driving-point and transfer admittances, here we solve for node voltages in terms of *driving-point* and *transfer impedances*. The driving-point impedances are z_{AA}, z_{BB}, and other coefficients of equations 4-2 with like subscripts; the transfer impedances are z_{AB}, z_{CB}, and other coefficients with unlike subscripts.

Driving-point and transfer impedances are defined by equations 4-3. Their *physical meaning* can be deduced from equations 4-2. Assume a network; the network of Figure 4a may serve as a simple example. The current I_A is

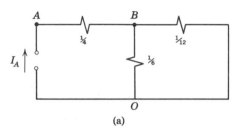

(a)

FIGURE 4a   A network with two independent nodes. Conductance values are shown in mhos.

produced by a source between node A and the reference node of the network. There are no other sources, so I_B, I_C, and so on, in equations 4-2 are all zero. Therefore $V_A = z_{AA} I_A$, from which $z_{AA} = V_A/I_A$. This means that if a passive network has one pair of terminals A and O, the ratio of voltage to current at those terminals is the driving-point impedance z_{AA}; it is the impedance that would be measured by a bridge or other measuring device connected to those terminals. This is the meaning of driving-point impedance.

The transfer impedance between B and A is similarly the ratio of the voltage at node B to the source current at node A, there being no other sources in the network. This follows from the second of equations 4-2 when I_B, I_C, and so on, are made zero, for then $V_B = z_{BA} I_A$. From this, $z_{BA} = V_B/I_A$.

It is evident from this physical interpretation that the driving-point impedance (such as z_{AA}) at a pair of terminals and the driving-point admittance (such as y_{11}) at the same pair of terminals are reciprocal quantities, for one is the ratio of voltage to current and the other is the ratio of the same current to voltage. (This assumes that the terminals of a one-terminal-pair network are at nodes A and O, and that they are in loop 1 only.) However, no such simple relation exists between transfer impedances and transfer admittances; these are *not* reciprocals of each other.†

---

† It is hardly necessary to give warning again that z_{AA} and the other lower case z's are totally different from Z_{11} and other capital Z's.

5. EXAMPLE OF NODE SOLUTION

Let us find node voltages in the network of Figure 5a. First, write the three node equations. The general form, from equations 4-1, is

$$Y_{AA}V_A + Y_{AB}V_B + Y_{AC}V_C = I_A$$
$$Y_{BA}V_A + Y_{BB}V_B + Y_{BC}V_C = I_B \qquad (5\text{-}1)$$
$$Y_{CA}V_A + Y_{CB}V_B + Y_{CC}V_C = I_C$$

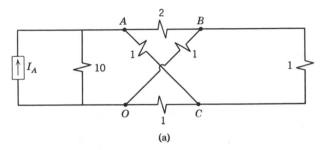

(a)

FIGURE 5a   A lattice network. Conductances are shown in mhos.

Putting in the numerical values for this particular network:

$$13V_A - 2V_B - 1V_C = I_A$$
$$-2V_A + 4V_B - 1V_C = 0 \qquad (5\text{-}2)$$
$$-1V_A - 1V_B + 3V_C = 0$$

Referring to equation 4-4, the first step in the solution is to write

$$\Delta = \begin{vmatrix} Y_{AA} & Y_{AB} & Y_{AC} \\ Y_{BA} & Y_{BB} & Y_{BC} \\ Y_{CA} & Y_{CB} & Y_{CC} \end{vmatrix} = \begin{vmatrix} 13 & -2 & -1 \\ -2 & 4 & -1 \\ -1 & -1 & 3 \end{vmatrix} = 123 \qquad (5\text{-}3)$$

The cofactor of Y_{AA} is

$$\Delta_{AA} = \begin{vmatrix} Y_{BB} & Y_{BC} \\ Y_{CB} & Y_{CC} \end{vmatrix} = \begin{vmatrix} 4 & -1 \\ -1 & 3 \end{vmatrix} = 11 \qquad (5\text{-}4)$$

The cofactors of Y_{AB} and Y_{AC} are, respectively,

$$\Delta_{AB} = -\begin{vmatrix} Y_{BA} & Y_{BC} \\ Y_{CA} & Y_{CC} \end{vmatrix} = -\begin{vmatrix} -2 & -1 \\ -1 & 3 \end{vmatrix} = 7 \qquad (5\text{-}5)$$

$$\Delta_{AC} = \begin{vmatrix} Y_{BA} & Y_{BB} \\ Y_{CA} & Y_{CB} \end{vmatrix} = \begin{vmatrix} -2 & 4 \\ -1 & -1 \end{vmatrix} = 6 \qquad (5\text{-}6)$$

Then, as in equations 4-3,

$$z_{AA} = \frac{\Delta_{AA}}{\Delta} = \frac{11}{123} \qquad (5\text{-}7)$$

$$z_{BA} = \frac{\Delta_{AB}}{\Delta} = \frac{7}{123} \qquad (5\text{-}8)$$

$$z_{CA} = \frac{\Delta_{AC}}{\Delta} = \frac{6}{123} \qquad (5\text{-}9)$$

Equations 4-2 give the formal solution for the voltages. It will be seen that there are nine coefficients (the small z's) in the general solution, but our present problem is simplified by the fact that I_B and I_C are both zero; this means that six of the coefficients are multiplied by zero and hence do not need to be computed. Using equations 4-2,

$$V_A = z_{AA}I_A + z_{AB}I_B + z_{AC}I_C = \tfrac{11}{123}I_A + 0 + 0$$
$$V_B = z_{BA}I_A + z_{BB}I_B + z_{BC}I_C = \tfrac{7}{123}I_A \qquad (5\text{-}10)$$
$$V_C = z_{CA}I_A + z_{CB}I_B + z_{CC}I_C = \tfrac{6}{123}I_A$$

These are the voltages of the network. If currents are desired, they are easily found from voltage difference and admittance. For instance, the current through the 1-mho element that represents the load on the lattice network is

$$I_{BC} = (V_B - V_C)(1) = \tfrac{1}{123}I_A \qquad (5\text{-}11)$$

This is the fraction of the input current that reaches the output of the lattice.

This example is the same, actually, as that of Section 11, Chapter 8, although the diagram is drawn here as a lattice and there as a bridge. Also, the nodes are lettered differently to give a more convenient choice of reference node in the present problem. The numerical answer is, of course, the same.

It is clear that any node may be the reference node. Indeed, it is important to realize that a reference node is not essential. It is essential that N equations be written for current at N independent nodes in terms of voltages at N independent node pairs. For simplicity and symmetry these node pairs *may* all have a reference node in common; however, in later work with two-port networks, the use of a common reference node will not always be found feasible.

6. EQUALITY OF TRANSFER ADMITTANCES AND IMPEDANCES

The transfer admittances y_{12} and y_{21} which are used with loop solution are rather similar to each other. It would not be unreasonable to wonder if they might be equal.

By definition, as given in equations 2-1, $y_{21} = \Delta_{12}/D$ and $y_{12} = \Delta_{21}/D$. Clearly, y_{21} and y_{12} are equal if Δ_{12} and Δ_{21} are equal. Let us see whether we can show that they are.

Δ_{21} is a determinant, given in equation 1-6 as:

$$\Delta_{21} = - \begin{vmatrix} Z_{12} & Z_{13} & Z_{14} & \cdots & Z_{1L} \\ Z_{32} & Z_{33} & Z_{34} & \cdots & Z_{3L} \\ Z_{42} & Z_{43} & Z_{44} & \cdots & Z_{4L} \\ \cdot & \cdot & \cdot & \cdots & \cdot \\ Z_{L2} & Z_{L3} & Z_{L4} & \cdots & Z_{LL} \end{vmatrix} \qquad (6\text{-}1)$$

A mathematical theorem (see Appendix 1) says that the value of a determinant remains the same if rows and columns of the determinant are interchanged; we can therefore write:

$$\Delta_{21} = - \begin{vmatrix} Z_{12} & Z_{32} & Z_{42} & \cdots & Z_{L2} \\ Z_{13} & Z_{33} & Z_{43} & \cdots & Z_{L3} \\ Z_{14} & Z_{34} & Z_{44} & \cdots & Z_{L4} \\ \cdot & \cdot & \cdot & \cdots & \cdot \\ Z_{1L} & Z_{3L} & Z_{4L} & \cdots & Z_{LL} \end{vmatrix} \qquad (6\text{-}2)$$

We now compare this expression for Δ_{21} with the determinant given in equation 1-8 for Δ_{12}, and we find that the two are the same in form but differ in the order of the subscripts. They would be identical if $Z_{12} = Z_{21}$, $Z_{32} = Z_{23}$, and in general

$$Z_{mn} = Z_{nm} \qquad (6\text{-}3)$$

Hence the determinants will be identical if *mutual impedances* between loops (as in equation 6-3) are the same.

We have seen a number of examples of networks in which mutual impedances are obviously the same. In Figure 2a of this chapter, or Figures 3a, 6a, or 9a of Chapter 8, the mutual impedances are simply the impedances of the common elements. It is less obvious but it is still true, as will be shown in Chapter 11, that mutual impedances are equal if circuits are connected by a transformer (even though the number of turns in the transformer windings might be unequal). Indeed, equation 6-3 fails only for certain networks employing transistors, vacuum tubes, and other devices that carry a signal more readily in one direction than the other. These devices are unilateral or, perhaps, unsymmetrically bilateral. If equation 6-3 applies, the network and the elements that comprise it are *reciprocal*, as are all ordinary passive elements including resistors, capacitors, inductors, and transformers.

It is concluded that

$$y_{pq} = y_{qp} \qquad (6\text{-}4)$$

for all linear, reciprocal networks. It can similarly be shown that the transfer impedances used for *node* solutions are likewise equal under the same restrictions; that is, for linear, reciprocal networks,

$$z_{PQ} = z_{QP} \tag{6-5}$$

These are conclusions of considerable importance. For one thing, the number of parameters required to characterize a network is reduced from four to three. For another, a foundation is laid for the reciprocity theorem (see Chapter 10). Indeed, it can be shown that the directional characteristics of a radio antenna when receiving signals are the same as its directional characteristics when transmitting [see, for instance, Terman (2)].

7. MATRICES

The sets of equations that we have been using in this chapter can be quite long if a network is complicated, and for the general consideration of an unspecified network they are of indeterminate length. A great deal of time and tiresome writing can be saved by a shorthand system of writing equations, and the use of matrices has been found particularly helpful.

In a matrix, the essential quantities are displayed in an orderly manner. For instance, the loop equations in equations 1-1 relate impedances and currents to voltages. The voltages on the right-hand side of the equations can be set out in a *column matrix*:

$$\begin{bmatrix} V_1 \\ V_2 \\ V_3 \\ \vdots \\ V_L \end{bmatrix} \tag{7-1}$$

These voltages are put between brackets to indicate that this is a display of voltages as if in a showcase. (Some authors omit the left-hand bracket.†)

The impedances are coefficients of terms on the left-hand side of equations 1-1. We may take them out, also, and set them in a showcase:

$$\begin{bmatrix} Z_{11} & Z_{12} & Z_{13} & \cdots & Z_{1L} \\ Z_{21} & Z_{22} & Z_{23} & \cdots & Z_{2L} \\ Z_{31} & Z_{32} & Z_{33} & \cdots & Z_{3L} \\ \cdot & \cdot & \cdot & \cdot & \cdot \\ Z_{L1} & Z_{L2} & Z_{L3} & \cdots & Z_{LL} \end{bmatrix} \tag{7-2}$$

---

† The orthogonal components of a vector can be expressed in a column matrix, and the column matrix is sometimes called a *vector*.

These are the impedances that tell how the network will behave when voltages are applied to it. They show a kind of abstract blueprint of the network. This array gives the form of the loop equations. The brackets indicate that the Z's enclosed constitute a related group, arranged in a systematic way.

Although this array looks a little like a determinant (like equation 1-3), it is *not* a determinant. It is simply a display of the coefficients of the terms of equations 1-1 in a systematic order. It is not to be expanded or evaluated like a determinant. Indeed, there is *nothing* that we can do with it at the moment except look at it. It is interesting because it gives the form of equations 1-1.

Such a display as 7-1 or 7-2 is called a *matrix*.† The former is called a column matrix, the latter a square matrix.

The economy of matrix notation comes from representing a matrix by a single letter. Thus the array of 7-2 can be symbolized as $[Z]$. (Notations vary; some authors use a bold-face Z and omit the brackets.) Symbolically, then,

$$[Z] = \begin{bmatrix} Z_{11} & Z_{12} & \cdots & Z_{1L} \\ Z_{21} & Z_{22} & \cdots & Z_{2L} \\ \cdot & \cdot & \cdot & \cdot \\ Z_{L1} & Z_{L2} & \cdots & Z_{LL} \end{bmatrix} \tag{7-3}$$

8. MATRIX EQUATIONS

With the understanding that a single symbol represents an array of impedances, currents, or voltages, the whole set of loop equations, which are written out fully in equations 1-1, can be indicated symbolically as

$$[Z][I] = [V] \tag{8-1}$$

Then the solutions for loop currents, which are given in equations 2-2, can be abbreviated as

$$[I] = [y][V] \tag{8-2}$$

In a similar way the node equations 4-1 can be written

$$[Y][V] = [I] \tag{8-3}$$

and their solutions in equations 4-2 take the form

$$[V] = [z][I] \tag{8-4}$$

Such equations as 8-1 and 8-2 suggest the usefulness of deriving one matrix from another in a way that parallels algebraic operations on the

---

† Matrix, pl. matrices: that which gives form. . . . . *Webster's Directory.*

equations they represent. Matrix operations of this kind have been devised, and certain rules will now be found helpful. Our present interest is in the facility these operations give to work with matrix equations, but it may also be found that their systematic nature makes them adaptable to numerical solutions on electronic computers.

For example, the $[y]$ matrix of equation 8-2 can be derived from the $[Z]$ matrix of equation 8-1 by the application of certain rules, and the resulting matrix $[I]$ gives the desired loop currents. Similarly, if node equations are used, $[z]$ of equation 8-4 can be found from $[Y]$ of equation 8-3, and the node voltages appear from $[V]$. The necessary methods will now be shown.

9. MATRIX OPERATIONS

Let us consider the node equations of Section 5 in matrix notation. By what rules can equation 8-3 be interpreted to mean the same as the set of equations 5-1? First equation 8-3 is expanded by writing out the matrices represented by $[Y]$, $[V]$, and $[I]$. Following the pattern of equations 7-1 and 2, we expand $[Y][V] = [I]$ to

$$\begin{bmatrix} Y_{AA} & Y_{AB} & Y_{AC} \\ Y_{BA} & Y_{BB} & Y_{BC} \\ Y_{CA} & Y_{CB} & Y_{CC} \end{bmatrix} \begin{bmatrix} V_A \\ V_B \\ V_C \end{bmatrix} = \begin{bmatrix} I_A \\ I_B \\ I_C \end{bmatrix} \qquad (9\text{-}1)$$

Here two matrices, a square matrix and a column matrix, are written side by side. The following rule is applied: When two such matrices are written side by side they are combined into a single matrix (in this case a column matrix) by the following process:

Multiply the first term in the first row of the $[Y]$ matrix by the first term in the column of the $[V]$ matrix. Multiply the second term in the first row of $[Y]$ by the second term in the column of $[V]$. Multiply the third term in the first row of $[Y]$ by the third term in the column of $[V]$. The sum of the three is then written as the first term in the column matrix:

$$[Y][V] = \begin{bmatrix} Y_{AA}V_A + Y_{AB}V_B + Y_{AC}V_C \\ Y_{BA}V_A + Y_{BB}V_B + Y_{BC}V_C \\ Y_{CA}V_A + Y_{CB}V_B + Y_{CC}V_C \end{bmatrix} \qquad (9\text{-}2)$$

In the same manner, multiply each term of the second row of $[Y]$ by the corresponding term of the column of $[V]$ and put the sum as the second term in the column matrix 9-2. Finally, multiply each term of the third row of $[Y]$ by the corresponding term of $[V]$ and enter as the third term of 9-2.

Clearly, 9-2 is a column matrix for, when the indicated multiplications and additions are performed, there will remain three numbers arrayed in a column.

10. MATRIX MULTIPLICATION

This process is called matrix multiplication, but it is not multiplication in the ordinary sense of the word. Ordinary multiplication is something that applies to two quantities, not to two arrays of quantities. It would probably be better if the word "multiplication" were not used, and some such word as "juxtaposition" were employed instead. But this is not done; the process is called multiplication. However, it is not surprising that many of the rules of ordinary multiplication do not apply to this process. For instance, the rule for matrix multiplication gives 9-2 as the matrix product $[Y][V]$. We should not be surprised that the matrix product $[V][Y]$ turns out to be something quite different.

11. MATRIX EQUALITY

Since the left-hand side of equation 9-1 is to be interpreted as the column matrix of equation 9-2, we equate 9-2 to the right-hand side of 9-1, giving:

$$\begin{bmatrix} Y_{AA}V_A + Y_{AB}V_B + Y_{AC}V_C \\ Y_{BA}V_A + Y_{BB}V_B + Y_{BC}V_C \\ Y_{CA}V_A + Y_{CB}V_B + Y_{CC}V_C \end{bmatrix} = \begin{bmatrix} I_A \\ I_B \\ I_C \end{bmatrix} \qquad (11\text{-}1)$$

What is meant by *equality* of matrices? A definition of equality is needed. A matrix is a mathematical showcase; the display in one showcase is the same as that in another if every item in one is identical with every item in the other. So with matrices: *one matrix equals another if, and only if, each element of one equals the corresponding element of the other.*

That being true, equation 11-1 says that $Y_{AA}V_A + Y_{AB}V_B + Y_{AC}V_C$ must equal I_A, that the second element of the left-hand matrix must equal I_B, and that the third element must equal I_C. But when these three are set equal, the result is precisely the set of equations 5-1.

12. PRODUCT OF SQUARE MATRICES

We have not yet encountered the need, but we shall soon, so let us generalize the definition of matrix multiplication. We know that a square matrix "multiplied" by a column matrix equals a column matrix, as in equation 9-1. What results when a square matrix is multiplied by another square matrix?

The definition of "multiplication" must be extended to cover an operation involving two square matrices. The following rule is adopted: the product matrix is a square matrix, and each column in the product matrix is obtained by "multiplying" the first matrix by a column of the second matrix. The following purely abstract matrices will illustrate the process. They are 2 × 2 matrices, square matrices of the second order. They are large enough to illustrate the process of matrix multiplication:

$$\begin{bmatrix} a_{11} & a_{12} \\ a_{21} & a_{22} \end{bmatrix} \begin{bmatrix} b_{11} & b_{12} \\ b_{21} & b_{22} \end{bmatrix} = \begin{bmatrix} (a_{11}b_{11} + a_{12}b_{21}) & (a_{11}b_{12} + a_{12}b_{22}) \\ (a_{21}b_{11} + a_{22}b_{21}) & (a_{21}b_{12} + a_{22}b_{22}) \end{bmatrix}$$

(12-1)

With larger matrices the process is merely continued. For a precise definition, matrix multiplication can be summed up in one formula: if $[a][b] = [c]$, then the elements of $[c]$ are

$$c_{pq} = \sum_{r=1}^{n} a_{pr} b_{rq}$$

(12-2)

and this is the law and the prophets.

13. CERTAIN MATRIX RELATIONS

The Commutative Law Is Not Valid. Multiply two random matrices, such as:

$$\begin{bmatrix} 4 & 3 \\ 2 & 1 \end{bmatrix} \begin{bmatrix} 2 & 1 \\ 1 & 0 \end{bmatrix} = \begin{bmatrix} 8+3 & 4+0 \\ 4+1 & 2+0 \end{bmatrix} = \begin{bmatrix} 11 & 4 \\ 5 & 2 \end{bmatrix}$$

(13-1)

But if the order of the same matrices is reversed:

$$\begin{bmatrix} 2 & 1 \\ 1 & 0 \end{bmatrix} \begin{bmatrix} 4 & 3 \\ 2 & 1 \end{bmatrix} = \begin{bmatrix} 8+2 & 6+1 \\ 4+0 & 3+0 \end{bmatrix} = \begin{bmatrix} 10 & 7 \\ 4 & 3 \end{bmatrix}$$

(13-2)

Clearly these two matrix products are not the same. This proves that *the commutative law of algebra does not hold* for matrix multiplication.† That is, $[a][b] \neq [b][a]$.

The Unit Matrix and Other Definitions. Another theorem is illustrated by the following matrix multiplications:

$$\begin{bmatrix} 1 & 0 \\ 0 & 1 \end{bmatrix} \begin{bmatrix} a & b \\ c & d \end{bmatrix} = \begin{bmatrix} a & b \\ c & d \end{bmatrix} \quad \text{and} \quad \begin{bmatrix} a & b \\ c & d \end{bmatrix} \begin{bmatrix} 1 & 0 \\ 0 & 1 \end{bmatrix} = \begin{bmatrix} a & b \\ c & d \end{bmatrix}$$

(13-3)

It makes no difference what values a, b, c, and d may have; the product is the same as the original matrix. This statement can be generalized by extending the multiplication to nth-order square matrices, with the conclusion that *a*

---

† The associative and distributive laws of algebra do apply to matrix multiplication. That is, $[A][B][C] = ([A][B])[C] = [A]([B][C])$; and $([A] + [B])[C] = [A][C] + [B][C]$, and so forth.

matrix either premultiplied or postmultiplied by the unit matrix remains unchanged. The words of this statement require a number of definitions:

1. The *unit matrix* is a square matrix in which all the elements of the principal diagonal are 1 and the other elements are 0. The symbol for the unit matrix is $[U]$, and $[A][U] = [U][A] = [A]$. (Some authors use $[I]$ or I rather than $[U]$, but I is here reserved for current.)
2. The *principal diagonal* is the line of elements extending from the upper left corner to the lower right corner of a matrix.
3. *Pre*multiplication and *post*multiplication have the meanings illustrated in equations 13-3; they must be distinguished because the commutative law does not apply and the products are, in general, different. (Premultiplication and postmultiplication by the *unit* matrix give the same product; this is an exception to the general rule.)

Addition of Matrices.   Addition merely means adding the corresponding terms of matrices. It is defined by saying that if $[A] + [B] = [C]$, then

$$c_{pq} = a_{pq} + b_{pq} \tag{13-4}$$

Product of a Matrix by a Number.   It follows from the rule for addition that if k identical matrices are added, every element in the matrix sum is k times the corresponding element in the original matrix. Multiplication of a matrix by a number is indicated by writing the number beside the matrix, as in equation 14-9. Each element of the matrix is to be multiplied by the number.

Division.   Division of one matrix by another is not defined. Instead of dividing, a matrix is multiplied by the *inverse* of another matrix as will be considered in the next section. If there were division, two kinds would be required, one corresponding to premultiplication by the inverse, the other to postmultiplication by the inverse. It is simpler to keep the inverse form and never speak of division.

Transposition.   Transposition of a matrix means interchanging rows and columns. $[A]_t$ is the transpose of $[A]$, and if

$$[A] = \begin{bmatrix} a & b & c \\ d & e & f \\ g & h & i \end{bmatrix} \quad \text{then} \quad [A]_t = \begin{bmatrix} a & d & g \\ b & e & h \\ c & f & i \end{bmatrix} \tag{13-5}$$

A *row* matrix is the transpose of a *column* matrix, but the row matrix is not used in this chapter.

Much could be said about nonsquare or *rectangular* matrices. In this chapter it is enough to say that the rules given above can be applied to nonsquare matrices by adding rows or columns of zeros until the rectangular matrices become square. Adding zeros does not change a matrix.

It is not the intention to present matrix algebra in this book. The aim, rather, is to give a few simple examples of how matrix methods are applied to network problems. Current electrical literature is full of references to matrix algebra for which an elementary understanding is enough. Those who feel the need to know more will find the subject treated fully in a number of good reference books.†

14. THE INVERSE MATRIX

Given a matrix $[Y]$, let us *define* another matrix, which we shall call the inverse of $[Y]$ and denote by $[Y]^{-1}$. By definition,

$$[Y]^{-1} = \frac{1}{\Delta} \begin{bmatrix} \Delta_{AA} & \Delta_{BA} & \Delta_{CA} \\ \Delta_{AB} & \Delta_{BB} & \Delta_{CB} \\ \Delta_{AC} & \Delta_{BC} & \Delta_{CC} \end{bmatrix} \qquad (14\text{-}1)$$

where Δ, a number, is the determinant of the matrix $[Y]$, and Δ_{AB} is the cofactor of the term Y_{AB} in that determinant. (This definition is expressly for a third-order matrix, but the extension to matrices of other orders is evident.) With $[Y]^{-1}$ defined in this way, it can be shown (and this is the reason for calling it the *inverse*) that

$$[Y][Y]^{-1} = [U] \quad \text{and also} \quad [Y]^{-1}[Y] = [U] \qquad (14\text{-}2)$$

That is, *the product of a matrix and its inverse is the unit matrix.*

Equations 14-2 are easily proved for a 2×2 matrix. Let $[M]$ be the matrix

$$[M] = \begin{bmatrix} M_{AA} & M_{AB} \\ M_{BA} & M_{BB} \end{bmatrix} \qquad (14\text{-}3)$$

First, the inverse is to be found. To compute the inverse, the determinant Δ of the matrix $[M]$ is needed:

$$\Delta = \begin{vmatrix} M_{AA} & M_{AB} \\ M_{BA} & M_{BB} \end{vmatrix} = M_{AA}M_{BB} - M_{AB}M_{BA} \qquad (14\text{-}4)$$

The cofactors of this determinant are simply

$$\Delta_{AA} = M_{BB} \qquad \Delta_{AB} = -M_{BA} \qquad \Delta_{BA} = -M_{AB} \qquad \Delta_{BB} = M_{AA}$$

$$(14\text{-}5)$$

---

† For example, Guillemin (1) or, better, (2), or Pipes and Hovanessian, or DeRusso, Roy, and Close.

Then, by equation 14-1, the inverse of $[M]$ is

$$[M]^{-1} = \frac{1}{\Delta}\begin{bmatrix} M_{BB} & -M_{AB} \\ -M_{BA} & M_{AA} \end{bmatrix} \tag{14-6}$$

Second, we prove that equations 14-2 are true for this matrix by multiplying:

$$[M]^{-1}[M] = \frac{1}{\Delta}\begin{bmatrix} M_{BB} & -M_{AB} \\ -M_{BA} & M_{AA} \end{bmatrix}\begin{bmatrix} M_{AA} & M_{AB} \\ M_{BA} & M_{BB} \end{bmatrix}$$

$$= \frac{1}{\Delta}\begin{bmatrix} M_{BB}M_{AA} - M_{AB}M_{BA} & M_{BB}M_{AB} - M_{AB}M_{BB} \\ -M_{BA}M_{AA} + M_{AA}M_{BA} & -M_{BA}M_{AB} + M_{AA}M_{BB} \end{bmatrix} \tag{14-7}$$

Two of the elements of the product matrix are zero, and when the other two elements are divided by Δ (as given in equation 14-4) the result is simply 1. Hence

$$[M]^{-1}[M] = \begin{bmatrix} 1 & 0 \\ 0 & 1 \end{bmatrix} = [U] \tag{14-8}$$

Thus a matrix premultiplied by its inverse gives the unit matrix. Postmultiplication by the inverse also gives the unit matrix, as can be shown in the same way.

The foregoing proof applies to a 2×2 square matrix. The same verification of equations 14-2 can be obtained for matrices of third or higher order by the patient student.

As a numerical example of finding an inverse matrix, suppose that we are given

$$[A] = \begin{bmatrix} 4 & 3 \\ 2 & 2 \end{bmatrix}$$

and wish to compute its inverse. First, we write the determinant

$$\Delta = \begin{vmatrix} 4 & 3 \\ 2 & 2 \end{vmatrix} = 8 - 6 = 2$$

Next we write the cofactors:

$$\Delta_{AA} = 2 \qquad \Delta_{AB} = -2 \qquad \Delta_{BA} = -3 \qquad \Delta_{BB} = 4$$

Then the inverse matrix is

$$[A]^{-1} = \frac{1}{2}\begin{bmatrix} 2 & -3 \\ -2 & 4 \end{bmatrix} = \begin{bmatrix} 1 & -\frac{3}{2} \\ -1 & 2 \end{bmatrix} \tag{14-9}$$

We now show that $[A][A]^{-1} = [U]$ by multiplying:

$$[A][A]^{-1} = \begin{bmatrix} 4 & 3 \\ 2 & 2 \end{bmatrix} \begin{bmatrix} 1 & -\frac{3}{2} \\ -1 & 2 \end{bmatrix} = \begin{bmatrix} 4-3 & -6+6 \\ 2-2 & -3+4 \end{bmatrix} = \begin{bmatrix} 1 & 0 \\ 0 & 1 \end{bmatrix}$$

This completes the numerical illustration.†

15. MATRIX SOLUTION OF THE NODE EQUATIONS

The theorems of the last few pages make it possible to solve simple matrix equations in two or three easy steps—and one step, inversion, that is not so easy. The process will be illustrated by solving the node equations that were given in matrix form as equation 8-3:

$$[Y][V] = [I] \tag{15-1}$$

We wish to solve for $[V]$. If this were an algebraic equation involving numbers, each side could be divided by $[Y]$; however, since $[Y]$ is a matrix, and division by matrices is not recognized, another plan must be followed: we premultiply each side by $[Y]^{-1}$. This gives

$$[Y]^{-1}[Y][V] = [Y]^{-1}[I] \tag{15-2}$$

Since $[Y]^{-1}[Y] = [U]$, and $[U][V] = [V]$, the left-hand side is merely $[V]$, and

$$[V] = [Y]^{-1}[I] \tag{15-3}$$

This is the required solution for the voltage matrix. Formally, equation 15-3 is the solution of the node equations represented by 15-1.

Equation 15-3 can be compared with equation 8-4 which gives the solution of the same node equations as

$$[V] = [z][I] \tag{15-4}$$

The comparison indicates that

$$[z] = [Y]^{-1} \tag{15-5}$$

That this is indeed true can be seen by comparing the derivation of the values of z as given in equations 4-3 with the rule for obtaining the inverse of a matrix as given in equation 14-1. The rule for matrix inversion is of course designed to be the rule for simultaneous solution.

---

† It is sometimes asked, Why is the inverse matrix defined in the peculiar way given in equation 14-1? As a matter of fact, the inverse is defined in this way in order to make the product of a matrix and its inverse equal the unit matrix. That is, equation 14-1 is designed to make equations 14-2 valid.

Thus, by matrix inversion, the *driving-point and transfer impedances* expressed as $[z]$ can be found from the *self and mutual admittances* that comprise $[Y]$.

16. A TWO-NODE EXAMPLE

Given the network of Figure 16a, with two independent current sources as shown, and three purely resistive elements for each of which Y is indicated on the diagram, find voltages at the two nodes A and B relative to the reference node O.

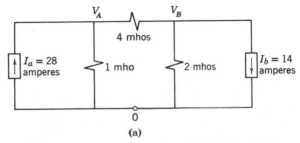

(a)

FIGURE 16a   A network of two independent nodes.

First, write the node equations by inspection:

$$[Y][V] = [I] = \begin{bmatrix} 5 & -4 \\ -4 & 6 \end{bmatrix} \begin{bmatrix} V_A \\ V_B \end{bmatrix} = \begin{bmatrix} 28 \\ -14 \end{bmatrix} \tag{16-1}$$

Then

$$[V] = [Y]^{-1}[I] \tag{16-2}$$

To find the inverse of $[Y]$ we need the determinants:

$$\Delta = \begin{vmatrix} 5 & -4 \\ -4 & 6 \end{vmatrix} = 30 - 16 = 14$$

$$\Delta_{AA} = 6 \qquad \Delta_{BA} = \Delta_{AB} = 4 \qquad \Delta_{BB} = 5 \tag{16-3}$$

These give the column matrix:

$$[V] = \frac{1}{14} \begin{bmatrix} 6 & 4 \\ 4 & 5 \end{bmatrix} \begin{bmatrix} 28 \\ -14 \end{bmatrix} = \begin{bmatrix} 12 - 4 \\ 8 - 5 \end{bmatrix} = \begin{bmatrix} 8 \\ 3 \end{bmatrix} \tag{16-4}$$

so $V_A = 8$ volts and $V_B = 3$ volts, the answer.

Let us, for comparison, solve the same problem by the method of superposition (see Section 1 of Chapter 10), without using either node equations or matrices. First, we require one of the source currents to be zero:†

Let $I_a = 28$, $I_b = 0$

$$V_A = \frac{I_a}{y_{aa}} = \frac{28}{1 + 8/6} = 12 \tag{16-5}$$

$$V_B = \tfrac{4}{6}12 = 8 \tag{16-6}$$

Next we require the other source to be zero:

Let $I_a = 0$, $I_b = 14$

$$V_B = \frac{-I_b}{y_{bb}} = \frac{-14}{2 + 4/5} = -5 \tag{16-7}$$

$$V_A = \tfrac{4}{5}(-5) = -4 \tag{16-8}$$

Finally we superimpose these two solutions, adding the two values of V_A and the two of V_B to get:

$$V_A = 12 - 4 = 8 \text{ volts} \qquad V_B = 8 - 5 = 3 \text{ volts} \tag{16-9}$$

This is the same answer that was obtained in equation 16-4 and thus we verify the answer as well as gaining a solution that is easier to visualize. Moreover, we have here an illustration of the method of superposition.

17. MATRIX SOLUTION OF THE LOOP EQUATIONS

A set of loop equations for a network can also be solved by matrix inversion—as, indeed, can any set of simultaneous linear equations. In matrix notation the loop equations 1-1 are written

$$[Z][I] = [V] \tag{17-1}$$

We wish to solve for the loop currents. Premultiplying each side of this equation by $[Z]^{-1}$ gives

$$[I] = [Z]^{-1}[V] \tag{17-2}$$

Thus the desired currents can be obtained by an inversion of the impedance matrix and a matrix multiplication.

---

† The method is to treat one source at a time, all other sources then being zero. Note that when an independent current source is required to be zero, the current through it is zero (as if it were an open circuit); on the other hand, when an independent voltage source is required to be zero, the voltage across it is zero (as if it were a short circuit).

This answer has also been obtained by nonmatrix methods, and equations 2-2 can be written

$$[I] = [y][V] \qquad (17\text{-}3)$$

where each y has the meaning indicated in equation 2-1. Comparison of equation 17-2 with 17-3, or study of the derivation of $[y]$, shows that

$$[y] = [Z]^{-1} \qquad (17\text{-}4)$$

Thus the *driving-point and transfer admittances* of a network can be found by inversion of the matrix of *self- and mutual impedances* of the loops.

For illustration, let us consider numerical values in a three-loop network.

18. A THREE-LOOP EXAMPLE

Our problem is to find currents and voltages in the network of Figure 18a, the impedance and the driving electromotive force being given. Network impedances are purely resistive; if there were also reactances and if the

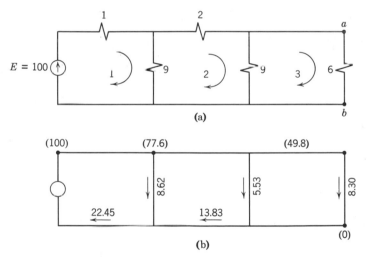

(a)

(b)

FIGURES 18a,b   Electromotive force and resistances are shown in (a); currents and potentials (in parentheses) are shown in (b).

electromotive force were alternating the arithmetic would be longer but the nature of the problem would be the same. The loop equations are

$$
\begin{aligned}
10I_1 - 9I_2 \quad\quad &= 100 \\
-9I_1 + 20I_2 - 9I_3 &= 0 \\
-9I_2 + 15I_3 &= 0
\end{aligned}
\qquad (18\text{-}1)
$$

After a little experience with this kind of problem has been gained, it will be unnecessary to write these equations at all; solution of the problem will be begun by writing the matrix:

$$[Z] = \begin{bmatrix} 10 & -9 & 0 \\ -9 & 20 & -9 \\ 0 & -9 & 15 \end{bmatrix} \qquad (18\text{-}2)$$

The determinant, being of the same form as the matrix, need not be written out. Its value is

$$\Delta = 3000 - (810 + 1215) = 975 \qquad (18\text{-}3)$$

Each cofactor is a second-order determinant in this example. For instance (striking out the third row and the first column),

$$\Delta_{31} = \begin{vmatrix} -9 & 0 \\ 20 & -9 \end{vmatrix} = 81 - 0 = 81 \qquad (18\text{-}4)$$

Thus the nine cofactors are computed and listed in the inverse matrix:

$$[Z]^{-1} = \frac{1}{975} \begin{bmatrix} 219 & 135 & 81 \\ 135 & 150 & 90 \\ 81 & 90 & 119 \end{bmatrix} \qquad (18\text{-}5)$$

To find currents, the inverse impedance is postmultiplied by the electromotive force matrix, obtained from the loop equations:

$$[I] = [Z]^{-1}[E] = \frac{1}{975} \begin{bmatrix} 219 & 135 & 81 \\ 135 & 150 & 90 \\ 81 & 90 & 119 \end{bmatrix} \begin{bmatrix} 100 \\ 0 \\ 0 \end{bmatrix}$$

$$= \frac{100}{975} \begin{bmatrix} 219 \\ 135 \\ 81 \end{bmatrix} = \begin{bmatrix} 22.45 \\ 13.83 \\ 8.30 \end{bmatrix}$$

The resulting current matrix gives the three loop currents. These are indicated in Figure 18b, and other branch currents are found as differences. Node voltages are computed as ZI drops in the branches. These provide a check on the computed currents, and the check shows satisfactory slide-rule accuracy.

It will be seen that only three of the cofactors were actually needed in the inverse matrix, for there is electromotive force in only one loop. The other cofactors need not have been computed but were, in this example, for completeness. They would have been needed if there had been other electromotive forces.

Notice the symmetry about the principal diagonals of the matrices for both $[Z]$ and $[Z]^{-1}$. Any square matrix with this kind of symmetry (with $a_{pq} = a_{qp}$) is called *symmetric*.† $[Y]$ and $[Z]$ matrices of reciprocal networks are symmetric (if formed in the usual manner), and should symmetry not appear in an ordinary network problem, a mistake is indicated.

19. NETWORKS EQUIVALENT AT ALL FREQUENCIES

An example of matrix operations is seen in the fascinating study of networks that are equivalent to each other at all frequencies. It is easy to find different networks that have the same admittance at a particular frequency; series and parallel combinations, for instance, as in Chapter 4, do this. To find different networks that have the same admittance at *all* frequencies is quite another matter.

Equivalence at all frequencies implies that if two such equivalent networks were enclosed in boxes with only the two terminals of each exposed, it would be impossible to determine which was which by any electrical tests, either steady-state or transient. Such equivalence is more than theoretically interesting; it may be highly practical, as in band-pass filter design.

Let us first state the problem. It is required that y_{11}, the driving-point admittance of the given network, be equal at all frequencies to y_{11}', the driving-point admittance of the network that is to be found. These admittances, y_{11} and y_{11}', will both be functions of $j\omega$ (or, what is the same thing, of the variable s), and they must both be the *same* function of $j\omega$ (or of s).

The solution of this problem will now be stated. An example will then be given. Proof of the validity of the solution will be given for a network of two loops. A general proof is not given here; it can be found in Guillemin's works.‡

To transform from one two-terminal network to another with the same input impedance, the impedance matrix $[Z]$ (as in equation 7-2) is postmultiplied by a *transformation* matrix and premultiplied by the transpose of the transformation matrix. Let us call the transformation matrix $[T]$. Then the impedance matrix of the new, equivalent network $[Z']$ is

$$[Z'] = [T]_t[Z][T] \qquad (19\text{-}1)$$

---

† A matrix in which $a_{pq} = -a_{qp}$ is called skew symmetric. These matrices appear in problems concerning rotating machinery. (See Kron.)

‡ Guillemin (3), Vol. II, pages 229 *et seq.*

The transformation matrix is of the form

$$[T] = \begin{bmatrix} 1 & 0 & \cdots & 0 \\ a_{21} & a_{22} & \cdots & a_{2n} \\ \cdot & \cdot & \cdot & \cdot \\ a_{n1} & a_{n2} & \cdots & a_{nn} \end{bmatrix} \tag{19-2}$$

The elements of $[T]$, the a's, may be any real numbers. Mathematically, they can be chosen quite arbitrarily; we can set down any numbers we please. Physically, however, care must be used to avoid negative resistance, inductance, or capacitance in the new network corresponding to $[Z']$.

20. AN EXAMPLE OF EQUIVALENCE

A simple example will show how the transformation works. Figure 20a shows a network for which the impedance matrix is

$$[Z] = \begin{bmatrix} 3 + j\omega 1 & -2 \\ -2 & 4 + j\omega 1 \end{bmatrix} \tag{20-1}$$

Let us write a transformation matrix $[T]$. The first row is specified by equation 19-2, but the second can be anything; we choose to write

$$[T] = \begin{bmatrix} 1 & 0 \\ \frac{1}{2} & -\frac{1}{2} \end{bmatrix} \tag{20-2}$$

and, transposing rows and columns:

$$[T]_t = \begin{bmatrix} 1 & \frac{1}{2} \\ 0 & -\frac{1}{2} \end{bmatrix} \tag{20-3}$$

The matrix multiplication proceeds as follows:

$$\begin{aligned} [Z'] &= \begin{bmatrix} 1 & \frac{1}{2} \\ 0 & -\frac{1}{2} \end{bmatrix} \begin{bmatrix} 3 + j\omega 1 & -2 \\ -2 & 4 + j\omega 1 \end{bmatrix} \begin{bmatrix} 1 & 0 \\ \frac{1}{2} & -\frac{1}{2} \end{bmatrix} \\ &= \begin{bmatrix} 1 & \frac{1}{2} \\ 0 & -\frac{1}{2} \end{bmatrix} \begin{bmatrix} 2 + j\omega 1 & 1 \\ j\omega\frac{1}{2} & -2 - j\omega\frac{1}{2} \end{bmatrix} \\ &= \begin{bmatrix} 2 + j\omega\frac{5}{4} & -j\omega\frac{1}{4} \\ -j\omega\frac{1}{4} & 1 + j\omega\frac{1}{4} \end{bmatrix} \end{aligned} \tag{20-4}$$

All that remains is to see whether a network can be found that is the physical realization of this $[Z']$ matrix. The upper right and lower left terms of the matrix are the mutual impedance between loops 1 and 2, and clearly these require the mutual element to be a $\frac{1}{4}$-henry inductance. This is shown in Figure 20b. The self-impedance of loop 1 consists of 2 ohms of resistance and $\frac{5}{4}$-henry inductance, including the $\frac{1}{4}$-henry mutual element; loop 1 is therefore completed by putting in the diagram 2 ohms of resistance and 1 henry

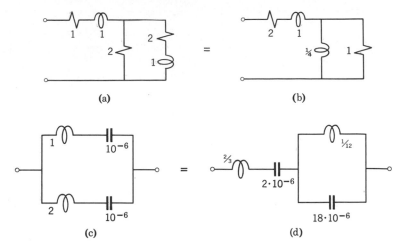

(a)   (b)

(c)   (d)

FIGURE 20 (a,b) Equivalent networks; resistances are shown in ohms, inductances in henrys. (c,d) Reactive networks equivalent at all frequencies. Inductances are given in henrys, and capacitances in farads.

of inductance. Similarly, loop 2 requires just 1 ohm of resistance in addition to the mutual element. Thus the equivalent network of Figure 20b is designed to fit the matrix $[Z']$.

We are fortunate in our choice of transformation matrix, for the resulting $[Z']$ is physically possible of realization. It is not easy to avoid getting $[Z']$ matrices that require negative R, L, or C in the equivalent network. This is particularly true if all three kinds of elements are present in the network.

Figures 20c and d show another pair of networks that are equivalent at all frequencies. These are composed of purely reactive elements, and equivalents of this kind are of practical importance in the design and construction of such networks as electric filters.

21. PROOF FOR TWO-LOOP NETWORK

The discussion will conclude with a proof that equation 19-1 is valid for all two-loop networks. The new network is equivalent to the original network if its input or driving-point admittance is the same. That is, it is equivalent if

$$y_{11} = y_{11}' \tag{21-1}$$

where (as in equations 2-1)

$$y_{11} = \frac{\Delta_{11}}{D} \quad \text{and} \quad y_{11}' = \frac{\Delta_{11}'}{D'} \tag{21-2}$$

Now, speaking quite generally,

$$[Z] = \begin{bmatrix} Z_{11} & Z_{12} \\ Z_{21} & Z_{22} \end{bmatrix} \quad \text{so} \quad D = \begin{bmatrix} Z_{11} & Z_{12} \\ Z_{21} & Z_{22} \end{bmatrix} = Z_{11}Z_{22} - Z_{12}Z_{21}$$

$$(21\text{-}3)$$

and Δ_{11}, the cofactor of Z_{11} in D, is Z_{22}. Therefore

$$y_{11} = \frac{Z_{22}}{Z_{11}Z_{22} - Z_{12}Z_{21}} \qquad (21\text{-}4)$$

We must now find, for comparison, y_{11}'. The first step is to find the completely general $[Z']$ for a two-loop network. The transformation matrix can be written (from equation 19-2):

$$[T] = \begin{bmatrix} 1 & 0 \\ a_{21} & a_{22} \end{bmatrix} \qquad (21\text{-}5)$$

Transforming $[Z]$ of equation 21-3,

$$[Z'] = [T]_t[Z][T] = \begin{bmatrix} 1 & a_{21} \\ 0 & a_{22} \end{bmatrix} \begin{bmatrix} Z_{11} & Z_{12} \\ Z_{21} & Z_{22} \end{bmatrix} \begin{bmatrix} 1 & 0 \\ a_{21} & a_{22} \end{bmatrix} \qquad (21\text{-}6)$$

and performance of the matrix multiplication gives

$$[Z'] = \begin{bmatrix} Z_{11} + a_{21}(Z_{12} + Z_{21}) + a_{21}{}^2 Z_{22} & a_{22}Z_{12} + a_{21}a_{22}Z_{22} \\ a_{22}Z_{21} + a_{21}a_{22}Z_{22} & a_{22}{}^2 Z_{22} \end{bmatrix}$$

$$(21\text{-}7)$$

As in equation 21-2, y_{11}' requires D' and Δ_{11}'. The determinant D' has the same form as $[Z']$ and need not be rewritten; Δ_{11}' is equal to $a_{22}{}^2 Z_{22}$. Therefore

$$y_{11}' = \frac{a_{22}{}^2 Z_{22}}{\left| \begin{array}{l} [Z_{11} + a_{21}(Z_{12} + Z_{21}) + a_{21}{}^2 Z_{22}]a_{22}{}^2 Z_{22} \\ \quad - (a_{22}Z_{12} + a_{21}a_{22}Z_{22})(a_{22}Z_{21} + a_{21}a_{22}Z_{22}) \end{array} \right|}$$

$$= \frac{Z_{22}}{Z_{11}Z_{22} - Z_{12}Z_{21}} = y_{11} \qquad \text{Q.E.D.} \qquad (21\text{-}8)$$

Thus a network for which the matrix is $[Z']$ is equivalent at all frequencies to a network for which the matrix is $[Z]$, and the elements of the transformation matrix a_{21} and a_{22} can be anything, for they cancel out of the final expression for y_{11}'.

22. SUMMARY

This chapter gives formal *solutions* for the *loop equations* of a network (equations 2-2) and for the *node equations* of a network (equations 4-2). It introduces the ideas of *driving-point* and *transfer admittances* in the solution of the loop equations, and the ideas of *driving-point* and *transfer impedances* in the solution of the node equations. Examples are given.

Equality between *transfer admittances*, equation 6-4, and between *transfer impedances*, equation 6-5, is shown.

Matrix notation is introduced. Matrix *equality*, "*multiplication*," *addition*, *transposition*, and *inversion* are defined.

The *loop* equations and the *node* equations are solved in *matrix* notation. Examples are given.

A condition for networks to be *equivalent at all frequencies* is expressed in matrix terms. An example is given, and a proof for networks of two loops.

PROBLEMS

9-1. In the network of Figure 18a, (*a*) what is the self-impedance of loop 1? By means of the loop equations, compute (*b*) the driving-point admittance of loop 1, and (*c*) the transfer admittance between loops 1 and 3.   §3*

9-2. Write equations 6-2 of Chapter 8 in matrix form. Knowing that $E = 10$ volts, write the solution $[I]$ of this matrix equation as a column matrix of three currents (numerical values).   §8*

9-3. In the network of Figure 18a, compute y_{11}, y_{13}, and y_{31} by means of the "ladder" method of solution.   §2

9-4. For the *RC* coupling network of Problem 4-15, page 98, (*a*) write node equations, writing sC for capacitor admittance, (*b*) find the transfer impedance $z_{BA} = V_B/I_A$ where I_A is the transform of input current and V_B is the transform of voltage across the 200,000-ohm output resistor, expressed as a function of the complex frequency variable s, and (*c*), find the numerical value (magnitude and angle) of V_B, the output voltage, when the frequency of the 20-milliampere input current is 1000 Hz.   §5*

9-5. Look into the network of Figure 5a from the terminals of the source. At these terminals, find (*a*) the driving-point impedance, using the node method, and (*b*) the driving-point admittance, using the loop method. (*c*) Are these reciprocal quantities?   §5

9-6. Find the matrix product: §12*

$$\begin{bmatrix} 3 & 1 & 2 \\ 2 & 3 & 1 \\ 4 & 1 & 1 \end{bmatrix} \begin{bmatrix} 1 & 0 & 1 \\ 2 & 1 & -1 \\ 1 & 3 & 2 \end{bmatrix} .$$

9-7. Interchange the two matrices in Problem 9-6 and find the product.   §12*

9-8. Find the matrix product: §13*

$$\begin{bmatrix} 9 \\ 5 \\ 1 \end{bmatrix}_t \begin{bmatrix} 2 \\ 6 \\ 4 \end{bmatrix}$$

9-9. By what matrix must $\begin{bmatrix} 10 & 5 \\ 3 & 8 \end{bmatrix}$ be premultiplied to give, as the matrix product, $\begin{bmatrix} 2 & 1 \\ 0 & 6 \end{bmatrix}$?

9-10. Let the following matrix be called $[F]$:

$$[F] = \begin{bmatrix} 1 & 0 & 1 \\ 2 & 1 & -1 \\ 1 & 3 & 2 \end{bmatrix}$$

(a) Find $[F]^{-1}$. (b) Find $[F][F]^{-1}$; work this out in detail to show that it equals $[U]$. §14*

9-11. Repeat Problem 9-10 with any other matrix, devised at random. §14*

9-12. The Z matrix of a network is given below. Draw the network, labeling the element values. §18

$$[Z] = \begin{bmatrix} 1 & j1 & -1 \\ j1 & 1+j1 & -1 \\ -1 & -1 & 2-j1 \end{bmatrix}$$

9-13. From the loop equations of a two-loop network, the matrix formula $[Z] \times [I] = [E]$ has been written below. (a) Draw the circuit, identifying the circuit elements and currents. (b) Find $[y]$, the matrix of driving-point and transfer admittances. (c) Find the value of I_2 from $[I] = [y][E]$. §18*

$$\begin{bmatrix} 3-j4 & j4 \\ j4 & 3+j4 \end{bmatrix} \begin{bmatrix} I_1 \\ I_2 \end{bmatrix} = \begin{bmatrix} 100 \\ 0 \end{bmatrix}$$

9-14. In Figure 18a, add to the network, in series with the 9-ohm resistor that is common to loops 2 and 3, a source of 72 volts of electromotive force, directed upwards. Using where applicable the matrices of Section 18, solve for all currents. §18*

9-15. Find the matrix $[y] = [Z]^{-1}$ of driving point and transfer admittances for the network shown, in which impedances are given in ohms. §18

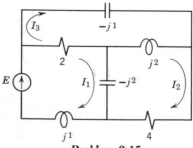

Problem 9-15.

9-16. Find the passive network whose matrix of driving point and transfer admittances is: §18

$$[y] = \begin{bmatrix} 1 - j1 & -j1 \\ 2 - j3 & 2 - j3 \\ -j1 & 2 - j1 \\ 2 - j3 & 2 - j3 \end{bmatrix}$$

9-17. Find a transformation matrix to change the network shown as (*a*) in the diagram for this problem to the network marked (*b*). Also, find a transformation matrix to change the network marked (*a*) to the network marked (*c*). Find the three driving-point admittances (using any method) to show that they are equal.   §20*

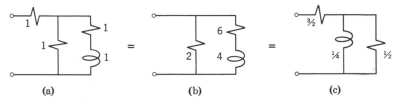

(a)   (b)   (c)

Problem 9-17   Resistances are in ohms and inductances in henrys.

9-18. Find a transformation matrix to change the network marked (a) in the diagram for Problem 9-17 to the network shown in the diagram for this Problem 9-18; n may have any value. What values of n are physically possible?   §20*

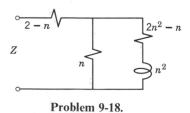

Problem 9-18.

9-19. Find the transformation matrix used to change the network of Figure 20c to that of Figure 20d. Show that the two networks are equivalent for all s.   §20

9-20. Draw any two-loop network that contains only elements of R and L. (*a*) Use any permissible transformation matrix $[T]$ to find another network that has the same input impedance at all frequencies. (*b*) Draw a network diagram for this new network; is it physically realizable, and if so over what range? (*c*) Compute the input impedance functions for both the original and the derived networks, and show that they are equal at all frequencies.   §20

9-21. Draw any two-loop network that contains only elements of R and C. Transform, repeating Problem 9-20.   §20

9-22. Draw any two-loop network that contains only elements of L and C. Transform, repeating Problem 9-20.   §20

CHAPTER
10

NETWORK THEOREMS

1. SUPERPOSITION

If two voltages are applied at the same time to the same network, it may be possible to use superposition. We first find the currents that would be produced in each element of the network by one of the voltages as if it were acting alone. Then we find the currents produced by the other voltage alone, and then finally add the two currents in each element. This superposition is permissible if the relations of the network are linear. If a network is composed of elements with *constant* values of R, L, and C the relations between currents and voltages are linear, but if the network contains elements with values of R, L, or C that are functions of the current or the voltage, the relations are not linear. Nonlinearity between current and voltage is likely to appear in networks containing transistors or other semiconductors or crystalline materials, vacuum tubes, coils with iron cores, or certain types of capacitors.

More generally, if a system is disturbed by two or more causes acting at the same time, we can find the disturbance produced by each of the causes acting alone and then, if the relation is linear, add the several individual disturbances to obtain the total disturbance.† The method of *superposition* can be used if the relations in the system are *linear*.

---

† *Linear* is defined in Section 8 of Chapter 1 (page 11) by this same statement put in mathematical terms; that is, a linear system is one in which superposition is correct.

234

The linearity of a coil with constant resistance R and constant inductance L is shown as follows. Voltage and current are related by:

$$v = Ri + L\frac{di}{dt} \tag{1-1}$$

The voltage produced by current i_1 is therefore

$$v_1 = Ri_1 + L\frac{di_1}{dt} \tag{1-2}$$

The voltage produced by i_2 is

$$v_2 = Ri_2 + L\frac{di_2}{dt} \tag{1-3}$$

The voltage produced by both together, by $i_1 + i_2$, is

$$v_3 = R(i_1 + i_2) + L\frac{d}{dt}(i_1 + i_2) \tag{1-4}$$

Since algebraic manipulation shows that $v_3 = v_1 + v_2$, superposition is justified and the relations are linear. But it is clear that if R and L were not constants (independent of i), the relations would no longer be linear; each voltage would be a nonlinear function of i, and since it would no longer be true that $v_3 = v_1 + v_2$, superposition would not be correct.

As an example of a relation that is *not* linear, the current through a crystalline material (such as Thyrite) can be approximated by $i = Kv^3$. This is a cubic, not a linear, relation and it does not meet the foregoing test for linearity. (See Problem 16 of Chapter 1.) Hence superposition cannot be used with a circuit containing such an element. Transistors, diodes, vacuum tubes, and many other devices are nonlinear through at least part of their ranges, and great care must be used when applying to such elements either superposition or any of the theorems or methods that require superposition or that assume linearity.

A relation that is *never* linear is that between current and power. Let us try the test for linearity on the equation:

$$p = Ri^2 \tag{1-5}$$

Then power p_1 due to current i_1 is $Ri_1{}^2$, and power p_2 due to current i_2 is $Ri_2{}^2$; the power due to both currents acting at the same time is $R(i_1 + i_2)^2$. This is *not* the sum of p_1 and p_2, so the relation between current and power is not linear (indeed it is obviously quadratic) and superposition is not permissible. You cannot add the power produced by each of two components of current in a resistor to find the total power.

In many of the examples and problems of the previous chapters there are linear networks with two or more sources, and any of these can be handled by superposition. The method is to compute the effects of one source at a time, and then to add the effects of the individual sources. While each source is considered, all other sources are required to be zero (independent voltage sources are required to have zero voltage; independent current sources to have zero current). An example is worked out in Section 16 of Chapter 9, where superposition is given as an alternative to solution by node equations.

The very concept of loop and node equations depends on linearity, for currents are added (or voltages are added) in mutual branches, and it is taken for granted that currents and voltages are linearly related. The concept of state-variable equations does not demand linearity, however, and this may prove to be important and valuable in the future, but state-variable analysis of nonlinear systems stumbles badly over inadequate techniques, and discussion in this book relates to linear systems only.

Some devices can be modeled as linear over a restricted range. Iron-cored inductance, for instance, may be treated as linear if the current is small enough to avoid magnetic saturation, particularly if there is an air gap in the magnetic core. Other devices, called piecewise-linear, can be treated in a number of ranges; a semiconductor diode, for instance, may be modeled as having one value of R for one range of current and a different value of R for another range.

Indeed, linearity is so valuable in analysis that systems are sometimes *linearized*, which means that a linear model is used when it is known to be inexact. This is convenient but it is likely to be dangerous. The principle of superposition seems natural, and probably more mistakes are made by using it where it does not apply than by failing to use it when correct.

2. THE SUBSTITUTION THEOREM

A number of useful theorems concern substitution. Part of a network can be removed, and something simpler can take its place.

If part of a network is connected to the rest of the network at one pair of terminals only, it can affect the rest of the network only by voltage and current at that pair of terminals. One such subnetwork can be replaced by another subnetwork if they have the same relation between voltage and current at their pair of terminals, and the rest of the network will not know the difference.

That is, if two subnetworks have the same voltages between terminals for all terminal currents, then one subnetwork may be substituted for the other and they are said to be *equivalent*.[†]

[†] A less restrictive requirement is called the *substitution theorem* in Skilling (7).

If there are more than two terminals, the condition for substitution is essentially the same: *two subnetworks are equivalent if the relations between terminal voltages and terminal currents are the same for both, and such equivalent subnetworks can be exchanged without affecting conditions in any part of the network external to them.*

This theorem may seem so obvious as to be almost simple minded, but it will be quite useful in the paragraphs to come.

3. EQUIVALENT SOURCES

An example of substitution is seen in the possible exchange of an independent current source and an independent voltage source. It makes no difference in the external load impedance Z_L whether it is connected to the independent voltage source V of Figure 3a with series impedance Z_V or to the independent current source I of Figure 3b with shunt impedance Z_I provided

$$I = \frac{V}{Z_V} \quad \text{and} \quad Z_I = Z_V \tag{3-1}$$

In the first case the current is

$$I_L = \frac{V}{Z_V + Z_L} \tag{3-2}$$

and in the second case it is

$$I_L = I \frac{Z_I}{Z_I + Z_L} = \frac{V}{Z_V} \frac{Z_V}{(Z_V + Z_L)} = \frac{V}{Z_V + Z_L} \tag{3-3}$$

which is the same. Terminal voltage in both cases is $V_L = Z_L I_L$.

This example can be made more general. The external network need not be either linear or passive; it need only be known that there is a load voltage V_L corresponding to every load current I_L. According to the substitution theorem, the source subnetworks of Figures 3c and 3d are equivalent if the relations between terminal voltage V_L and terminal current I_L are the same for both. For the independent voltage source of Figure 3c,

$$V_L = V - ZI_L \tag{3-4}$$

For the independent current source of Figure 3d,

$$V_L = Z(I - I_L) \tag{3-5}$$

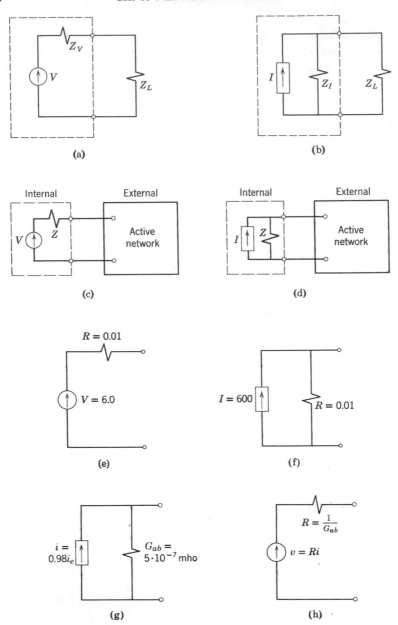

FIGURE 3 (a,b) Equivalent sources. (c,d) Equivalent sources connected to an active network. (e,f) Equivalent representations of a battery. (g,h) A transistor model and equivalent.

and when V is substituted for ZI as the necessary condition from equation 3-1 (the subscript of the Z has been dropped), equation 3-5 becomes identical with equation 3-4, thus demonstrating equivalence. This demonstration places no kind of limitation on the external network.

Example 1. A storage battery has a no-load terminal voltage of 6.0 volts. When current through the battery is 100 amperes, the terminal voltage drops to 5.0 volts.

Two models of the battery are shown in Figures 3e and 3f. These models are entirely equivalent in any external circuit to which the battery might be connected. However, they are by no means equivalent as representations of the actual battery. While it is probable that neither is an exact model, surely the former is much closer to reality than the latter for several apparent reasons.

Example 2. The output circuit of a common-base transistor with input i_e can be modeled as the controlled current source of Figure 3g; current is controlled by input to the transistor emitter which is not part of the subnetwork shown. The controlled voltage source of Figure 3h is equivalent provided $R = 2 \times 10^8$ ohms and

$$v = Ri = 1.96i_e \times 10^8 \text{ volts} \tag{3-6}$$

4. THÉVENIN'S THEOREM

The box in Figure 4a contains a linear network from which two terminals emerge. The linear network may include any configuration of impedances with constant values of R, L, or C, together with independent voltage sources or independent current sources or both. (Controlled voltage or controlled current sources are also permitted if their control is independent of or linear with respect to any current or voltage of the network in the box or its output.)

When all the sources within the box are operating normally, a voltage that we shall call V_θ appears between the terminals. This voltage can be computed or measured.

When all the sources within the box are reduced to zero,† no voltage appears between the terminals of the box. Impedance between these terminals, impedance looking into the box, can then be computed or measured, and this impedance is called Z_θ, the Thévenin impedance.

---

† *Sources reduced to zero* means that the *voltage* of each independent voltage source is reduced to zero, which can be conceived as a short circuit in place of the source, and that the *current* of each independent current source is reduced to zero, which can be conceived as an open circuit at the place of the source.

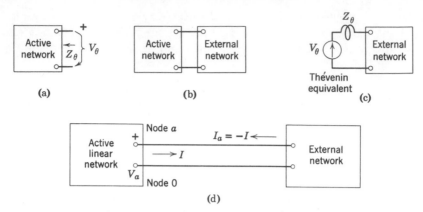

FIGURE 4 (a) An active network, (b) connected to an external network, and (c) its Thévenin equivalent. (d) With external network.

Thévenin's theorem, which is one of the most useful theorems of network analysis, says that the following substitution can be made. If the network within the box of Figure 4a is connected to some external network as in Figure 4b, a single source of voltage V_θ in series with impedance Z_θ can be substituted for the box, as in Figure 4c, and all results in the external network will remain unchanged.

The network for which substitution is made must be linear. It may contain sources that are time-varying, so it can include dc, ac, or transient sources. Impedance is the appropriate value or function of $j\omega$ or s. The external network can be active or passive. If it is linear, solution with Thévenin's theorem is straightforward. If it is nonlinear, so that superposition is not permissible, the usual difficulties are encountered.

Proof of Thévenin's theorem is based on the node equations of Chapter 8 and their solution in Chapter 9. Let us assume, as in Figure 4d, that the two terminals coming from the box are arbitrarily called the node pair a and o. The node equations for the linear network can be solved for V_a as in equations 4-2 of Chapter 9. The solution is

$$V_a = z_{aa} I_a + z_{ab} I_b + z_{ac} I_c + \cdots \qquad (4\text{-}1)$$

where I_a is the current injected into node a from the external network as shown in Figure 4d. I_b, I_c, and so on, are independent source currents within the active linear network. If the active linear network contains independent voltage sources, as it probably does, their values may be written into the node equations or, what is surely simpler in concept, they may all be changed to independent current sources by the theorem of Section 3. In any case, I_b, I_c, and so on, represent sources within the linear network.

Equation 4-1 contains one driving-point impedance and a number of transfer impedances. The driving-point impedance z_{aa} is (from Chapter 9) that seen looking in at the terminals ao, and is therefore the same as the Thévenin impedance Z_θ.

It is convenient to have a symbol for the current *from* node a, the current from the linear network to the external network; this is simply the negative of I_a. Let us call it I, and write $-I = I_a$. With this change, and the change of symbol for impedance, equation 4-1 becomes:

$$V_a = -Z_\theta I + z_{ab}I_b + z_{ac}I_c + \cdots \tag{4-2}$$

If we now disconnect the external network (as in Figure 4a) so that I (and I_a) are forced to be zero, we can see from equation 4-2 (or 4-1) that the terminal voltage under these conditions, which we have already called V_θ, must be

$$V_\theta = V_{a(\text{open})} = z_{ab}I_b + z_{ac}I_c + \cdots \tag{4-3}$$

Now this open-circuit voltage can be substituted for the several latter terms of equation 4-2, giving:

$$V_a - V_\theta - Z_\theta I \tag{4-4}$$

But this, the terminal voltage for our network as in Figure 4b, is also the terminal voltage for the Thévenin equivalent of Figure 4c, and so the foregoing statement of the theorem is proved.

5. THÉVENIN–NORTON EQUIVALENCE

It is sometimes convenient to express Thévenin's theorem in terms of open-circuit voltage and short-circuit current, with no explicit mention of impedance. To do so, merely short circuit the terminal a to o, as in Figure 5a, thereby making $V_a = 0$, and write from equation 4-4 that

$$0 = V_\theta - Z_\theta I_\theta \tag{4-5}$$

where I_θ is the current in the short circuit, the value of I that applies when $V_a = 0$. From this, the passive Thévenin impedance of the network is

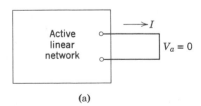

(a)

FIGURE 5a   With short circuit.

$Z_\theta = V_\theta/I_\theta$, the ratio of open-circuit voltage to short-circuit current. This permits a restatement of Thévenin's theorem in a form that will be given at the end of this section.

But an alternative statement may also be included. Norton's theorem is derived from Thévenin's theorem by substituting an independent current source for the independent voltage source of the Thévenin equivalent, using Section 3. The result is shown in Figure 5d, and both Norton's and Thévenin's theorems can be stated as follows.

Open-circuit voltage V_θ and short-circuit current I_θ are measured or computed at a pair of terminals of an active linear network. The active linear network is equivalent at these terminals to either an independent voltage source V_θ in series with an impedance $Z_\theta = V_\theta/I_\theta$, or alternatively to an independent current source I_θ in parallel with the same Z_θ. See Figures 5b,c, and d.

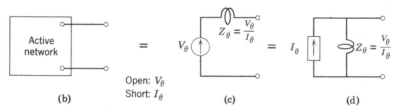

(b)   Open: V_θ
             Short: I_θ (c)   (d)

FIGURES 5b,c,d   Thévenin's and Norton's theorems.

6. A TRANSISTOR EXAMPLE

Figure 6a shows a transistor with the collector connected to two sources. Operation of the transistor, a nonlinear device, can best be determined graphically from collector curves and load line, and the solution is greatly aided by reducing all of the network except the transistor, everything to the right of terminals a and o, to the Thévenin equivalent.†

Figure 6b shows the equivalent. With the transistor removed, the open-circuit voltage at ao is V_θ:

$$V_\theta = 10 + \tfrac{1}{6}60 = 20 \text{ volts} \tag{6-1}$$

Reducing both sources to zero voltage, the impedance looking into the linear network is Z_θ:

$$Z_\theta = \frac{(1000)(5000)}{(1000) + (5000)} = 833 \text{ ohms} \tag{6-2}$$

---

† Adapted from Nelson Hibbs, "Understanding and Using Thévenin's Theorem," *Service Scope*, No. 40, October 1966, Tektronix, Inc.

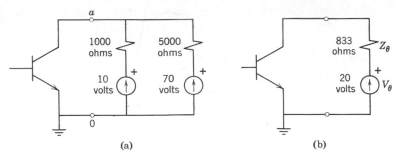

FIGURES 6a,b   Transistor with a split collector load.

From the Thévenin equivalent of Figure 6b and the necessary transistor curves, operating conditions are easily found. It is evident that this same kind of reduction can be useful with any nonlinear device connected to a linear network.

7. A BRIDGE EXAMPLE

Thévenin's and Norton's theorems are often useful when several values are to be given to one part of a network while keeping the rest of the network unchanged. For instance, which of a number of galvanometers gives the greatest sensitivity in the Wheatstone bridge of Figure 7a? Sensitivity

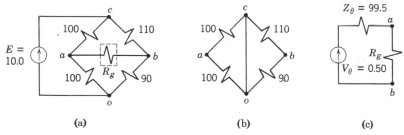

FIGURE 7 (a) A bridge circuit, (b) its passive impedance between terminals a and b, and   (c) its Thévenin equivalent.

depends on the deflection of the galvanometer, and hence on both its resistance and the delicacy of its movement. We need to find the current through the galvanometer when R_g has values of 0, 1, 10, and 100 ohms.

First, find the Thévenin equivalent of the network, omitting R_g. To find Z_θ, reduce E to zero, leaving a short circuit from c to o, and combine the remaining four resistances in parallel and series as indicated in Figure 7b:

$$Z_\theta = \frac{100}{2} + \frac{90 \cdot 110}{90 + 110} = 50 + 49.5 = 99.5$$

V_θ is found by computing the voltage between a and b with R_g removed:

$$V_\theta = V_{ao} - V_{bo} = \tfrac{100}{200}10.0 - \tfrac{90}{200}10.0 = 0.50$$

Hence the equivalent circuit is that of Figure 7c.

Current through the galvanometer is now found by substituting various values of R_g in

$$I_g = \frac{V_\theta}{R_g + Z_\theta} = \frac{0.50}{R_g + 99.5}$$

$R_g =$	0	1	10	100
$R_g + 99.5 =$	99.5	100.5	109.5	199.5
$I_g \cdot 10^3 =$	5.02	4.97	4.56	2.50

The third row gives galvanometer current in milliamperes.

Since galvanometer sensitivity in scale divisions per milliampere is related to the number of turns in the moving coil, it is also related to galvanometer resistance. It appears that there will be some best value of R_g. Clearly, a galvanometer of infinite resistance would be useless as it would carry no current; a galvanometer of zero resistance would be equally useless as it would have no sensitivity. A choice of instrument can be made when the resistance, sensitivity, and price of available galvanometers are known.

It is worth noticing that solution of this problem without Thévenin's theorem would be very much less easy.

8. A TRANSIENT EXAMPLE

The theorems of Thévenin and Norton can be applied in transients problems with sources that are discontinuous. Superposition is required, so all elements should be linear.

Proof of the validity of the theorems comes most readily by superposition of Laplace components as considered in Chapter 14, but a simple example shows the application of Thévenin's theorem to a transients problem in an exceptionally simple case. (For a more typical example, see Problem 15-6.)

In Figure 8a, a switch is closed at $t = 0$, when there is no current in either inductance. Current i_2 in the right-hand branch is to be found as a function of time after the switch is closed.

First, break the network at nodes a and b and apply Thévenin's theorem to the network on the left. V_θ is the voltage between these nodes with the network broken; because the two impedances shown in the diagram are equal, $V_\theta = 12$ volts at all times after closing the switch. This voltage is entered in Figure 8b.

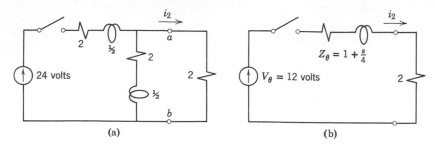

FIGURES 8a,b   A network and its Thévenin equivalent; resistances and inductances are in ohms and henrys.

Z_θ is the passive impedance looking in at terminals a and b. Two equal impedances in parallel are each $Z(s) = 2 + s/2$, so $Z_\theta = 1 + s/4$. This also is entered in Figure 8b.

Now there is but one loop for current, and simply from our knowledge of current in an inductive loop (or, if necessary, using the four rules of Chapter 6) we can write the current:

$$i_2 = \frac{V_\theta}{R}(1 - e^{-(R/L)t}) = 4(1 - e^{-12t}) \text{ amperes} \qquad (8\text{-}1)$$

This same current was found for the same network without Thévenin's theorem as equation 10-11 in Chapter 6. Admittedly, the equality of impedances makes this solution especially easy.

9. THREE-TERMINAL NETWORKS

A passive linear network in the form of a Y, Figure 9a, can be interchangeable with a passive linear network in the form of a Δ, Figure 8b, and either one can be substituted for the other without affecting the rest of a connected network. The necessary condition is that the three terminal currents for the Δ must be the same as the three terminal currents for the Y, given the same three terminal voltages.

Current entering terminal a of the Δ is

$$I_a = I_{ab} - I_{ca} = \frac{V_{ab}}{Z_{ab}} - \frac{V_{ca}}{Z_{ca}} \qquad (9\text{-}1)$$

The other two terminal currents, being exactly symmetrical, do not have to be written; note the symmetry of symbols in both the Y and Δ of Figures 9a and b.

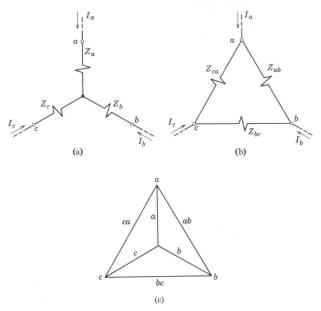

FIGURE 9 **(a,b)** Y and Δ networks that can be equivalent. **(c)** Diagram to show the pattern of subscripts used in equations.

Terminal voltages of the Y of Figure 9b are:

$$V_{ab} = I_a Z_a - I_b Z_b \tag{9-2}$$

$$V_{ca} = I_c Z_c - I_a Z_a \tag{9-3}$$

Since $I_a + I_b + I_c = 0$, I_c can be removed from equation 9-3:

$$V_{ca} = -I_a(Z_c + Z_a) - I_b Z_c \tag{9-4}$$

Equations 9-2 and 9-4 are now solved simultaneously (quite easily by determinants) to give the entering terminal current to the Y:

$$I_a = \frac{V_{ab} Z_c - V_{ca} Z_b}{Z_a Z_b + Z_b Z_c + Z_c Z_a} \tag{9-5}$$

Finally, the necessary conditions for current entering the Δ to equal current entering the Y are seen by comparing equations 9-1 and 9-5. It is necessary that

$$Z_{ab} = \frac{Z_a Z_b + Z_b Z_c + Z_c Z_a}{Z_c} \tag{9-6}$$

and that the similar expressions for Z_{bc} and Z_{ca} be exactly symmetrical.

This statement is perhaps more easily appreciated in words: *The impedance of any side of the equivalent Δ is equal to the sum of the three products* of Y impedances multiplied together in pairs, divided by the impedance of the *opposite* branch of the Y.

Figure 9c superimposes a Δ and a Y to show the pattern of subscripts used in the equations.

Equation 9-6 is used to change a Y to a Δ. Its inverse for changing a Δ to a Y is equally useful, and possibly even more so. The three symmetrical equations of which equation 9-6 is one are solved simultaneously to obtain the Y impedances in terms of the Δ impedances, and the results are

$$Z_a = \frac{Z_{ab} Z_{ca}}{Z_{ab} + Z_{bc} + Z_{ca}} \tag{9-7}$$

and two other equations for Z_b and Z_c.

This relation is conveniently expressed in words, also. *The impedance of any branch of the Y is equal to the product of the two adjacent sides of the Δ divided by the sum of the three Δ impedances.*

10. Y-Δ EXAMPLES

The substitution of Y for Δ or Δ for Y can be made with three of the resistances in a resistive network, or with three impedances if the network is reactive. With a network of impedances and with alternating voltage and current, the substitute Y or Δ can be computed to have a numerical value at a given frequency or, more generally, the impedances of the three substitute branches can be determined as functions of frequency. As functions of frequency, either $Z(j\omega)$ or $Z(s)$ can be used and they are, of course, interchangeable.

If the network is purely resistive, an actual physical substitution of a Y for a Δ or a Δ for a Y can be made, with wire-cutters and a soldering iron. If an impedance network is for use at a *single* sinusoidal frequency it is usually possible to make such a physical substitution for use at that frequency alone, although some rather extreme networks lead to substitutes with impractical negative values of resistance. But the substitution found as a *function* of frequency is not usually realizable over a *range* of frequencies, though it is perfectly permissible to use the function for computation of either steady or transient values.

Two examples will be shown, one of an easy substitution of resistances and one of a transient-current computation. Other possibilities have obvious similarity to these.

Example 1. What impedance is seen by the source in the network of Figure 10a? Resistances are given in ohms.

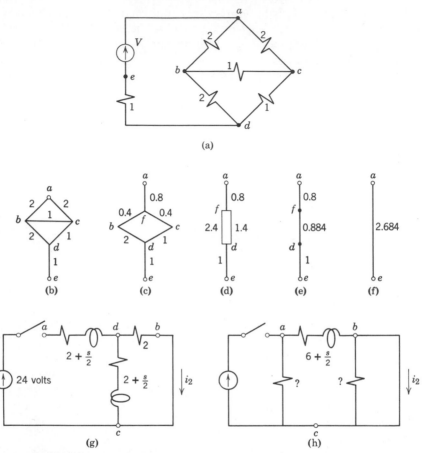

FIGURE 10   (a) A bridge circuit.   (b,c,d,e,f) Reduction involving a Δ-Y substitution. Numbers are resistances in ohms.   (g,h) Transient current found by substitution.

Solution: In this bridge network there are no branches to combine in series or in parallel. A solution of loop equations could be used, but a substitution is easier.

An even more diagrammatic sketch is given in Figure 10b, and it is evident that by substituting a Y for a Δ we can obtain a network that then reduces easily to a single resistance by series and parallel combinations. Substitution could be made for the Δ*abc*, or for *bcd*, or for the Y centered at *b*, or the Y centered at *c*. Let us do the first of these; the result is Figure 10a. The new branch *af* is the product of adjacent sides over the sum, which is $(2 \cdot 2)/5 = 4/5 = 0.8$ ohm as shown in the diagram. The other branches, *bf* and *cf*, are both $(2 \cdot 1)/5 = 0.4$ ohm. This completes the Δ-Y substitution, and the series and parallel combinations shown in Figures 10d, 10e, and 10f lead to the answer, 2.684 ohms.

Example 2. Figure 10g shows a network of 2-ohm resistors and $\frac{1}{2}$-henry inductors. The impedance of each branch (as a function of s) is given on the diagram. Find the current i_2 after the switch is closed, there being no initial currents in the network.

Solution: Substitute a Δ for the Y in Figure 10g, giving Figure 10h. Impedance of the new branch ab is the sum of the products by pairs divided by the opposite side:

$$Z_{ab} = \frac{2(2 + s/2) + 2(2 + s/2) + (2 + s/2)^2}{2 + s/2} = 6 + \frac{s}{2} \qquad (10\text{-}1)$$

The impedances of the other two sides, Z_{ac} and Z_{bc}, do not matter and are not computed, except to note that they are not zero. The value of Z_{ac} does not matter because the voltage at its terminals is forced by the source to be 24 volts. Z_{bc} does not matter because it is short circuited and carries no current.

Hence the current in which we are interested is that produced by the sudden application of 24 volts to an inductive circuit of 6 ohms and $\frac{1}{2}$ henry (there being no initial current), and we know that the answer to this simple problem is

$$i_2 = \frac{24}{6}(1 - e^{-12t}) \qquad (10\text{-}2)$$

This same answer was obtained for this same problem by Thévenin's theorem in equation 8-1 of this chapter, and in equation 10-11 of Chapter 6 by the four rules of that chapter.

As an example of solution by Y-Δ substitution, this problem is highly non-random. Equality of impedances of two branches of the Y results in the great simplification seen in equation 10-1. More typically, Z_{ab} might be obtained as a rational function. Such an example will be seen in Problem 15-7. The present example illustrates the substitution while avoiding complications.

11. THE RECIPROCITY THEOREM

Unlike the other theorems of this chapter, the reciprocity theorem is not one with many obvious uses. Nevertheless, it is an elegant theorem and seems to be one that every educated man is expected to know.

It is not a substitution theorem, at least not in the sense of those that have been discussed, but it serves to turn a network inside out. Stated very crudely, the theorem says that if there are a battery and an ammeter in different parts of a network, the battery and ammeter can be interchanged and the meter will read the same.

Expressed with more care, it can be put as follows: if an independent voltage source in one branch of a linear network that is otherwise passive produces a certain current in any second branch, *the same source if placed in the second branch will produce that same current in the first branch.*

Proof is easy. Equations 2-2 of Chapter 9 give the current in loop 2 of a network when the only applied voltage is in loop 1:

$$I_2 = y_{21} V_1 \qquad\qquad (11\text{-}1)$$

Since the numbers of loops are quite arbitrary, this equation can describe our first connection. Now, for our second connection, let there be a source in loop 2 but none in loop 1; by the same equations from Chapter 9:

$$I_1 = y_{12} V_2 \qquad\qquad (11\text{-}2)$$

It is specified that V_2 is now equal to the previous V_1, so I_1 will now equal the previous I_2 provided $y_{12} = y_{21}$. But Section 6 of Chapter 9 was devoted to proving that $y_{12} = y_{21}$ in all linear networks; hence the reciprocity theorem is proved.

The dual of this theorem states that: if an independent current source connected to one node of a linear network that is otherwise passive produces a certain voltage at some second node, *the same source if connected to the second node will produce that same voltage at the first node.* This dual theorem can be put crudely to say that an independent current source and a voltmeter with infinite resistance can be interchanged in a network. The dual is proved by reference to the node equations and the transfer impedances of Chapter 9.

Example. Figure 11a shows a network with E_1 applied at the left and I_2 read at the right. Using the numerical values of resistances shown, I_2 can be found by computation to be $E_1/244$.

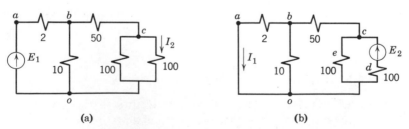

(a)   (b)

FIGURES 11a,b   Illustration of the reciprocity theorem.

To demonstrate the reciprocity theorem, interchange the source and the ammeter, as in Figure 11b, and again compute the current. This time it is found that $I_1 = E_2/244$, as one would expect from the theorem.

It does not follow that any other voltages or currents in the network are the same after the interchange is made. In the first case, for example, the generator current is 22 times the measured current, whereas in the second the generator current is only 1.82 times the measured current. In one case V_{bo} is 200 times the current; in the other it is 2 times the current.

Reciprocity applies, as seen, between a voltage and a current, either being the excitation and the other the response. It does not apply, however, between two voltages in different parts of a network or between two currents.

12. MAXIMUM POWER TRANSFER THEOREM

One of the most important requirements in transmitting information is to deliver as much power as possible. When an independent voltage source, modeled as the electromotive force E in Figure 12a, supplies power to a load of impedance Z_l through a fixed value of series impedance Z_s (which may include internal impedance of the source), maximum power is received by the load if its impedance is the conjugate of Z_s.

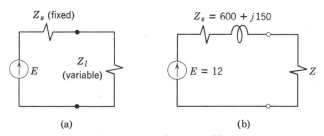

(a)   (b)

FIGURE 12 (a) Maximum power transfer. (b) Maximum power to Z is required.

If Z_s were a pure resistance, Z_l would be an equal resistance in order to receive maximum power. If Z_s were somewhat inductive, Z_l would need to be somewhat capacitive, their respective reactances being equal and opposite. That is, for maximum power transfer,

$$Z_l = Z_s^* \qquad R_l = R_s \qquad X_l = -X_s \qquad (12\text{-}1)$$

Proof: Power to the load, to Z_l, is

$$P = |I|^2 R_l = \left(\frac{E}{|Z_s + Z_l|}\right)^2 R_l = \frac{E^2 R_l}{(R_s + R_l)^2 + (X_s + X_l)^2} \qquad (12\text{-}2)$$

We want the maximum value of this expression. R_l and X_l can be varied independently and can have any desired values. E, R_s, and X_s cannot be varied. Let us first consider the result of varying X_l while holding R_l constant. Mere inspection of the formula shows that P will be greatest when $X_l = -X_s$, for this makes the denominator minimum, whatever the value of R_l may be.

Let us therefore make $X_l = -X_s$; then

$$P = \frac{E^2 R_l}{(R_s + R_l)^2} \tag{12-3}$$

Now we study the effect of varying R_l. To find the value of R_l that gives maximum P, differentiate and set the derivative equal to zero:

$$\frac{dP}{dR_l} = E^2[(R_s + R_l)^{-2} - 2R_l(R_s + R_l)^{-3}] = 0 \tag{12-4}$$

$$2R_l = R_s + R_l \tag{12-5}$$

$$R_l = R_s \tag{12-6}$$

Maximum power to the load is therefore obtained by the conditions of equation 12-1. Such a load receives half the power output of the source, the other half going into the resistance R_s.

By Thévenin's theorem, the source E and the source impedance Z_s may be the equivalent of a more complicated active network. Thus Thévenin's theorem gives the maximum power transfer theorem extremely general application. Note, however, that a complicated network may consume more power than does its Thévenin equivalent; hence the load that is found by applying the maximum power transfer theorem to the Thévenin equivalent will receive the maximum possible power, but this may be less than half of the power actually generated.

It is clear that a load can be adjusted to receive maximum power only if its resistance and reactance can both be varied independently. If the *magnitude* of Z_l could be changed, but the ratio of X_l to R_l were necessarily constant (constant phase angle, constant power factor, in the load), it would in general be impossible to attain as great a transfer of power as if each were adjustable separately. Maximum power transfer, under the limiting condition that phase angle is constant, is obtained when the magnitude of load impedance equals the magnitude of Z_s:

$$|Z_l| = |Z_s| \tag{12-7}$$

Proof of equation 12-7 comes most easily by differentiating $P = \frac{1}{2}(VI^* + V^*I)$, equation 26-4 of Chapter 4, with respect to $|Z_l|$. Before differentiating, V_l and I are expressed in terms of E, $|Z_s|$, $|Z_l|$, and the angles of Z_s and Z_l. The only variable is $|Z_l|$, and after a lengthy but not at all difficult manipulation of the algebra this turns out equal to $|Z_s|$ as in equation 12-7.

Equation 12-7 finds frequent practical use in matching a load to receive as much power as possible from a source if phase angles of the impedances are disregarded. If the angles of Z_l and Z_s are small, it is probably not worth while to add reactance to a load for the purpose of making the impedances conjugate, as the following example illustrates.

Example. Figure 12b shows a 12-volt source feeding a load of impedance Z through $600 + j150$ ohms. Find the load impedance to obtain maximum power, and the power received. Find also the best resistance if the load must be nonreactive, and find the power received.

Solution: For maximum power transfer, by equation 12-1:

$$Z_s = 600 + j150$$

$$Z = 600 - j150$$

$$I = \frac{12}{600 + j150 + 600 - j150} = \frac{12}{1200}$$

$$= 0.01 \text{ ampere or 10 milliamperes}$$

$$P = |I|^2 R = (0.01)^2 600 = 0.06 \text{ watt or 60 milliwatts}$$

For maximum power to a purely resistive load, using equation 12-7:

$$Z_s = 600 + j150 = 618\underline{/-14.03^\circ}$$

$$Z = R = 618$$

$$|I| = \frac{12}{|600 + j150 + 618|} = \frac{12}{1225}$$

$$= 9.80 \cdot 10^{-3} \text{ ampere or 9.80 milliamperes}$$

$$P = |I|^2 R = (9.80)^2 (618) 10^{-6} = 59.3 \cdot 10^{-3} \text{ watt or 59.3 milliwatts}$$

It is hard to imagine a situation in which the gain of the additional fraction of a milliwatt that is theoretically possible would justify the expense of adding a capacitor at the load.

Impedance matching, largely for the purpose of obtaining maximum power to the load, is important in practically all communication engineering. A conjugate match is the ideal, but in practice both source impedance and load impedance are likely to be principally resistive, and often it is enough to make their impedances equal in magnitude. Although a close match is not usually necessary, the general principle of impedance matching must never be forgotten.

In power engineering, on the other hand, the condition of maximum power transfer is not wanted. It is clear from the condition for maximum power transfer, and it is illustrated by the example, that when the load is receiving the greatest possible power from a given source, an equal amount of power is being lost in Z_s. This means that the efficiency of power transmission is only 50 per cent. It also means that terminal voltage drops to half when maximum load is applied. Neither of these is tolerable on a power system.

In communication engineering, the objective is to receive a useful signal from a given transmitter; it makes no difference how much of the power is lost if the received signal is good. In power engineering, the objective is to get a large proportion of the generated power to the consumer and to provide him with reasonably constant terminal voltage.

13. SUMMARY

The *superposition* theorem (or principle), Section 1, is used when there are several sources in a linear network; it permits considering the sources separately.

By the *substitution* theorem, Section 2, two subnetworks are equivalent if the relations between terminal voltages and terminal currents are the same for both, and such equivalent subnetworks can be exchanged without affecting conditions in any part of the network external to them. This theorem is used to prove several of the following theorems.

Equivalent sources can be exchanged as in Section 3 if an independent voltage source with series impedance takes the place of an independent current source with the same impedance in parallel, or vice versa.

Thévenin's theorem and Norton's theorem, Sections 4 and 5, permit substituting a single source and impedance in place of any linear section of a network, however complicated the section may be.

Three-terminal networks, shaped Y or Δ, if linear and passive, are equivalent under the conditions of Section 9 and may be exchanged for each other.

The *reciprocity* theorem permits exchanging a source and a point of measurement in a linear network that is passive apart from the source under consideration.

The *maximum power transfer* theorem says that power to a load is maximum if load impedance is the conjugate of impedance looking toward the generator from the load terminals. If the magnitude of load impedance can be varied but its phase angle cannot (the practical case being that of a load of pure resistance), then the magnitude of load impedance should equal the magnitude of impedance looking toward the generator. If the network looking toward the generator is complicated, its Thévenin equivalent can be considered.

It is emphasized by precept and example that substitutions are valid for direct or alternating current in the steady state, or for transient current; that is, for $Z(0)$, $Z(j\omega)$, or $Z(s)$.

PROBLEMS

10-1. Find currents in the branches of Figure 3a of Chapter 8 by superposition.
§2*

10-2. In the network shown, use the superposition theorem to find I.   §2

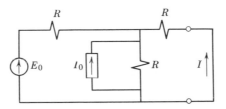

Problem 10-2.

10-3. In the diagram, use superposition to find the voltage V_{AB}, across the 8-ohm resistor. The diagram gives voltages and resistances in volts and ohms.   §2

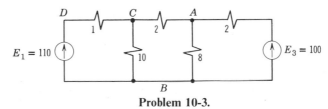

Problem 10-3.

10-4. In the circuit shown, $V_1 = 24$ volts and the source marked S_2 is a constant-current source carrying always 5 amperes. Each resistance is 6 ohms. Find I_c.
§3*

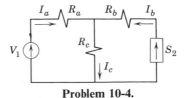

Problem 10-4.

10-5. (a) Find V' and Z' to make these two networks equivalent at terminals a-b. (b) Repeat, considering $R_1 = 0$ §3*

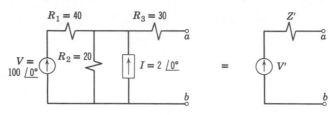

Problem 10-5.

10-6. The diagram shows a network (as encountered in power transmission) with two generators. Find the power P being delivered from one to the other, and tell which is producing and which absorbing electrical power. Draw phasor diagrams for your own guidance. Solve by substitution of equivalent sources.   §3

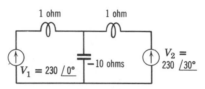

Problem 10-6.

10-7. (a) Given circuit A in the figure, find I and R_3 to make circuit B equivalent. (b) Find V_1 and Z_1 to make circuit C equivalent to circuits A and B. Express these all in terms of the quantities of circuit A.   §3

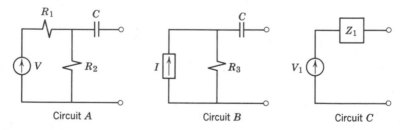

Circuit A Circuit B Circuit C

Problem 10-7.

10-8. Solve Problem 10-3 by use of Thévenin's theorem.   §7*

10-9. A farm draws its electricity from a transformer whose secondary voltage is low, partly because of losses in the insulation, which has deteriorated, and partly because of overload due to severe winter weather. An open circuit develops in the line which supplies the power for heating an incubator, and the farmer improvises a long emergency line and plugs in a 100-watt lamp, to be used for heating. The lamp does not get hot enough, and a check reveals that when the lamp is removed from its socket only 90 volts can be measured. Resistors of various values are available, but it is absolutely essential to dissipate exactly 100 watts in the incubator. The farmer takes a reading on an ammeter with which he briefly short-circuits the terminals and, quickly applying Thévenin's theorem, selects a 36-ohm wire-wound resistor to replace the lamp. What did the ammeter read?   §7

10-10. In the network shown, an independent-current source and an independent-voltage source are connected together through a highly inductive transmission line to supply power to a load, L. The whole arrangement may be replaced by a Thévenin equivalent source and impedance connected to L. Find the Thévenin equivalent source and impedance.   §7*

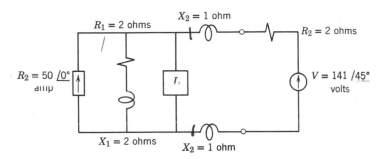

Problem 10-10.

10-11. The diagram shows a "pad" as used in telephone systems to reduce voltage and current. Resistances are given in ohms. A load R is to be connected between terminals A and B. Find power to the load if R equals (a) 200, (b) 283, (c) 350 ohms.   §7

Problem 10-11.

10-12. It is desired to substitute for the circuit shown its Thévenin equivalent between terminals A and B. $R_1 = R_2 = 5$ ohms, $L = 25$ millihenrys, and $C = 0.25$ microfarad. $E = 50$ millivolts. (a) Compute the equivalents as functions of $j\omega$ where ω is the frequency, and (b) compute numerical values for 5KHz.   §7*

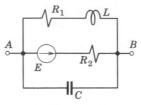

Problem 10-12.

10-13. Find, by any method, the Thévenin and Norton equivalent circuits for the network shown in the diagram.   §7

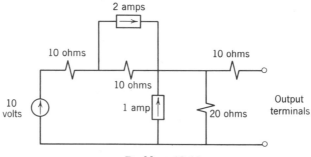

Problem 10-13.

10-14. The diagram shows a type of circuit useful for a purpose that the solution makes evident. Find current through the resistor if R is (a) 10, (b) 20, (c) 30 ohms. Use Norton's theorem, and also verify the result by some different means of solution.   §7*

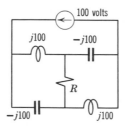

Problem 10-14.

10-15. The source of 100 volts in Problem 10-14 is replaced by an independent-current source of 4 amperes. Find voltage across the resistor as a function of its resistance R.   §7

10-16. Find Z_1, Z_2, and Z_3 that will make the two circuits equivalent at $f = 1600$ Hz. If you had the given circuit, could any of the elements be used to build the new T circuit?   §10*

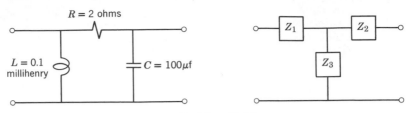

Problem 10-16.

10-17. In the network shown, $Z = 1 + j1$. What must be the value of X to make the three-terminal network abc equivalent to a purely resistive network? Is the resulting network resistive at all frequencies, or only at one frequency? §10*

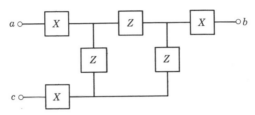

Problem 10-17.

10-18. The diagram shows a bridge circuit. If $V = 20$ volts, and each resistance is 2 ohms except R_e, which is 4 ohms, find (a) a single equivalent resistance between the generator terminals, and (b) current through R_f. §10*

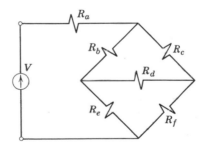

Problems 10-18 and 10-19.

10-19. In the bridge circuit shown, let $V = 10$ volts. $R_a = R_d = 1$ ohm; $R_b = R_c = R_e = 10$ ohms; $R_f = 20$ ohms. Find the current through R_d (magnitude and direction). §10*

10-20. Show that with appropriate choice of elements the circuit on the right can be made equivalent to the circuit on the left at all frequencies. Give the values of all elements. §10*

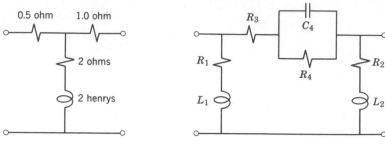

Problem 10-20.

10-21. Find the magnitude of the current through the two-ohm resistor. $V = 100$ volts, and $Z_1 = Z_2 = Z_3 = Z_4 = 8 + j12$ ohms.

Note: There is an easy way!   §11

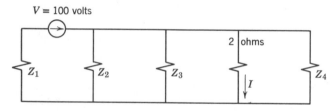

Problem 10-21.

10-22. In Figure 7a, let R_g be 10 ohms. By any method, find all branch currents. Then move the source E to be in series with R_g, and again find all branch currents. Does this illustrate the reciprocity theorem?   §11*

10-23. Prove equation 12-7.   §12

10-24. What load resistance R, connected at terminals A-B of Problem 10-11, will receive maximum power?   §12

10-25. From equation 12-3, show how load power varies with load resistance (that is, plot $P(R_s/E^2)$ as a function of R_1/R_s).   §12*

CHAPTER
11

MUTUAL INDUCTANCE

1. MAGNETIC COUPLING

Electric networks, whether for conveying power or information, often connect a number of *two-port* components in cascade, each component having a pair of input terminals and a pair of output terminals. One kind of two-port component is the transformer. A telephone conversation, for instance, is carried with the aid of transformers, and electric power to your home comes through transformers. The input circuit of any transformer affects the output circuit through magnetic induction.

Mutual inductance is one of the elementary parameters of electrical action and might logically have been introduced in Chapter 1 with self-inductance, capacitance, and resistance. However, the effect of one circuit on another is best considered with loop equations. Now, with loop equations ready to use, and with two-port networks coming in the next chapter, it is time to consider magnetic coupling.

Consider two circuits that are close together but not conductively connected, as suggested by Figure 1a. Current in either circuit produces *mutual* magnetic flux Φ_M that links both circuits, linkages with circuit 1 being $N_1 \Phi_M$ and linkages with circuit 2 being $N_2 \Phi_M$. Let there be current i_1 in circuit 1 but no current in circuit 2. A change of i_1 induces voltage in both circuits, and the voltage induced in circuit 2 is

$$v_2 = \frac{d}{dt}(N_2 \Phi_M) = N_2 \frac{d\Phi_M}{dt} = \frac{d}{dt}(L_{21}i_1) \qquad (1\text{-}1)$$

261

where L_{21}, called the mutual inductance, is by definition $L_{21} = N_2 \Phi_M / i_1$. Thus L_{21} is the number of flux linkages with circuit 2 produced by unit current in circuit 1. In a linear time-invariant system, in which L_{21} is constant, equation 1-1 can be written:

$$v_2 = L_{21} \frac{di_1}{dt} \tag{1-2}$$

Similarly a changing current in circuit 2 induces voltage in circuit 1. It is not obvious, but it will be shown in Section 6, that $L_{12} = L_{21}$.

A transformer is a device in which the mutual coupling is said to be *close*. There are typically two coils wound on a common core of steel or some other ferromagnetic material. When alternating voltage and current are supplied to one coil of the transformer, the other coil can supply voltage and current to a load. An actual transformer can sometimes be modeled as an *ideal* transformer (which will be defined in Section 5) with such close coupling that *all* magnetic flux is mutual; that is, all the flux produced by current in circuit 1 links circuit 2 also, and vice versa. As will be shown in Section 5, the *voltages* of the two coils of an ideal transformer are related by the ratio of turns in the coils, and *currents* by the reciprocal of the turn ratio:

$$\frac{v_1}{v_2} = \frac{N_1}{N_2} \quad \text{and} \quad \frac{i_1}{i_2} = \frac{N_2}{N_1} \tag{1-3}$$

As an example, suppose Figure 1a represents an ideal transformer. The secondary voltage, across N_2, is twice the primary voltage across N_1. But the secondary current is only half the primary current.

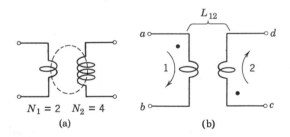

$N_1 = 2 \quad N_2 = 4$

(a)   (b)

FIGURES 1a,b   Inductive coupling.

On the other hand, many circuits are so *loosely* coupled by magnetic field that they cannot be modeled by ideal transformers. For these, a more detailed consideration of mutual inductance is required.

When, as in Figure 1b, there is magnetic coupling between two (or more) coils, the voltage across coil 1 results partly from a rate of change of current in coil 1 itself, and partly from a rate of change of current in coil 2 (and in other coils if there is coupling to others also). That is,

$$v_{ab} = \left(L_{11} \frac{di_1}{dt} \pm L_{12} \frac{di_2}{dt} \pm \cdots \right) \tag{1-4}$$

Correspondingly, the voltage across coil 2 is:

$$v_{cd} = \left(\pm L_{21} \frac{di_1}{dt} + L_{22} \frac{di_2}{dt} \pm \cdots \right) \tag{1-5}$$

L_{11} and L_{22} are the self-inductances of the coils; L_{12} and L_{21} are mutual inductances. $L_{21} = L_{12}$ in all cases.

Equations 1-4 and 5 are valid for currents that vary in any manner. If the currents are sinusoidal, the transformed equations are:

$$V_{ab} = j\omega(L_{11}I_1 \pm L_{12}I_2 \pm \cdots) \tag{1-6}$$

$$V_{cd} = j\omega(\pm L_{21}I_1 + L_{22}I_2 \pm \cdots) \tag{1-7}$$

2. POLARITY OF COILS

Should the $+$ or $-$ signs be used in these equations? This depends on the polarity with which coupled coils are wound compared with the reference directions of current indicated by arrows. Figure 2a shows a pair of coupled

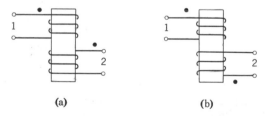

(a)   (b)

FIGURES 2a,b   Polarity of coils.

coils; so does Figure 2b. It is conventional to place dots on the diagram to indicate *corresponding* ends of coils; current entering the dotted end of one coil should produce magnetic flux linking that coil with the same polarity as the magnetic flux produced by current entering the dotted end of the other coil. Thus, in Figure 2a, current entering the dotted end of either coil produces magnetic flux upward in the core. The same is true of Figure 2b. Such use of dots, as in Figure 1b, makes a detailed picture or drawing of the coils unnecessary.

A diagram, then, shows arrows and dots. The rule for signs in the foregoing equations is as follows:

1. If both arrows enter dotted ends of coupled coils, or if both arrows enter undotted ends, the sign before the corresponding mutual inductance term is $+$.
2. If one arrow enters a dotted end, and the other an undotted end, the sign is $-$.

This rule rests on the relative directions of voltage induced in a coil by changing currents, and a brief consideration shows its validity.[†]

3. MUTUAL INDUCTANCE IN LOOP EQUATIONS

Two circuits with mutual inductance are shown in Figure 3a. Since there are only two magnetically coupled circuits, and since $L_{21} = L_{12}$, we follow the common practice of designating either of these mutual inductances by the letter M.

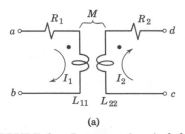

(a)

FIGURE 3a   Loop equations include M.

In view of the foregoing sections, it is easy to write the loop equations:

$$(R_1 + j\omega L_{11})I_1 - j\omega M I_2 = V_{ab} \qquad (3\text{-}1)$$

$$(R_2 + j\omega L_{22})I_2 - j\omega M I_1 = V_{cd} \qquad (3\text{-}2)$$

If there is not much magnetic coupling between the two circuits, they are *loosely* coupled. If almost all the magnetic flux produced by one coil links the other also, the coupling is said to be *close*. The closeness of coupling is measured by the *coefficient of coupling k* which is defined as

$$k = \frac{M}{\sqrt{L_{11}L_{22}}} \qquad (3\text{-}3)$$

For close coupling, k can approach 1, but it can never be greater.

[†] The rule, if possible, is correct. However, in networks with more than two loops, it is not always possible to dot corresponding ends of all coils. For more complete discussion, see Skilling (7).

4. EQUIVALENT NETWORKS

It is often helpful in analysis, and even more so in visualization, to replace inductively coupled circuits by a network with conductive coupling. This substitute network is called *equivalent* because it has the same loop equations, although in other ways it may act differently.

Equations 3-1 and 3-2 were written for the inductively coupled circuits of Figure 3a. These same equations can be written for the conductively connected loops of Figure 4a provided the following conditions are met. The inductance of the common element in Figure 4a is M, the self-inductance of circuit 1 is L_{11} which must therefore equal $L_1 + M$, and the self-inductance of circuit 2 is L_{22} which must therefore equal $L_2 + M$. L_{11} and L_{22} are called self-inductances.

For many purposes this equivalent network is convenient. If the coils are closely coupled, however, and if at the same time the numbers of turns in the coils are quite different, as is usual with a transformer, it is possible for M to be greater than one of the self-inductances, and this results in either L_1 or L_2 being negative. A negative inductance can be handled perfectly well in the mathematics, but it is inconvenient for visualization and it would make physical realization of the model impossible. It is usual, therefore, to use a different model, a more generalized equivalent network, for transformers and other devices that give negative inductances in Figure 4a.

To consider the more general equivalent network, let us start with Figure 4b. Here the common element is not M but aM where a is a number. It can be any number, but there are two most convenient choices: We can let $a = 1$, which gives the special case shown in Figure 4a, or we can let $a = N_1/N_2$. With the latter choice, a is the *turn ratio*, a number of great significance in transformers.

Loop equations for Figure 4b are written in the usual way, and with a

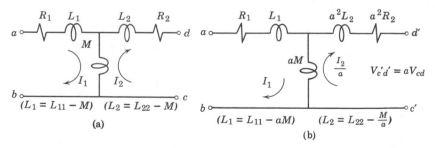

FIGURE 4 (a) A network with the same loop equations as the network of Figure 3a. (b) A more general equivalent network.

little algebraic juggling they can be put in the form:

$$(R_1 + j\omega L_{11})I_1 - j\omega M I_2 = V_{ab} \tag{4-1}$$

$$(R_2 + j\omega L_{22})I_2 - j\omega M I_1 = \frac{V_{c'd'}}{a} \tag{4-2}$$

Here equation 4-1 is identical with equation 3-1, which is as it should be. However, the output voltage from the network of Figure 4b, called $V_{c'd'}$, is not equal to V_{cd} but to aV_{cd}, while the output current from the network is not I_2 but I_2/a, and equation 4-2 is correspondingly different from equation 3-2.

If we could connect at the output terminals of Figure 4b, in cascade with the network, an ideal transformer to multiply the output voltage by $1/a$ while multiplying the current by a, as in Figure 4c, such a transformer would have an output voltage V_{cd} and an output current I_2. Then the equations 3-1 and 3-2 which were written for the loops with mutual inductance would apply exactly to the equivalent network of Figure 4c.

Thus an equivalent to magnetically coupled circuits is provided by the network of Figure 4b followed by an ideal transformer, and this equivalent circuit can be connected to any load, impedance, or system that might be connected to circuit 2 of the actual coupled circuits with identical results.

Another way to use the network of Figure 4b is to connect to its terminals a load, impedance, or system like the one actually connected to circuit 2 of the magnetically coupled circuits of Figure 3a but with every impedance

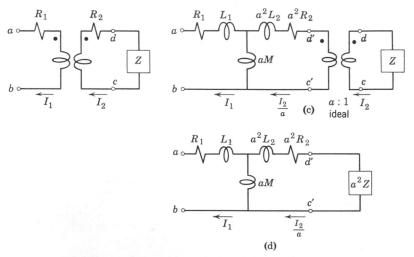

FIGURE 4c,d   Networks equivalent to coupled circuits: $L_1 - L_{11} - aM$, and $L_2 = L_{22} - M/a$.

multiplied by the factor a^2 as suggested in Figure 4d. At the same time, any source in the connected system must have its voltage multiplied by a and current multiplied by $1/a$. The result is a system that is fully equivalent to the original coupled circuits on the side of circuit 1, but with all voltages and currents on the side of circuit 2 altered by a factor of either a or $1/a$.

This latter interpretation of the equivalent circuit, as suggested in Figure 4d, is quite widely useful when a is made equal to the turn ratio N_1/N_2 of the coupled coils. When applied to power systems it leads to such concepts as the "secondary voltage (or current, or impedance) referred to the primary side." In systems for the transmission of information it is quite common to speak of "impedance transformation." Clearly, from Figure 4d, the load on an incoming line is approximately a^2Z if power is supplied to an impedance Z through a transformer with turn ratio a. After a further discussion of transformers, these ideas can be illustrated numerically.

5. TRANSFORMERS

A transformer is a device that transmits power from one circuit to another through a magnetic field, often with no conductive connection. Transformers usually have quite close coupling.

A transformer in a power system has two (or more) coils wound together on a steel core, so interleaved that k may be as great as 0.98. A transformer in an information system is usually satisfactory with a lower k, perhaps between 0.6 and 0.9; it is very much smaller because both voltage and current are less. It is probably wound on a ferromagnetic core of some kind, although "air-core" transformers are possible if relatively low inductances are satisfactory.

The purpose of a power transformer is usually to change voltage, the voltage at the output terminals being either higher or lower than the input voltage in proportion to the ratio of turns in the primary and secondary windings. A transformer in an information system serves the same purpose, although the object may be spoken of as transforming impedances.

Any transformer, if it has magnetic coupling but no electrical connection between its two coils, may serve to isolate the output circuit from the input circuit. In an information system this isolation may permit a component of direct current in one winding with no corresponding direct current in the other, although input alternating current is reproduced in the output circuit. Power-system transformer banks often change three-phase connections from Y to Δ, or Δ to Y.

An actual transformer can be modeled, as in Figure 4c, by a T-shaped network cascaded with an ideal transformer. The T network provides its non-ideal characteristics, while the ideal transformer with the same turn ratio as the actual transformer provides only the turn-ratio transformation:

$$\frac{V_1}{V_2} = \frac{N_1}{N_2} = a \quad \text{and} \quad \frac{I_1}{I_2} = \frac{N_2}{N_1} = \frac{1}{a} \tag{5-1}$$

If the transformer being modeled were itself ideal, the T network would vanish. This happens for a transformer that has no resistance or power loss, that has such close coupling that $k = 1$, and that has such great mutual inductance that M is infinite. These are the specifications for an *ideal* transformer, and they can be approached but never attained by an actual transformer.

To show that these specifications do indeed make the T network vanish, consider that if the shunt element aM in Figure 4c becomes infinite in inductance it carries no current and can be eliminated. Also, R_1 and R_2 are both zero. Finally, since $k = 1$ and therefore $(L_1 + aM)(L_2 + M/a) = M^2$ it follows that when M becomes infinite, $L_1 + a^2 L_2 = 0$ and thus the series inductances in the T network vanish, leaving nothing.

A physical picture of an ideal transformer is easily seen in terms of magnetic flux. Since the same mutual flux links both coils, voltage induced is proportional to the number of turns, and the voltage ratio of equations 5-1 follows. Since mutual inductance is so great, magnetomotive force of one coil equals that of the other for any finite flux, and $N_1 I_1 = N_2 I_2$, giving the current ratio of equations 5-1.

To design an actual transformer to behave more like an ideal transformer, copper conductors are made larger, coils are wound closer together and perhaps interleaved, and ferromagnetic cores are made of better material with larger cross section. All these improvements are expensive. It is the designer's job to minimize the undesirable aspects without too greatly increasing the cost. Energy transmitted by a large transformer in a power system is so valuable that the design of the transformer can be close to ideal, but a transformer to supply a few watts to a home radio receiver from a wall outlet is probably designed for cheapness rather than efficiency.

For numerical illustration, Figure 1a can represent a small, cheap transformer. The primary coil will surely not have 2 turns; 400 is more likely. The secondary winding can then have 800 turns, and the turn ratio $N_1/N_2 = a = 1/2$. This is a step-up transformer, increasing voltage from, say, 120 to 240 volts. The laminated steel core will perhaps be a rectangular frame of four legs, one leg having half the primary coil wound upon it with half the secondary coil wound as closely as possible on the same leg. On the opposite leg of the rectangular frame the other half of the primary coil is wound, with the other half of the secondary coil wound closely upon it.

If the transformer is supplied from an independent source of 120 volts (which may be our model of a wall outlet), and if the transformer were ideal, its output voltage would be 240 volts independent of load.

Let the load be 100 ohms of resistance. Secondary current is 2.4 amperes. Primary current is (still assuming ideal relations) 4.8 amperes. Power to the load is $V_2 I_2 = I_2{}^2 R_{load} = 576$ watts. Power to the primary winding of the transformer is $V_1 I_1 = 576$ watts.

The power engineer would say that secondary voltage "referred to the primary side" was 120 volts, this being aV_2. The secondary current "referred to the primary side," I_2/a, is 4.8 amperes. The load impedance "referred to the primary side," $a^2 Z$ (see Figure 4d) is 25 ohms.

The communication engineer might say that he is transforming a load impedance of 100 ohms to 25 ohms, or that the two-to-one transformer matches 100 ohms to 25 ohms. See *impedance matching* in Chapter 10.

Considering that the actual transformer is not, in fact, ideal, the resistances and self-inductances of the coils and their mutual inductance can be taken into account as in Figures 4c or d if this is considered to be necessary.

6. EQUALITY OF MUTUAL INDUCTANCES

It can be shown experimentally that $L_{21} = L_{12}$, but experimental demonstration is not necessary. If conservation of energy is accepted this equality follows. The demonstration can be given in several ways, and perhaps the following is well matched to previous work in this book.†

The average energy output from mutual inductance cannot be greater than the energy supplied to it. Consider two circuits with mutual as well as self inductance, as in Figure 6a. Resistance is taken to be zero, for simplicity. Voltages and currents are sinusoidal. Loop equations, following the pattern of equations 3-1 and 3-2, are written:

$$j\omega L_{11} I_1 - j\omega L_{12} I_2 = V_1 \tag{6-1}$$

$$j\omega L_{22} I_2 - j\omega L_{21} I_1 = -V_2 \tag{6-2}$$

The latter voltage is negative because V_2 is measured (in the diagram) from top to bottom.

Equation 26-4 of Chapter 4 gives power (average over a cycle) as $P = \frac{1}{2}(VI^* + V^*I)$, and equation 6-1 yields the following by multiplication:

$$j\omega L_{11} I_1 I_1{}^* - j\omega L_{12} I_2 I_1{}^* = V_1 I_1{}^* \tag{6-3}$$

$$(j\omega L_{11} I_1)^* I_1 - (j\omega L_{12} I_2)^* I_1 = V_1{}^* I_1 \tag{6-4}$$

---

† For others see Ramo, Whinnery and Van Duzer; or Skilling (7).

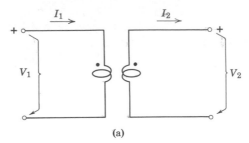

(a)

FIGURE 6a   To show that $L_{21} = L_{12}$.

The sum of these expressions is twice the power *input* to the transformer of Figure 6a. Note that $(jI_1)^* = -jI_1{}^*$ as the conjugate of $j1$ is $-j1$. Then:

$$2P_{in} = j\omega L_{11}(I_1 I_1{}^* - I_1{}^*I_1) - j\omega L_{12}(I_2 I_1{}^* - I_2{}^*I_1) \qquad (6\text{-}5)$$

The first term on the right is obviously zero.

The equation from Chapter 4 is now used again to find the power output from circuit 2. By an exactly similar process applied to equation 6-2, twice the power *output* from the transformer is found to be:

$$2P_{out} = -j\omega L_{22}(I_2 I_2{}^* - I_2{}^*I_2) + j\omega L_{21}(I_1 I_2{}^* - I_1{}^*I_2) \qquad (6\text{-}6)$$

Again the first term of the right-hand member is zero.

Now, average power out cannot be greater than average power in,

$$P_{out} \ngtr P_{in} \qquad (6\text{-}7)$$

We compare the non-zero parts of equations 6-6 and 6-5, and find that

$$L_{21} \ngtr L_{12} \qquad (6\text{-}8)$$

Hence, when average power flows from left to right, L_{21} cannot be greater than L_{12}.

But when average power flows from right to left the same demonstration shows that L_{12} cannot be greater than L_{21}. Therefore the mutual inductances can only be equal, and

$$L_{12} = L_{21} \qquad (6\text{-}9)$$

Since the value of mutual inductance is not affected by the wave form of voltage or current in a linear system, this proof of equality is not limited to sinusoids.

7. SUMMARY

Change of current in one circuit can induce voltage in a *nearby* circuit. The parameter describing this induction is called *mutual inductance*. Mutual inductance is ascribed to *mutual flux*.

Two circuits can be closely coupled through a *transformer*. The limit of close coupling is given by the *ideal* transformer in which all flux links both circuits. In an ideal transformer (as defined in Section 5) voltages are directly proportional and currents inversely proportional to turn ratio.

Two circuits can be *loosely* coupled, in which case loop equations are convenient. Expressions are given for *voltages* induced in circuits, with rules for the *signs* to be used.

Loop equations are written for circuits with mutual inductance between them. The *coefficient of coupling* is defined.

An *equivalent* T network is developed that is subject to three interpretations. There is a parameter *a*.

1. If coupling is loose, let $a = 1$; then the loop equations of the network are identically those of the actual coupled circuits.
2. If coupling is close and the turn ratio N_1/N_2 is quite different from 1, let $a = N_1/N_2$. The actual coupled circuits can then be modeled by a T network of impedances followed by an ideal transformer.
3. The T network can be used without the ideal transformer, but, to keep identical loop equations, all elements connected to circuit 2 must be changed: impedances must be multiplied by a^2, voltages by a, and currents by $1/a$.

Some of the uses and purposes of *transformers* are discussed. An *ideal* transformer is *defined*. A practical model of a transformer can often be an ideal transformer plus a T network, with all non-ideal characteristics of the actual transformer appearing in the T network.

It is shown that $L_{21} = L_{12}$, for if this were not true the power output from mutual inductance could be greater than the power supplied to it, contrary to the principle of *conservation of energy*.

PROBLEMS

11-1. Two coils are in fixed positions relative to each other, but connections to them may be changed. Equipment for measuring inductance (presumably an inductance bridge) is available. Describe an experiment for determining the mutual inductance with the least number of measurements, and explain how the desired mutual inductance is to be computed from the measurements.

§3

11-2. If the self-inductance of a coil is $L_1 = 0.5$ henry and its mutual inductance to another coil is $L_{12} = 1.2$ henrys, and current in the coil is $i_1 = 3 \cos \omega t$ and current in the other coil is $i_2 = 2 \sin \omega t$, and $\omega = 500$, find voltage v_1 across the coil. (See Figure 1b.) What item of information is missing? Two answers are consistent with the data; give both.   §3*

11-3. Find an expression for the impedance of the circuit shown. The loop method is suggested.   §3*

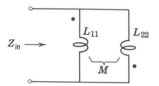

Problems 11-3 through 11-6.

11-4. In Problem 11-3, let $L_{11} = 100$, $L_{22} = 100$, and $M = 90$ millihenrys. Find the coefficient of coupling k. Find the equivalent input inductance of the circuit.   §3*

11-5. The network is the same as for Problem 11-4, but the dot on the right-hand coil in the diagram is now moved from the bottom to the top. What is physically different? Find the new equivalent input inductance.   §3

11-6. Let each coil of Problem 11-4 have 10.0 ohms of resistance as well as the indicated inductance. Frequency is 100 Hz. Find the input impedance.   §3

11-7. Mutual inductance between two identical coils is to be found. Bridge measurement gives 137 millihenrys for the two coils in series. After reversing connections to one of the coils the measurement is 43 millihenrys. Find (a) the mutual inductance, (b) the self-inductance of each coil, and (c) the coefficient of coupling.   §3*

11-8. In the circuit shown, each of the three coils has mutual inductance with each of the others in the amounts shown. (a) Write the loop equations for these inductively coupled circuits. (b) Draw an equivalent network similar to Figure 4a, giving the values of each of the three equivalent inductances in terms of the actual self- and mutual inductances.   §4*

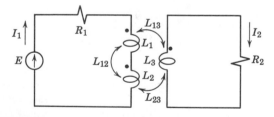

Problem 11-8.

11-9. In Figures 3a and 4a, $V_{ab} = 100\underline{/0°}$ and $V_{dc} = 1000\underline{/0°}$ volts; $L_{11} = 0.30$ henry, $L_{22} = 30.0$ henrys, and $M = 3.00$ henrys; $R_1 = 0.5$ ohm and $R_2 = 50$ ohms; frequency is 60 Hz. Write the loop equations and solve for I_1.   §4*

11-10. Devise an equivalent circuit for the coupled circuits of Problem 11-9, using $a = 1$. Repeat, using $a = 10$.   §4*

11-11. A transformer has $L_{11} = 5.00$, $L_{22} = 21.0$, $M = 9.80$ henrys; $R_1 = 1.5$, $R_2 = 5.1$ ohms. When 100 volts, 60 Hz, is applied to the primary terminals, find (a) primary current and secondary voltage, the secondary circuit being open; and (b) primary current and secondary current, the secondary terminals being short-circuited. Use the equivalent circuit of Figure 4a. §4*

11-12. A test on the low-voltage winding of a 230- to 2300-volt power transformer (60 Hz, 20 kva) gives 2.0 amperes, 0 watt, at 230 volts applied, with the high-voltage winding *open-circuited*. A test on the same winding with the high-voltage winding *short-circuited* gives 100 amperes, 0 watt, at 5.3 volts. Evaluate as many as possible of the inductances and resistances of Figure 4c, accepting reasonable approximations. Discuss the approximations involved. (These are known as the open-circuit and short-circuit tests of the transformer.)

11-13. In the circuit shown, find the value of the load impedance Z_L for maximum power transfer from the source to the load. It is known that $L_{11} = L_{22}$, that $k = 0.5$, and that $j\omega M = j100$ ohms. §5

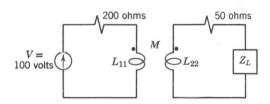

Problem 11-13.

11-14. Draw equivalent circuits of the type shown in Figures 4a and 4c for an *ideal* transformer. To do so, define an ideal transformer. Can the loop equations such as equations 8-1 and 8-2 be written to apply to an ideal transformer? Are your definitions and results consistent with equations 1-3? §5

CHAPTER
12

TWO-PORT NETWORKS

1. USEFUL FORMS

A network that has two pairs of terminals is often called a *two-port* network; a signal enters the net at one port and leaves at the other. A number of two-port nets may be cascaded to form a communication channel. Ladder networks, transformers, transmission lines, amplifiers, and filters are examples of two-port nets often cascaded in a channel.

Several of our early chapters emphasized networks with one pair of terminals, or one port. A network with two pairs of terminals may consist of several such one-port networks, as shown in Figure 1a. Each box in this figure may contain anything from a single element to a complicated one-port network.

Two-port networks are used to transmit signals and also to distribute power. The first part of this chapter applies equally to either use, but the latter sections discuss image-impedance termination which is used for information systems only.

2. LOOP EQUATIONS

Networks with two pairs of terminals are readily considered as special cases of the networks described by loop and node equations in Chapter 8. Figure 2a shows such a network; it is specified that the two terminals at one end carry equal but opposite currents, and so also do the two terminals at the

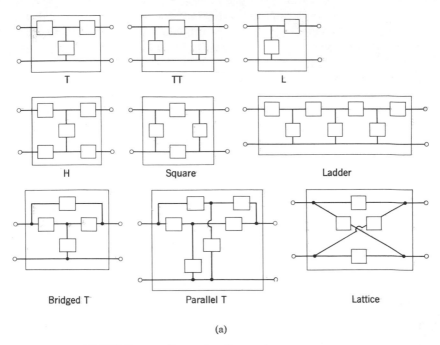

(a)

FIGURE 1a   Illustrative forms of two-port networks.

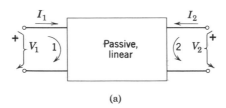

(a)

FIGURE 2a   Node and loop analysis of the two-port net.

other end. (This requirement distinguishes the two-terminal-pair network from a more general four-terminal network.) Calling the input loop 1 and the output loop 2 (loop numbers are entirely arbitrary), and specifying that there are no independent sources within the box, equations 2-2 of Chapter 9 give us

$$I_1 = y_{11} V_1 + y_{12} V_2$$
$$I_2 = y_{21} V_1 + y_{22} V_2 \qquad (2\text{-}1)$$

It will be remembered that y_{11} is the driving-point admittance at the terminals in loop 1, which is the ratio of current to voltage at these terminals when voltages at all other input terminals are zero; that is, it is the driving-point admittance when all other pairs of input terminals are short circuited. Now, specifically for two-port networks, y_{11} is the admittance at terminal-pair 1 when there is a short circuit at terminal-pair 2, and y_{11} is correspondingly called the *short-circuit driving-point admittance* at port 1. Similarly, y_{22} is the short-circuit driving-point admittance at port 2, which can be computed or measured when there is a short circuit at port 1.

These equations use also the *short-circuit transfer admittances* y_{21} and y_{12}, which are equal. These can be computed as the ratio of the short-circuit current at the farther terminals to voltage applied at the nearer terminals.

Certain implicit limitations should perhaps be mentioned. The network must be *linear*. Also, it must be *reciprocal*, which means that y_{12} must equal y_{21}. Note that these limitations apply throughout all this chapter.

Equations 2-1 are convenient for finding current at either end of a two-port net if both voltages are known. A very simple numerical example will show how they work.

3. AN EXAMPLE: FINDING CURRENTS

Each resistance in Figure 3a is 4 ohms. Short circuiting port 2, the impedance seen at port 1 is 6 ohms; hence $y_{11} = 1/6$. Similarly, short circuiting port 1, $y_{22} = 1/6$. Since (in either case) the current at the short-circuited port

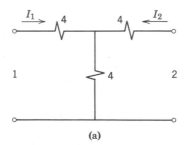

(a)

FIGURE 3a   Each resistance is 4 ohms.

is half the entering current but (note the direction of arrows) of the opposite sign, $y_{21} = y_{12} = -1/12$. With these values, equations 2-1 are:

$$I_1 = \tfrac{1}{6}V_1 - \tfrac{1}{12}V_2$$
$$I_2 = -\tfrac{1}{12}V_1 + \tfrac{1}{6}V_2 \tag{3-1}$$

As an example, let a 36-volt battery be connected at port 1, and a 24-volt battery at port 2. Find the currents. Using equations 3-1,

$$I_1 = \frac{36}{6} - \frac{24}{12} = 4 \text{ amperes}$$

$$I_2 = -\frac{36}{12} + \frac{24}{6} = 1 \text{ ampere}$$

(3-2)

The result can be checked by superposition or otherwise.

This example uses batteries and direct current, but equations 2-1 and 3-1 apply equally well with alternating voltages and currents. V and I are then the *transforms* of voltage and current, and the admittances are complex quantities, functions of $j\omega$ or of s.

4. NODE EQUATIONS

Equations 2-1 give the terminal *currents* of two-port nets in terms of applied voltages. If it happens that one wants the *voltages* in terms of terminal currents it is convenient to start with the node equations 4-2 of Chapter 9. If there are no sources within the box of Figure 2a, and if the current entering port 1 is I_1 while the current entering port 2 is I_2, these node equations become:†

$$V_1 = z_{11}I_1 + z_{12}I_2$$

$$V_2 = z_{21}I_1 + z_{22}I_2$$

(4-1)

Here z_{11} is the driving point impedance at port 1, the ratio of voltage to current at that port, when I_2 is zero and hence when port 2 is open circuited. It is called the *open-circuit driving-point impedance* at port 1. Similarly, z_{22} is the open-circuit driving-point impedance at port 2, and it can be computed or measured when there is an open circuit at port 1.

The *open-circuit transfer impedances* z_{12} and z_{21} are equal (in all reciprocal networks). They can be computed as the ratio of open-circuit voltage at one port to the entering current at the other port.

These node equations, like the loop equations, are valid for either direct or alternating current. With sinusoidally alternating currents and voltages, I and V are transforms, and each Z is a complex quantity. In the special case of direct current and voltage, I and V are currents and voltages, and the impedances are merely resistances.

---

† In Chapter 8 node voltages were measured from each "independent" node to one "reference" node. This is convenient but not necessary. In the present work, N independent node pairs are so selected that one such pair is terminal pair 1 and another is terminal pair 2. They may or may not have a common "reference" node.

5. ANOTHER EXAMPLE: FINDING VOLTAGES

Again a simple numerical example may be helpful. Let us use the same network of Figure 3a, each element having 4 ohms of resistance.

With open circuit at port 2, the impedance seen at port 1 is 8 ohms, so the open-circuit driving-point impedance at this end is $z_{11} = 8$. Similarly, with open circuit at port 1, we find that $z_{22} = 8$. To find the open-circuit transfer impedance, let the current into port 1 be 1 ampere while port 2 is open circuited; the voltage across port 2 is then 4 volts, so $z_{12} = 4$, and this is also the value of z_{21}. Hence:

$$V_1 = 8I_1 + 4I_2$$
$$V_2 = 4I_1 + 8I_2 \tag{5-1}$$

As an example, suppose an independent current supply forces 4 amperes to enter port 1, while another supply forces 1 ampere to enter port 2 (both in the arrow directions). What are the voltages at the two ports? Equations 5-1 give the answers:

$$V_1 = (8)(4) + (4)(1) = 36 \text{ volts}$$
$$V_2 = (4)(4) + (8)(1) = 24 \text{ volts} \tag{5-2}$$

6. SETS OF PARAMETERS

We have now developed two sets of parameters, the *short-circuit admittances* and the *open-circuit impedances*. With the admittances, the y's, we can compute currents at the two ports in terms of voltages; with the impedances, the z's, the voltages can be computed in terms of currents. There is of course a relation between the z's and the y's but it is not simple or obvious.†

---

† Relations between y's and z's are found as follows. Simultaneous solution of equations 2-1 for V_1 gives

$$V_1 = \frac{\begin{vmatrix} I_1 & y_{12} \\ I_2 & y_{22} \end{vmatrix}}{\begin{vmatrix} y_{11} & y_{12} \\ y_{21} & y_{22} \end{vmatrix}} = \frac{y_{22}}{\Delta \text{ of } y\text{'s}} I_1 - \frac{y_{12}}{\Delta \text{ of } y\text{'s}} I_2 \tag{6-1}$$

But by equations 4-1,

$$V_1 = z_{11}I_1 + z_{12}I_2 \tag{6-2}$$

Therefore

$$z_{11} = \frac{y_{22}}{\Delta \text{ of } y\text{'s}} \quad \text{and} \quad z_{12} = -\frac{y_{12}}{\Delta \text{ of } y\text{'s}} \tag{6-3}$$

The other relations are found in a similar manner. It is interesting to try these relations on the parameters of Sections 3 and 5.

There are three independent parameters in either of these sets, the y's or the z's. If the network is also *symmetrical* (as well as being reciprocal) there are only two independent parameters in each set, for by definition a network is symmetrical if $y_{11} = y_{22}$ or, the same thing, if $z_{11} = z_{22}$.

A network that is symmetrical can be turned end for end in a system without altering currents or voltages elsewhere in the system. Thus the T of Figure 3a is symmetrical. This is obvious from its symmetry of structure. However, a network can be symmetrical in the electrical sense without having structural symmetry. Networks in communication systems are often symmetrical, as we shall see in the latter part of this chapter.

Two-port problems can be solved with the aid of either the y's or the z's (or with some mixture, for a number of so-called *hybrid* sets of parameters are in use). Nevertheless, we shall find it convenient to introduce still another set. These, *transmission parameters*, are called the *ABCD* parameters, or the general circuit constants, or more properly the general network functions. They are convenient because they give voltage and current at one port in terms of voltage and current at the other port. Thus they give the quantities that we usually want in terms of the quantities that we usually know.

7. THE TRANSMISSION PARAMETERS

When signals or power always pass along a system in one direction, it is convenient to speak of the sending end and the receiving end. Thus Figure 7a uses the subscripts s and r, with signals presumed to pass from left to right, and the arrow direction of I_r is correspondingly *out* of the two-port net.

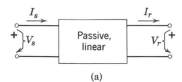

(a)

FIGURE 7a   Notation used with transmission parameters A, B, C, and D.

Convenient equations for working with transmission problems could have the form:

$$V_s = AV_r + BI_r \qquad (7\text{-}1)$$

$$I_s = CV_r + DI_r \qquad (7\text{-}2)$$

The necessary information to find A, B, C, and D can come from the loop equations 2-1. When these loop equations are rewritten with subscripts s and r, and I_2 is changed to $-I_r$, we have

$$I_s = y_{11} V_s + y_{12} V_r \qquad (7\text{-}3)$$

$$-I_r = y_{21} V_s + y_{22} V_r \qquad (7\text{-}4)$$

It is now easy to solve equations 7-3 and 7-4 simultaneously for V_s and I_s in terms of V_r and I_r, and when this is done we find that we obtain equations 7-1 and 7-2 with

$$A = -\frac{y_{22}}{y_{21}} \qquad B = -\frac{1}{y_{21}} \qquad C = y_{12} - \frac{y_{11} y_{22}}{y_{21}} \qquad D = -\frac{y_{11}}{y_{21}} \qquad (7\text{-}5)$$

Thus A, B, C, and D are found in terms of the *short-circuit admittances* of the network. They can also be found in terms of the *open-circuit impedances*. The node equations 4-1, rewritten with the subscripts and signs of Figure 7a, can be solved simultaneously for V_s and I_s. The results give, for use in equations 7-1 and 2, the values:

$$A = \frac{z_{11}}{z_{21}} \qquad B = \frac{z_{11} z_{22}}{z_{21}} - z_{12} \qquad C = \frac{1}{z_{21}} \qquad D = \frac{z_{22}}{z_{21}} \qquad (7\text{-}6)$$

We have seen that among the four y parameters, only three are independent, and likewise only three of the four z parameters are independent. If our network is *symmetrical* only two are independent. We may expect some similar kind of independence among the $ABCD$ functions. From equations 7-6 it is always true that

$$AD - BC = 1 \qquad \text{or} \qquad \begin{vmatrix} A & B \\ C & D \end{vmatrix} = 1 \qquad (7\text{-}7)$$

and this relation permits finding the fourth parameter if any three are known. This is another way of saying that only three are independent. Further, if the network is symmetrical, so that $z_{22} = z_{11}$, equations 7-6 tell us that $D = A$, and this reduces to two the independent parameters for a *symmetrical* network.

It is often easier to find the values of A, B, C, and D directly, without having to find either the y's or the z's first. Suppose the network in question is open circuited at the receiving end, forcing I_r to be zero. Then, by equation 7-1, $V_s = AV_r$, and, by equation 7-2, $I_s = CV_r$. It is now only necessary to determine V_r with the receiving end open, and we may then compute:

$$A = \frac{V_s}{V_r}\bigg|_{I_r = 0} \qquad\qquad C = \frac{I_s}{V_r}\bigg|_{I_r = 0} \qquad (7\text{-}8)$$

Similarly, with the receiving end short circuited, so that $V_r = 0$, we obtain from the same two equations:

$$B = \frac{V_s}{I_r}\bigg|_{V_r=0} \qquad D = \frac{I_s}{I_r}\bigg|_{V_r=0} \qquad (7\text{-}9)$$

8. AN EXAMPLE OF TRANSMISSION PARAMETERS

The T network of Figure 3a with 4 ohms in each branch offers an exceedingly simple illustration of the $ABCD$ parameters. We found for this network in Section 3 that:

$$y_{11} = \tfrac{1}{6} \qquad y_{12} = y_{21} = -\tfrac{1}{12} \qquad y_{22} = \tfrac{1}{6} \qquad (8\text{-}1)$$

When these values of *short-circuit admittances* are put into equations 7-5 they give:

$$A = 2 \qquad B = 12 \qquad C = \tfrac{1}{4} \qquad D = 2 \qquad (8\text{-}2)$$

As another example, we found for this same network in Section 5 that the *open-circuit impedances* are:

$$z_{11} = 8 \qquad z_{12} = z_{21} = 4 \qquad z_{22} = 8 \qquad (8\text{-}3)$$

When these are substituted into equations 7-6 the $ABCD$ values prove to be identical with those already determined in equations 8-2.

It was stated in equation 7-7 that for all networks (linear and reciprocal), $AD - BC = 1$, and this is seen to be true for our example in which $AD = 4$ and $BC = 3$. Further, our T network is symmetrical and hence $A = D = 2$.

To illustrate the direct method, not using the y's or z's, we look at Figure 3a and assume the receiving end, end number 2, to be open. Then the receiving-end voltage, V_r, is half the sending end voltage, and $V_s/V_r = A = 2$. With the receiving end still open, if I_s, the current into the network, equals 1, then V_r, the receiving-end voltage, equals 4, and $I_s/V_r = C = 1/4$.

Next we must short circuit the receiving end. Entering current now divides so that I_s, the current entering the sending end, is twice I_r, the current *out* at the receiving end, and $D = I_s/I_r = 2$. Finally, if we assume that the short-circuited $I_r = 1$ (arbitrarily), then $V_s = 4 + 8 = 12$, and $B = V_s/I_r = 12$. The four parameters thus computed are the same as those found in equation 8-2, as of course they should be. This direct method of computation is often the easiest to use.

It must be carefully remembered that this network of pure resistances is a highly special case. In general the admittances and impedances are complex and therefore the values of A, B, C, and D are complex also. Moreover, they are functions of frequency, and hence may be written in terms of $j\omega$ for some purposes, or in terms of s for other purposes. The most general form is in terms of s, for the parameters can then be interpreted to apply to either dc or ac problems, either steady or transient. See Section 25 for an example.

9. RATIOS OF TRANSMISSION PARAMETERS

Certain ratios of the $ABCD$ parameters that will soon be needed can now be derived.

From equations 7-5,

$$\frac{D}{B} = \frac{-(y_{11}/y_{21})}{-(1/y_{21})} = y_{11} \tag{9-1}$$

and from equations 7-6,

$$\frac{A}{C} = \frac{(z_{11}/z_{21})}{(1/z_{21})} = z_{11} \tag{9-2}$$

When these are multiplied, and then divided, we obtain:

$$z_{11}y_{11} = \frac{AD}{BC} \quad \text{and} \quad \frac{z_{11}}{y_{11}} = \frac{AB}{CD} \tag{9-3}$$

These results will be useful later, and since we shall use them for symmetrical networks, for which $A = D$, we can now write that for a symmetrical network:

$$z_{11}y_{11} = \frac{A^2}{BC} \quad \text{and} \quad \frac{z_{11}}{y_{11}} = \frac{B}{C} \tag{9-4}$$

10. PARAMETERS FOR TYPICAL NETWORKS

Let us find $A, B, C,$ and D for several two-port networks. The basic equations are

$$V_s = AV_r + BI_r \tag{10-1}$$

$$I_s = CV_r + DI_r \tag{10-2}$$

From these, $A, B, C,$ and D are computed (with the aid of equations 7-8 and 9), as follows.

Series Impedance. In Figure 10a the "network" is merely an impedance leading to some load. First, let there be no load, so $I_r = 0$. Then $V_s = V_r$, so from equation 10-1, $A = 1$. From the diagram, $I_s = 0$, so from equation 10-2, $C = 0$.

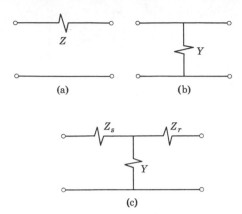

FIGURES 10a,b,c Finding *ABCD* parameters for series and shunt elements, and a T network.

With the receiving end short-circuited, $V_r = 0$. I_r is now equal to V_s/Z, so $B = Z$ and $D = 1$. Thus we have found:

$$A = 1 \qquad B = Z \qquad C = 0 \qquad D = 1 \qquad (10\text{-}3)$$

Note that $AD - BC = 1$, as it must, and because there is obvious symmetry $A = D$.

Shunt Admittance. Now let the network be the shunt admittance of Figure 10b. With the receiving end open, $I_r = 0$, and we see that $A = 1$ and $C = Y$. With a short circuit at the receiving end, $V_r = 0$, and $I_s = I_r$, so clearly $D = 1$. Since V_s and $V_r = 0$, $B = 0$. Thus:

$$A = 1 \qquad B = 0 \qquad C = Y \qquad D = 1 \qquad (10\text{-}4)$$

Again, $AD - BC = 1$, and $A = D$ because of symmetry.

The Unsymmetrical T. A less obvious example is the T network of Figure 10c. When the terminals at the receiving end of the network are open circuited, and $I_r = 0$, we see from Figure 10a that $V_r = I_s/Y$. Then, by equation 7-8, $C = Y$. Also, $V_s = V_r + Z_s I_s = V_r + Z_s Y V_r$ and by equation 7-8 we have $A = 1 + Z_s Y$.

Now we make $V_r = 0$ by short circuiting the right-hand end, and from the diagram we have:

$$I_s = \frac{Z_r + (1/Y)}{1/Y} I_r = (Z_r Y + 1)I_r \qquad (10\text{-}5)$$

from which equation 7-9 gives $D = Z_r Y + 1$. Also, with the short circuit,

$$V_s = Z_r I_r + Z_s I_s = Z_r I_r + Z_s(Z_r Y + 1)I_r$$

$$B = \frac{V_s}{I_r} = Z_r + Z_s + Z_r Z_s Y \qquad (10\text{-}6)$$

Thus we have obtained:

$$A = 1 + Z_s Y \qquad B = Z_s + Z_r + Z_s Z_r Y \qquad C = Y \qquad D = 1 + Z_r Y$$
$$(10\text{-}7)$$

Here, also, $AD - BC = 1$, as it must. However, A does not equal D because the network is not symmetrical unless $Z_s = Z_r$, which gives the following special case.

The Symmetrical T. To make the T symmetrical, let $Z_s = \frac{1}{2}Z$ and $Z_r = \frac{1}{2}Z$, so that the total series impedance of the T is $Z_s + Z_r = Z$. Equations 10-7 then tell us that for such a symmetrical T network the transmission parameters are:

$$A = 1 + \tfrac{1}{2}ZY \qquad B = Z + \tfrac{1}{4}Z^2 Y \qquad C = Y \qquad D = 1 + \tfrac{1}{2}ZY$$
$$(10\text{-}8)$$

Now it is indeed true that $A = D$.

Series impedance and shunt admittance are clearly special cases of a T. If $Y = 0$ in equations 10-8, the $ABCD$ values of equation 10-3 result. If $Z = 0$ in equations 10-8, equations 10-4 appear.

The L Network. The L network of Figure 1a is obtained by letting $Z_s = 0$ in the unsymmetrical T. Then $Z_r = Z$, and:

$$A = 1 \qquad B = Z \qquad C = Y \qquad D = 1 + ZY \qquad (10\text{-}9)$$

These parameters are for an L as shown, with the series impedance Z downstream from the shunt Y; for an L network with Z upstream, keep B and C but interchange A and D.

Parameters for a Π network could be computed by any of three methods that have now been shown but to illustrate another method the computation will be done on page 286 in Section 12.

Balanced Networks. An H network (Figure 1a) has the same $ABCD$ parameters as a T; it does not matter whether the series impedance is in the upper conductor or the lower conductor, or both, since it is stipulated that the terminals are in pairs that carry equal but opposite currents. Similarly, a square network is the same as a Π. The H and the square are called *balanced* networks if the upper and lower conductors have equal impedances and so can be electrically balanced to ground. This is important in avoiding *cross-talk* on telephone systems and for other such purposes.

11. CASCADED NETWORKS

One of the conveniences of the $ABCD$ parameters is the ease of finding overall parameters for several cascaded two-port nets. Figure 11a shows two cascaded networks, and one that is the equivalent of both. The transmission equations 10-1 and 10-2 are conveniently and neatly written in matrix form:

$$\begin{bmatrix} V_s \\ I_s \end{bmatrix} = \begin{bmatrix} A & B \\ C & D \end{bmatrix} \begin{bmatrix} V_r \\ I_r \end{bmatrix} \tag{11-1}$$

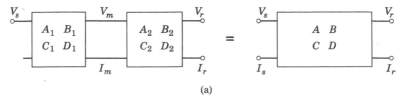

(a)

FIGURE 11a   **Equivalent of two cascaded networks.**

This, when interpreted by the rules of Section 10 of Chapter 9, clearly gives equations 10-1 and 2. To make our work even easier, let us use $[S]$ for the sending-end matrix, $[K]$ for the $ABCD$ matrix, and $[R]$ for the receiving-end matrix, writing equation 11-1 as

$$[S] = [K][R] \tag{11-2}$$

In this very abbreviated form, manipulation is quick and requires a minimum of writing.

Figure 11a shows two two-port networks cascaded. Let us represent the $ABCD$ matrix of the first by $[K_1]$ and that of the second by $[K_2]$. Let us represent the voltage-and-current matrix at the sending end by $[S]$, at the midpoint by $[M]$, and at the receiving end by $[R]$.

Now, looking at the right-hand half only,

$$[M] = [K_2][R] \tag{11-3}$$

Looking at the left-hand half,

$$[S] = [K_1][M] \tag{11-4}$$

Combining the two equations,

$$[S] = [K_1][K_2][R] \tag{11-5}$$

Thus we are able to relate sending-end voltage and current to receiving-end voltage and current by means of the matrix product $[K_1][K_2]$.

If the parameters for the two cascaded networks together are represented by the matrix $[K]$, then

$$[S] = [K][R] \qquad (11\text{-}6)$$

and, by comparison with equation 11-5,

$$[K] = [K_1][K_2] \qquad (11\text{-}7)$$

K_1 and K_2 being known, matrix multiplication gives $[K]$, and the components of $[K]$ are the overall A, B, C, and D as in equation 11-1.

If three or more networks are cascaded, the overall $[K]$ is the product of the individual matrices:

$$[K] = [K_1][K_2][K_3] \cdots \qquad (11\text{-}8)$$

12. THE SYMMETRICAL Π NETWORK

It is often easy to obtain the $ABCD$ parameters of an unknown network by cascading known networks. For instance, think of the Π network of Figure 12a as composed of two L networks, each having shunt admittance of $\frac{1}{2}Y$

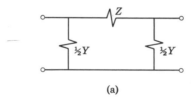

(a)

FIGURE 12a   A symmetrical Π network.

and series impedance of $\frac{1}{2}Z$. Let one of these face to the left and the other to the right. These two in cascade form the Π. For the upstream one, equation 10-9 gives

$$[K_1] = \begin{bmatrix} 1 & \frac{1}{2}Z \\ \frac{1}{2}Y & 1 + \frac{1}{4}ZY \end{bmatrix} \qquad (12\text{-}1)$$

For the downstream one (interchanging A and D):

$$[K_2] = \begin{bmatrix} 1 + \frac{1}{4}ZY & \frac{1}{2}Z \\ \frac{1}{2}Y & 1 \end{bmatrix} \qquad (12\text{-}2)$$

For the Π, formed by cascading the two:

$$[K] = [K_1][K_2] = \begin{bmatrix} 1 + \frac{1}{2}ZY & Z \\ Y + \frac{1}{4}ZY^2 & 1 + \frac{1}{2}ZY \end{bmatrix} \qquad (12\text{-}3)$$

Thus, *for the symmetrical* Π *of Figure 12a,*

$$A = 1 + \tfrac{1}{2}ZY \qquad B = Z \qquad C = Y + \tfrac{1}{4}ZY^2 \qquad D = 1 + \tfrac{1}{2}ZY$$

$$(12\text{-}4)$$

Note that A and D are equal for this symmetrical Π, though each of the components was unsymmetrical. Also, as always, $AD - BC = 1$.

13. INVERSE EQUATIONS

The general equations 10-1 and 10-2 give sending-end voltage and current in terms of receiving-end voltage and current. In matrix form these are equation 11-2,

$$[S] = [K][R] \tag{13-1}$$

where $[S]$ is the matrix of sending-end voltage and current, $[R]$ the matrix of receiving-end voltage and current, and $[K]$ the $ABCD$ matrix of the network.

It is just as likely that we shall want to know the receiving-end quantities in terms of the sending-end values, so let us express the inverse general equations. This is done by premultiplying each side of equation 13-1 by $[K]^{-1}$. The unit matrix results when $[K]$ is multiplied by $[K]^{-1}$, giving

$$[K]^{-1}[S] = [R] \quad \text{or} \quad [R] = [K]^{-1}[S] \tag{13-2}$$

This, then, is the desired result, but to be useful it is necessary to find $[K]^{-1}$.
$[K]^{-1}$ is obtained from equation 14-6 of Chapter 9:

$$[K]^{-1} = \frac{1}{\begin{vmatrix} A & B \\ C & D \end{vmatrix}} \begin{bmatrix} D & -B \\ -C & A \end{bmatrix} \tag{13-3}$$

By equations 7-7, the determinant is equal to 1, so $[K]^{-1}$ is just the matrix shown. Equations 13-2 can now be expanded to

$$\begin{bmatrix} V_r \\ I_r \end{bmatrix} = \begin{bmatrix} D & -B \\ -C & A \end{bmatrix} \begin{bmatrix} V_s \\ I_s \end{bmatrix} \tag{13-4}$$

or finally, if it is desired to write the equations out fully,

$$V_r = DV_s - BI_s \tag{13-5}$$

$$I_r = -CV_s + AI_s \tag{13-6}$$

These equations will be as useful as the original pair.

14. IMAGE IMPEDANCE

When information is being sent along a system of cascaded two-port networks that follow each other like links of a chain, the maximum power transfer theorem suggests that it would be well to match each network to the ones that precede and follow.

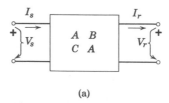

(a)

FIGURE 14a   A symmetrical, passive, linear network.

Figure 14a shows a symmetrical two-port net. It is characterized by the functions A, B, and C; since it is symmetrical, $D = A$. Thus:

$$V_s = AV_r + BI_r \qquad\qquad (14\text{-}1)$$

$$I_s = CV_r + AI_r \qquad\qquad (14\text{-}2)$$

The input impedance to this network is $z_1 = V_s/I_s$. The external impedance connected to the output terminals is $z_2 = V_r/I_r$. Obviously the input impedance depends on what is connected to the output terminals, and hence the value of z_1 depends on the value of z_2. Changing z_2 will result in changing z_1 also.

Is there some value of terminating impedance for which the input impedance will be equal to the terminating impedance? Can we adjust z_2 until $z_1 = z_2$? It will be shown that we can, and a special name is given to the impedance that makes this true. It is called the *image impedance*,[†] the symbol z_0 is used, and for a network terminated in its image impedance

$$z_1 = z_2 = z_0 \qquad\qquad (14\text{-}3)$$

[†] For a symmetrical network, image impedance and iterative impedance are the same. Since networks for carrying information are most commonly symmetrical, and since the mathematics of symmetrical networks is a good deal simpler, we shall here give attention mainly to symmetrical networks with asymmetrical networks mentioned from time to time in footnotes.

Iterative impedance is defined as "the impedance that, when connected to one pair of terminals, produces a *like* impedance at the other pair of terminals." Image impedances are: "the impedances that will simultaneously terminate ... inputs and outputs in such a way that at each ... the impedances in both directions are equal. *Note*: The image impedances of a four-terminal transducer are in general not equal to each other, but for any symmetrical transducer the image impedances are equal and are the same as the iterative impedances." (IEEE Standard Dictionary of Electrical and Electronics Terms, 1972.)

The value of z_0 can be found in terms of A, B, and C, which characterize the network. Dividing equation 14-1 by 14-2 gives z_1:

$$z_1 = \frac{V_s}{I_s} = \frac{AV_r + BI_r}{CV_r + AI_r} \tag{14-4}$$

Substituting $z_2 I_r$ for V_r,

$$z_1 = \frac{Az_2 I_r + BI_r}{Cz_2 I_r + AI_r} = \frac{Az_2 + B}{Cz_2 + A} \tag{14-5}$$

But, since z_1 and z_2 must be equal, and we are to call them z_0, the latter can be substituted for both z_1 and z_2 in equation 14-5:

$$z_0 = \frac{Az_0 + B}{Cz_0 + A} \tag{14-6}$$

We now solve for z_0, and the result is

$$z_0 = \sqrt{\frac{B}{C}} \tag{14-7}$$

This is the image impedance, expressed in terms of B and C; it is the value of terminating impedance that makes the input impedance equal the terminating impedance.

If B and C were known, z_0 could thus be found. Commonly, however, and conveniently, z_0 is expressed in terms of two impedances that are easier to measure or to calculate. One of these is z_{op}, the impedance looking into one port of the network in question when the other port is open circuited. But this is the impedance that was called, in previous sections, the open-circuit driving-point impedance of the network. Thus at port 1, $z_{op} = z_{11}$.

The other impedance that we shall now use is z_{sh}, the impedance looking into one port of the network in question when the other port is short circuited. But this is the reciprocal of the short-circuit driving-point admittance, and therefore (if measured at port 1) $z_{sh} = 1/y_{11}$.

Now we note from equation 14-7 that $z_0{}^2 = B/C$, and from equation 9-4 that $B/C = z_{11}/y_{11} = z_{op} z_{sh}$, and we combine these to obtain:

$$z_0 = \sqrt{z_{op} z_{sh}} \tag{14-8}$$

That is, the image impedance is simply the geometric mean, the square root of the product, of the open-circuit input impedance and the short-circuit input impedance.†

---

† A network that is *not* symmetrical has different image impedances at the two ends: $z_{01}{}^2 = z_{1(op)} z_{1(sh)}$ and $z_{02}{}^2 = z_{2(op)} z_{2(sh)}$. In terms of $ABCD$ functions, $z_{01}{}^2 = AB/CD$ and $z_{02}{}^2 = BD/AC$.

Note that z_0 is in general a function of frequency. The purely resistive network that is used for an example in the next section has a purely resistive image impedance that is independent of frequency, but this is not general. Hence the impedance of the load connected to the output terminals of a network cannot usually be equal to the image impedance of the network at more than one frequency. Over a band of frequencies (as, for example in a telephone channel) the equality is usually only approximate.

15. AN EXAMPLE OF IMAGE IMPEDANCE

A simple example will show that it is indeed possible to have the input impedance to a network equal the load impedance. The T network with 4 ohms in each branch which has been used so many times before is shown in Figure 15a. The problem is to find z_0. We use equation 14-8. Disconnect

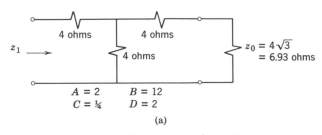

(a)

FIGURE 15a   A T network terminated in its z_0.

the load, and the impedance looking into the open-circuited T (at either end) is 8 ohms. Then short circuit one pair of terminals, and the impedance looking in at the other pair is 6 ohms. This gives:

$$z_0 = \sqrt{z_{op} z_{sh}} = \sqrt{(8)(6)} = 4\sqrt{3} = 6.93 \text{ ohms} \qquad (15\text{-}1)$$

Alternatively, this same result can be obtained by using equation 14-7.

To show that this is the right answer, we can compute the input impedance to the network of Figure 15a when complete with load. Series and parallel combinations give:

$$z_1 = 4 + \frac{(10.93)(4)}{(10.93) + (4)} = 4 + 2.93 = 6.93 \text{ ohms} \qquad (15\text{-}2)$$

Thus it comes out as we desired; the input impedance to the network is the same as the impedance of the load alone.

16. IMAGE IMPEDANCE MATCHING

Figure 16a shows a communication system consisting of a source, a load, and two intermediate two-port nets. If each network is symmetrical, with image impedance of z_0 at each pair of terminals, and load impedance and generator impedance are both equal to z_0, then the system is said to have impedances matched on an image-impedance basis. The impedance "looking in either direction" at any intermediate point is z_0, as indicated by the arrows.

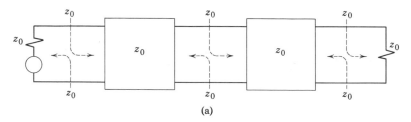

FIGURE 16a   A system matched on image-impedance basis.

It will be recalled (Section 12 of Chapter 10) that maximum power is received by a load if the impedance looking into the load is the conjugate of the impedance looking away from the load into the network. This is not the same as the condition for image-impedance matching *unless* the image impedance is purely resistive. But it is usual to design practical systems to have image impedance that *is* purely resistive, and the consequence is that a system matched on an image-impedance basis also provides maximum power transfer. So, in practice, some convenient value† of impedance level is chosen, and all cascaded networks are designed with this image impedance. In actual construction a perfect match is not necessary, and an approximate match is readily obtained.

Sometimes it is convenient to have different impedance levels in different parts of the same system. In a telephone system, for instance, terminal equipment may have 900 ohms as standard impedance whereas the line impedance is 600 ohms. Impedance-matching transformers can then be used (see Sections 4, 5, and 6 of Chapter 11). As an alternative means of matching, one of the two-port nets can be asymmetrical. The two-port net is then designed to match different impedances at its two ports and so serves as an impedance-matching network in addition to whatever other purpose it may have.

---

† The value is sometimes dictated by associated transmission lines, as 75 ohms for coaxial, 600 ohms for open-wire, 900 ohms for switching; see Skilling (3).

The basic need for maximum power transfer in every communication or signal system is probably the most persuasive reason for designing the system with image impedances matched. Matching on an image-impedance basis is not the only method of design that will give maximum power transfer, but it has a number of advantages including simplicity.

Since image impedance is in general a function of frequency, it is not ordinarily possible to design a system with perfect matching at more than one frequency. It is often desirable to have good matching over a narrow or even over a wide band of frequencies, but the designer must be satisfied with some degree of approximation.

17. IMAGE TRANSFER FUNCTION

As a sinusoidal signal passes from the input terminals to the output terminals of a two-port network it loses energy and changes phase angle. The amount of change can be found from the $ABCD$ functions, and especially easily if the network is terminated in its image impedance. Then

$$V_{in} = AV_{out} + BI_{out} = AV_{out} + B\frac{V_{out}}{z_0} \tag{17-1}$$

Substituting $\sqrt{B/C}$ for z_0 (equation 14-7) we obtain the transfer function:

$$\frac{V_{in}}{V_{out}} = A + \sqrt{BC} \tag{17-2}$$

In general, $BC = AD - 1$. Let us require our network to be symmetrical, with $A = D$; then $BC = A^2 - 1$, and

$$\frac{V_{in}}{V_{out}} = A + \sqrt{A^2 - 1} \tag{17-3}$$

Here we have the transfer function for this symmetrical two-port net terminated in its image impedance, given in terms of the parameter A alone. It is in general a complex number, giving change of both magnitude and angle. Note that only one parameter is necessary when symmetry and termination are both specified.

Equation 17-3 is good for actual computation provided A is easily found. We have considered several ways of computing A, and perhaps the easiest is to note that $A = V_{in}/V_{out}$ for a network when that network is open-circuited so that I_{out} is forced to be zero (see equation 17-1).

A numerical illustration comes from the T network of Figure 15a that has 4 ohms in each branch. We know that $A = 2$, and this can indeed be seen from the diagram. By equation 17-3, then,

$$\frac{V_{in}}{V_{out}} = 2 + \sqrt{4 - 1} = 3.73 \qquad (17\text{-}4)$$

Thus the input voltage is 3.73 times the output voltage in magnitude and the phase difference is zero. Because of image-impedance termination, the input current is also 3.73 times the output current.

It is easy to check the answer. Let the output voltage be 1 volt. The output current is then $1/6.93 = 1/(4\sqrt{3})$ amperes. Using the ladder method, voltage across the 4-ohm shunt at the midpoint is $1 + 1/\sqrt{3}$. Input current is $1/4 + 1/(2\sqrt{3})$. Input voltage is then $2 + \sqrt{3}$, and since this equals 3.73 the ladder method gives the same answer as equation 17-4.

18. PROPAGATION

When a signal is propagated along a channel of many cascaded two-port networks, there is a certain analogy to propagation along a transmission line. This concept is used in speaking of a *propagation function*.

For a symmetrical two-port net of lumped parameters, let us arbitrarily write

$$\frac{V_{in}}{V_{out}} = e^{\gamma} \qquad (18\text{-}1)$$

in which we call γ the propagation function. Now two trigonometric identities are used; these are borrowed from pure mathematics, and can easily be derived or found in handbooks (e.g., Eshbach, Rybner) if they are not familiar. First, $e^{\gamma} = \cosh \gamma + \sinh \gamma$. Also, $\sinh \gamma = \sqrt{\cosh^2 \gamma - 1}$. These we use in equation 18-1 to get:

$$\frac{V_{in}}{V_{out}} = e^{\gamma} = \cosh \gamma + \sinh \gamma = \cosh \gamma + \sqrt{\cosh^2 \gamma - 1} \qquad (18\text{-}2)$$

Now we compare this equation with 17-3, and find that

$$\cosh \gamma = A \quad \text{or} \quad \gamma = \text{inv cosh } A \qquad (18\text{-}3)$$

Here we might stop, and use equation 18-3 to compute γ from A. But perhaps we do not know A. Then equation 18-3 can be changed to

$$\tanh \gamma = \frac{\sinh \gamma}{\cosh \gamma} = \frac{\sqrt{A^2 - 1}}{A} = \frac{\sqrt{BC}}{A} \qquad (18\text{-}4)$$

Now from equation 9-4, $BC/A^2 = 1/z_{11} y_{11}$, and in Section 14 we see that $1/z_{11} y_{11} = z_{sh}/z_{op}$, so finally equation 18-4 becomes:

$$\tanh \gamma = \sqrt{\frac{z_{sh}}{z_{op}}} \quad \text{or} \quad \gamma = \text{inv tanh } \sqrt{\frac{z_{sh}}{z_{op}}} \qquad (18\text{-}5)$$

Thus the propagation function can be expressed in terms of the open and short-circuit impedances of a network.† In general the propagation function γ will be a complex quantity.‡ (See Skilling (3) regarding the use of γ with transmission lines.)

If we try this on the familiar T network of Figure 15a, we find (as we have done before) that $z_{op} = 8$ ohms and $z_{sh} = 6$ ohms. Then $\tanh \gamma = \sqrt{6/8} = \frac{1}{2}\sqrt{3} = 0.866$, and from a table of hyperbolic functions, $\gamma = 1.32$. Here we have the propagation constant for this simple resistive network, and if we want the transfer function V_s/V_r, it is, by equation 18-1, $e^{1.32}$; which we find from a table of exponentials to be 3.73.

Actually, we have a check on this result, for the same answer was obtained by a much easier method in equation 17-4.

19. CURRENT RATIO

Equations 17-3 and 18-1 give the voltage ratio between input and output terminals. Since voltage and current are related at each pair of terminals by the same impedance z_0, they will necessarily be in the same ratio, so

$$\frac{I_{in}}{I_{out}} = \frac{V_{in}}{V_{out}} = A + \sqrt{A^2 - 1} = e^{\gamma} \tag{19-1}$$

(The definitions given by the IEEE are based on *current* ratios.)

---

† The derivation as given is for a symmetrical two-port network. For an *asymmetrical* network,

$$\frac{V_{in}}{V_{out}} = \sqrt{\frac{z_{01}}{z_{02}}} e^{\gamma} \tag{18-6}$$

instead of equation 18-1, and γ must be computed from equation 18-5, or from $\cosh \gamma = \sqrt{AD}$.
‡ Since the radical of equation 18-5 may in general be a complex quantity, it follows that γ will in general be complex also. To find complex γ from equation 18-5, trigonometry gives the following method (Eshbach's *Handbook*). Since γ is a complex quantity, it may be written $\gamma = \alpha + j\beta$. Then, if $\tanh \gamma = \tanh(\alpha + j\beta) = p + jq$, we can express α and β in terms of p and q as follows:

$$\alpha = \frac{1}{2} \tanh^{-1} \frac{2p}{1 + p^2 + q^2} = 0.5756 \log_{10} \frac{(1+p)^2 + q^2}{(1-p)^2 + q^2}$$

$$\beta = \frac{1}{2} \tan^{-1} \frac{2q}{1 - p^2 + q^2} \tag{18-7}$$

The attenuation α can now be computed with the aid of a table of hyperbolic tangents or, perhaps more readily available, a table of common logarithms; and β, the phase angle, requires only a table of ordinary tangents. To avoid spurious values for β, however, it is necessary to note that if $p^2 + q^2 < 1$, the angle β can only have values that will make $\cos 2\beta$ positive, whereas if $p^2 + q^2 > 1$, $\cos 2\beta$ must be negative (see Rybner).

20. ATTENUATION AND PHASE FUNCTIONS

The propagation function γ is in general a complex quantity, and we can write it with real and imaginary components:

$$\gamma = \alpha + j\beta \qquad (20\text{-}1)$$

Then equation 19-1 becomes:

$$\frac{V_{\text{in}}}{V_{\text{out}}} = \frac{I_{\text{in}}}{I_{\text{out}}} = e^{\gamma} = e^{\alpha + j\beta} = e^{\alpha}e^{j\beta} = e^{\alpha}\underline{/\beta} \qquad (20\text{-}2)$$

By definition, α and β are real, e^{α} is real, and by equation 20-2:

$$e^{\alpha} = \left|\frac{V_{\text{in}}}{V_{\text{out}}}\right| = \left|\frac{I_{\text{in}}}{I_{\text{out}}}\right| \qquad (20\text{-}3)$$

Thus α shows the change of magnitude between input and output signals and is appropriately called the *attenuation function* (it is, in general, a function of frequency). A large α corresponds to a large loss of signal strength in the network. The phase difference between input and output signals is β, and β is called the *phase function*.

It is possible for β to be zero. An example is seen in the purely resistive network of Section 18 in which the network attenuates an applied signal; the input and output signals are in phase with each other.

It is possible for α to be zero. A network of purely reactive elements, for instance, may give an output signal of exactly the same magnitude as the input signal, but with a phase difference of β.

21. NEPERS

Like γ, α and β are dimensionless numbers; yet it is convenient to assign names to the units in which they are measured. Since β is interpreted as angle, its unit is naturally called the *radian*. In the final form of equation 20-2 (although never in the exponential form), β may be expressed in *degrees*. Whether in degrees or radians, β is a dimensionless value, angle being basically the ratio of two distances.

The unit of α is given a similar dimensionless name, the *neper*.[†] It is convenient to speak of attenuation of so many nepers. It is the natural logarithm of the magnitude of the voltage or current ratio:

$$\alpha = \ln\left|\frac{V_{\text{in}}}{V_{\text{out}}}\right| = \ln\left|\frac{I_{\text{in}}}{I_{\text{out}}}\right| \qquad (21\text{-}1)$$

[†] From the Latin form, *Neperus*, of Napier, the Scottish mathematician (sixteenth century) whose name is connected with logarithms.

Attenuation in nepers is a measure of the ratio of voltages or currents, under certain specified conditions of propagation. Another unit of attenuation which is called the decibel is defined to measure the ratio of values of power.

22. POWER RATIO

A unit of attenuation called the bel was invented some years ago. (Its immediate predecessor was the telephone man's "miles of standard cable.") It was defined as follows:

$$\text{Attenuation in bels} = \log_{10}\frac{P_{in}}{P_{out}} \tag{22-1}$$

A more practical unit of attenuation was one-tenth as much, and the decibel, abbreviated db, is now commonly used:

$$\text{Attenuation in decibels} = 10 \log_{10}\frac{P_{in}}{P_{out}} \tag{22-2}$$

This decibel measure of a power ratio is by no means limited to attenuation; the ratio of any two quantities of power may be so measured. It is quite as common to measure gain in db as to measure loss or attenuation.

In any cascaded system the gain (or loss) of the whole system is the sum of the separate gains (or losses) of the cascaded sections; adding the db gains (and subtracting the db losses) gives the total system gain.

23. DECIBELS

Table 12-23 gives power ratios corresponding to decibel values. It will be found convenient to remember a few of these. The whole system is based on the relation:

$$10 \text{ db gives a power ratio of } 10$$
$$10 \cdot n \text{ db gives a power ratio of } 10^n$$

from which

$$20 \text{ db gives a power ratio of } 100$$
$$30 \text{ db gives a power ratio of } 1{,}000$$
$$60 \text{ db gives a power ratio of } 1{,}000{,}000$$

and so on. For smaller differences:

$$1 \text{ db gives a power ratio of about } \tfrac{5}{4}$$
$$3 \text{ db gives a power ratio of about } 2$$
$$7 \text{ db gives a power ratio of about } 5$$

TABLE 12-23 Power Ratio in Decibels

Deci-bels	Power Ratio		Deci-bels	Power Ratio		Deci-bels	Power Ratio	
	Gain	Loss		Gain	Loss		Gain	Loss
0.1	1.02	.977	3.6	2.29	.437	7.1	5.13	.195
0.2	1.05	.955	3.7	2.35	.427	7.2	5.25	.191
0.3	1.07	.933	3.8	2.40	.417	7.3	5.37	.186
0.4	1.10	.912	3.9	2.45	.407	7.4	5.50	.182
0.5	1.12	.891	4.0	2.51	.398	7.5	5.62	.178
0.6	1.15	.871	4.1	2.57	.389	7.6	5.76	.174
0.7	1.17	.851	4.2	2.63	.380	7.7	5.89	.170
0.8	1.20	.832	4.3	2.69	.372	7.8	6.03	.166
0.9	1.23	.813	4.4	2.75	.363	7.9	6.17	.162
1.0	1.26	.794	4.5	2.82	.354	8.0	6.31	.159
1.1	1.29	.776	4.6	2.88	.347	8.1	6.46	.155
1.2	1.32	.759	4.7	2.95	.339	8.2	6.61	.151
1.3	1.35	.741	4.8	3.02	.331	8.3	6.76	.148
1.4	1.38	.724	4.9	3.09	.324	8.4	6.92	.145
1.5	1.41	.708	5.0	3.16	.316	8.5	7.08	.141
1.6	1.44	.692	5.1	3.24	.309	8.6	7.25	.138
1.7	1.48	.676	5.2	3.31	.302	8.7	7.41	.135
1.8	1.51	.661	5.3	3.39	.295	8.8	7.59	.132
1.9	1.55	.646	5.4	3.47	.288	8.9	7.76	.129
2.0	1.58	.631	5.5	3.55	.282	9.0	7.94	.126
2.1	1.62	.617	5.6	3.63	.275	9.1	8.13	.123
2.2	1.66	.603	5.7	3.72	.269	9.2	8.32	.120
2.3	1.70	.589	5.8	3.80	.263	9.3	8.51	.118
2.4	1.74	.575	5.9	3.89	.257	9.4	8.71	.115
2.5	1.78	.562	6.0	3.98	.251	9.5	8.91	.112
2.6	1.82	.550	6.1	4.07	.246	9.6	9.12	.110
2.7	1.86	.537	6.2	4.17	.240	9.7	9.33	.107
2.8	1.91	.525	6.3	4.27	.234	9.8	9.55	.105
2.9	1.95	.513	6.4	4.37	.229	9.9	9.77	.102
3.0	1.99	.501	6.5	4.47	.224	10.0	10.00	.100
3.1	2.04	.490	6.6	4.57	.219			
3.2	2.09	.479	6.7	4.68	.214			
3.3	2.14	.468	6.8	4.79	.209			
3.4	2.19	.457	6.9	4.90	.204			
3.5	2.24	.447	7.0	5.01	.200			

For larger db values, add db and multiply ratios. Thus for 36.5 db:

Read: 6.5 db, ratio is 4.47
add: 30.0 db, times 10^3

Hence: 36.5 db, ratio is 4470.

If only these are remembered, other values can be estimated as needed by adding decibels and multiplying power ratios. Thus 4 db equals 3 + 1 db, giving a power ratio of $2 \times \frac{5}{4}$ or $\frac{5}{2}$. Also, 27 db equals 20 + 7 db, giving a power ratio of 100×5 or (very nearly) 500.

24. VOLTAGE AND CURRENT RATIOS IN DECIBELS

Any two values of power can be compared by using the decibel expression:

$$Db = 10 \log \frac{P_1}{P_2} \tag{24-1}$$

It is often heard that since

$$P_1 = |I_1{}^2|R_1 \qquad \text{and} \qquad P_2 = |I_2{}^2|R_2 \tag{24-2}$$

we can therefore make the same comparison by using the two values of current in the following expression:

$$Db = 10 \log \frac{P_1}{P_2} = 10 \log \frac{|I_1{}^2|R_1}{|I_2{}^2|R_2} = 20 \log \frac{|I_1|}{|I_2|} = 20 \log \frac{|V_1|}{|V_2|} \tag{24-3}$$

This, however, is true only if $R_1 = R_2$. That is, equation 24-3 is valid if the current I_2 feeds into the same resistance as does the current I_1, or into an impedance with the same resistive component.

If I_1 and I_2 are the input and output currents of a symmetrical two-port net, equation 24-3 is valid if the load connected to the *output* terminals is the image impedance z_0, for then the input impedance to the net is also z_0 (or, for an asymmetrical net, the load would have to be the iterative impedance). If the net is terminated in any other way, the equation cannot be expected to hold.

Since equation 24-3 gives attenuation in decibels and equation 21-1 gives attenuation in nepers, both being in terms of current ratios, and under identical conditions of loading, it follows that these expressions give a relation between decibels and nepers. The result is:

$$\text{Attenuation in db} = 8.686 \times \text{attenuation in nepers} \tag{24-5}$$

Image-impedance (or iterative-impedance) loading permits equation 24-5 to be valid; more precisely, its limits of validity are those of equation 24-3.

As a practical matter, it is much more usual to specify attenuation in decibels than in nepers.

25. A COMPLEX EXAMPLE

A number of the concepts of this chapter have been illustrated by using as an example a T network of resistive elements. This simple example has served its purpose well, but the real world is not so simple. The various impedances and admittances, the $ABCD$ parameters, the transfer functions, the image impedance, and the propagation function are in general complex quantities. They are also, in general, functions of the frequency variable s. This uncomfortably complicated but usual situation is illustrated by the network of Figure 25a.

Here a reactive network is feeding into a load of resistance R. This reactive network is, indeed, a rather simple filter circuit,† and the following analysis shows what may be expected of it. The reactive network has the form of a T, a symmetrical T, and since this is a filter section the obvious problem is to find how the output of this reactive section compares with its input as a function of frequency. This comparison of V_{out}/V_{in} is given by the transfer function.

A convenient way to find V_{out}/V_{in} for this filter section is to begin by writing, in terms of the transmission parameters,

$$V_{in} = A V_{out} + B I_{out} \qquad (25\text{-}1)$$

I_{out} is now written in terms of the load resistance:

$$V_{in} = A V_{out} + B \frac{V_{out}}{R} \qquad (25\text{-}2)$$

Note that R is *not* equal to z_0 (except at one frequency) and hence the simpler equations of Sections 14 to 18 do not apply, except perhaps as approximations. Now the desired transfer function is

$$\frac{V_{out}}{V_{in}} = \frac{R}{AR + B} \qquad (25\text{-}3)$$

This is the transfer function that will tell us what we want to know about the filter, but it cannot be useful until we know A and B. To determine A and B, we note that the reactive filter section is a symmetrical T section and we have already obtained formulas for A and B in equations 10-8. For the network of Figure 25a, the Z and Y of the formulas are:

$$Y = \frac{1}{sL_b} \qquad (25\text{-}4)$$

---

† A chapter on filter circuits is given in Electrical Engineering Circuits, Skilling (7).

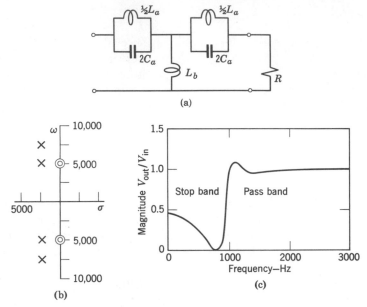

FIGURE 25   **(a)** $L_a = 0.2085$ henry; $L_b = 0.093$ henry; $C_a = 0.189$ microfarad; $R = 700$ ohms.   **(b)**   **Poles and zeros for filter section of Figure 25a.**   **(c)**   **Magnitude of the transfer function for the filter section of Figure 25a.**

and, taking twice the impedance of the elements $\frac{1}{2}L_a$ and $2C_a$ in parallel,

$$Z = \frac{sL_a}{s^2 L_a C_a + 1} \tag{25-5}$$

Hence, using A and B from equations 10-8, the denominator in the transfer function of equation 25-3 is:

$$AR + B = \left[1 + \frac{\frac{1}{2}(L_a/L_b)}{s^2 L_a C_a + 1}\right]R + \left[\frac{sL_a}{s^2 L_a C_a + 1}\left(1 + \frac{1}{4}\frac{L_a/L_b}{(s^2 L_a C_a + 1)}\right)\right] \tag{25-6}$$

With this denominator inserted, the transfer function of equation 25-3 can be written as a rational function, the ratio of two polynomials in s, and factoring of these polynomials puts the transfer function into the standard form of equation 13-1 of Chapter 5:

$$\frac{V_{out}}{V_{in}} = H\frac{(s - s_a)(s - s_b)(s - s_c)(s - s_d)}{(s - s_1)(s - s_2)(s - s_3)(s - s_4)} \tag{25-7}$$

The factoring that is here indicated is not easy; indeed, it is much the hardest part of the solution, and it is surely a job for an electronic computer if one is available. Both numerator and denominator are quartic functions. This means that the transfer function has four zeros, at s equals s_a, s_b, s_c, and s_d, and four poles, at s equals s_1, s_2, s_3, and s_4. When these zeros and poles are known, they can easily be plotted in the complex s plane, and this will be done in Figure 25b, but first there are certain general relations to observe.

It is easy to locate the zeros of the transfer function. These occur where the fraction of equation 25-3 is zero, and hence where either A or B is infinite. But A is the first bracket in equation 25-6 and B is the second bracket; either of these is infinite when

$$s^2 L_a C_a + 1 = 0 \qquad (25\text{-}8)$$

and hence the zeros of the transfer function are at

$$s = \pm j \frac{1}{\sqrt{L_a C_a}} \qquad (25\text{-}9)$$

Because the numerator of the transfer function contains the square of equation 25-8, it follows that there are second-order zeros at each of the roots of equation 25-9; hence the roots of this equation give all four of the zeros of the transfer function which have been called s_a, s_b, s_c, and s_d. These zeros, which come out to be at $\pm j5037$ when the numerical values given under Figure 25a are used, can now be plotted in Figure 25b.

Next, and not so easily, the poles are to be found. Poles are at values of s for which the transfer function is infinite, and hence where $AR + B = 0$. Setting equation 25-6 equal to zero and solving for s we obtain the positions of the poles. There turn out to be four poles, as of course there must because equation 25-6 is quartic, and they are in two conjugate complex pairs, as shown in Figure 25b.

The statement of the problem asks for V_{out}/V_{in} to show how well the filter section works. Such a transfer function is of course a function of frequency. We already know that if the input frequency is steady at $\omega = 5037$ radians per second (this is 800 Hz) the output is zero, for equation 25-3 gives zero for $s = \pm j5037$. Other numerical values of the transfer function are likewise found for other values of frequency by entering imaginary values of s into equation 25-3. The resulting values of the transfer function range from zero at 800 Hz to slightly over 1 at very high frequency, as shown in Figure 25c. Here we have the solution of our problem.

Two limiting values of the transfer function are easily verified; these are the values at high and low frequencies. Let us consider the behavior of the circuit of Figure 25a at infinite frequency; then the capacitors can be considered short circuits, and the inductances open, and whatever voltage is applied at the input is transmitted to the load R. Thus as frequency becomes high without limit, V_{out}/V_{in} approaches 1, as shown in Figure 25c. This fact can be verified mathematically by letting $s = j\infty$ in equation 25-6, whereupon $AR + B = R$, and in equation 25-3 the transfer function equals 1.

The other limiting value is found when ω approaches zero. Letting $s = 0$ in equation 25-6, and using the result in equation 25-3:

$$\frac{V_{out}}{V_{in}} = \frac{L_b}{L_b + \frac{1}{2}L_a} = 0.471 \tag{25-10}$$

This result also can be reached from Figure 25a, for if the frequency is approaching zero there is no appreciable current through the capacitors; hence input voltage divides between L_b and $\frac{1}{2}L_a$ in proportion to their inductances. The voltage across L_b, which is $V_{in}L_b/(L_b + \frac{1}{2}L_a)$, also appears across R because the inductive branch in series with R is practically a short circuit (compared with R); hence equation 25-10 appears from this reasoning also. By whichever means the result is obtained, the load voltage is 0.471 times the applied voltage at zero frequency.

Let us now return to a comparison of the pole-zero plot of Figure 25b with the curve in Figure 25c of the magnitude of the transfer function. Figure 25b suggests a rubber sheet above a table top (the analogy of Chapter 5, Section 14). The sheet is held down to the table top by tacks at the zeros, and is supported on tall vertical rods at each pole. The sheet is held 1 unit above the table top at infinity in all directions. The vertical distance from table top to rubber sheet along the ω axis, is then a rough indication of the magnitude of the transfer function and should suggest Figure 25c.

The analogy is reasonably good. The second-order zero at 800 Hz is evident. At lower frequency the rubber sheet is above the table top being supported by the nearer pole. Both poles are effective in causing an abrupt rise of the sheet at frequencies just above 800 Hz (where $\omega = 5037$ rn/sec). Then, supported by the upper pole, and retained by the fact that the height of the sheet is 1 at infinite frequency, the magnitude of the function remains near 1 at all high frequencies. Just above 1000 Hz ($\omega = 6280$) the amplitude of the transfer function rises a bit above 1, lifted by the nearer pole; it then falls again, rises again where the upper pole exerts more influence than the second-order zero, and finally approaches 1 at high frequencies. The reactive section is a high-pass filter.

The numerical values of Figure 25c are computed from equation 25-3 with the aid of 25-6. The pole-zero plot helps greatly with the understanding. The angle of the transfer function is given by the computation, as well as the magnitude, for the transfer function is a complex quantity.

26. PHYSICAL MEANING OF ZEROS AND POLES

One naturally asks, either here or earlier, whether the zeros and poles of a function have any physical meaning. Can you look at a network diagram and tell where a function of the network will have a zero or a pole? Sometimes you can.

Let us consider the transfer function V_{out}/V_{in} of a two-port network such as that of Figure 26a. For this example, there is a zero of the function at the

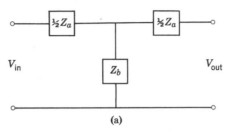

(a)

FIGURE 26a   There is a zero if $V_{out} = 0$ when $V_{in} \neq 0$, or a pole if $V_{out} \neq 0$ when $V_{in} = 0$.

value of s for which $V_{out} = 0$ when V_{in} has some finite value. Looking at Figure 26a, there are two arrangements of impedance that provide this condition:

1. Either the series impedance Z_a is infinite and the shunt impedance Z_b is not infinite,
2. Or the shunt impedance Z_b is zero and the series impedance Z_a is not zero.

To be specific, let us consider the filter network of Figure 25a. At what frequency is there no output to the load R even though voltage is applied to the input terminals? This happens when the parallel circuits in the series arms are resonant, for then the impedance of each series arm is infinite. Since at this frequency the shunt impedance ωL_b is not infinite, it is impossible for any voltage to reach the output terminals. Hence at this frequency, at this value of s, the transfer function has a *zero*. This agrees with the mathematical determination of the preceding section.

Note that the zeros fall directly on the imaginary axis of the s plane. This is because the filter network is assumed to be purely reactive; if any small resistance had been considered in the parallel circuits of the filter arms, the zeros would have appeared slightly off the ω axis in the left half-plane. Then the filter would not have exactly zero output for constant voltage of any frequency; the output would be exactly zero only for a voltage of slightly damped sinusoidal form. However (remembering the rubber sheet) there would be some value of steady alternating voltage for which the output would be very nearly zero.

Next, what of the poles? At a pole there must be output when there is no input, or infinite output when there is only finite input. Can this happen? Yes, it is related to resonance.

Let us again look at the network of Figure 25a. Let the input terminals be short circuited together, thus providing zero input voltage. What kind of current can now flow through the load; what kind of current, that is, can continue assuming that it is once started? The answer is that there are two modes of resonance for which this is possible, the resulting currents being damped sinusoids, corresponding to two conjugate pairs of poles. This is an interesting concept, but it does not in the least help our computations.

Zeros are often easy to recognize. Poles are harder.

27. SUMMARY

A *two-port network* has two pairs of terminals, and power or signals enter at one pair and leave at the other. The two terminal pairs, or two ports, are distinguished by *subscripts* 1 and 2, or *s* and *r*, or *in* and *out*.

Two-port networks are so widely used, both for power transmission and in communication channels, that it is important to find *parameters* to express their behavior conveniently. Three sets of parameters are derived in this chapter.

The *short-circuit driving point and transfer admittances* are useful for certain problems, and are easily obtained from the familiar loop equations of a network.

Also, the *open-circuit driving-point and transfer impedances* can be useful. These are obtained from the familiar node equations.†

The *transmission parameters*, or *general network functions*, or *ABCD* functions, give voltage and current at one port in terms of voltage and current at the other port. This is often a most useful form of expression. The *ABCD* parameters can be determined from the foregoing admittances and impedances, but it is often more convenient to compute them directly.

The *ABCD* parameters are computed in terms of impedances for several simple two-port *nets in common use*, including series impedance, shunt admittance, T networks and Π networks.

It is shown that always $AD - BC = 1$. For symmetrical networks, $A = D$. Thus an asymmetrical network has three *independent parameters*; a symmetrical network only two.

---

† In addition to those listed, certain *hybrid* sets can be derived that are useful.

Equations are written to give *receiving-end* voltage and current in terms of *sending-end* voltage and current.

Parameters for *cascaded networks* are computed.

If *image impedance* is connected at one port of a symmetrical network the input impedance at the other port is equal to the image impedance. (For an *asymmetrical* network, iterative impedance, not image impedance, is required.)

Communication networks are commonly *matched* on an *image impedance* basis. Useful parameters are the image impedance and the *image transfer function.*

The concept of signal propagation along a channel of cascaded two-port nets leads to the *propagation function* and the *attenuation and phase functions.*

Attenuation along a channel with image-impedance matching can be expressed in *nepers.*

Power ratios can be expressed in *decibels.* Power loss (or gain) in decibels $= 10 \log_{10} (P_{in}/P_{out})$.

Along a channel with image-impedance matching the same attenuation can be computed in decibels from a *voltage ratio* or a *current ratio*: Power loss (or gain) in decibels $= 20 \log_{10} (V_{in}/V_{out}) = 20 \log_{10} (I_{in}/I_{out})$. The voltage ratio or the current ratio can replace the power ratio and give the same decibel value only if the systems compared are feeding into equivalent *loads.*

Computation with *complex parameters* is illustrated in an example. A simple filter network with reactive elements is used; it is a symmetrical T network and the parameters are *complex functions of frequency.* Its transfer function is found, and the *poles and zeros* of the function are located in the complex s plane. It is concluded that the network is a fairly effective high-pass *filter.*

The physical meaning of *zeros and poles* is considered. It is fairly easy to recognize the physical conditions that provide zeros; more difficult to recognize poles.

PROBLEMS

12-1. Equations 6-3 give z_{11} and z_{12} of a network in terms of the y's of that network. Find y_{11}, y_{12}, and y_{22} (short-circuit admittances) of the network in terms of the z's (open-circuit impedances). §6

12-2. Equations 6-3 give relations between z_{11} and z_{12} of a network and the y's of that network. Find similar relations between z_{22} and z_{21} (open-circuit impedances) of the network and its y's (short-circuit admittances). §6

12-3. (a) The figure shows a two-port net in which a series resistance is 5 ohms and a shunt resistance is 3 ohms. Compute y_{11}, y_{22}, and y_{12}. (b) For the network

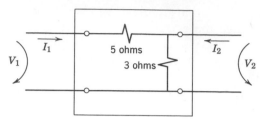

Problems 12-3 through 12-7.

shown, compute z_{11}, z_{22}, and z_{12}. (*c*) Show that these z's and y's are in agreement with equations 6-3. $§6*$

12-4. For the network of Problem 12-3, and using values computed in that problem, show that the general relations between y's and z's, the relations found in Problems 12-1 and 12-2, are satisfied. $§6$

12-5. In the network shown, use the parameters computed in Problem 12-3, to answer the following questions. (*a*) If V_1 and V_2 are each 10 volts, find I_1; find I_2. (*b*) If $V_1 = 10$ and $V_2 = 5$ volts, find I_1; find I_2. (*c*) If I_1 and I_2 are each 5 amperes, find V_1; find V_2. (*d*) If $I_1 = 5$ and $I_2 = -5$ amperes, find V_1 and V_2. How can part (*d*) be achieved, physically? $§6*$

12-6. Change the network so that the element shown in the figure as a 3-ohm resistance becomes, instead, a 3-henry inductance. The 5-ohm resistance is unchanged. Compute y_{11}, y_{22}, and y_{12}, each being a function of s. $§6$

12-7. Using the y's computed in Problem 12-6, find current in each battery if a 10-volt battery is suddenly connected (at $t = 0$) to the terminals marked V_1, and a 5-volt battery is connected at the same instant (at $t = 0$) to the terminals marked V_2. (Each applied voltage has a positive polarity.) Use any method desired. $§6*$

12-8. Figure 12a shows a symmetrical Π network. $Z = 25$ ohms and $Y = 0.02$ mho, both purely resistive. (*a*) Find short-circuit driving-point admittance and short-circuit transfer admittance. (*b*) Find open-circuit driving-point impedance and open-circuit transfer impedance. (*c*) Show that these satisfy equations 6-3. (Considering the reference arrows of Figure 2a, be specially careful of the signs of transfer functions.) $§6*$

12-9. Figure 10c shows a T network. $Z_s = 12$ ohms, $Z_r = 15$ ohms, and $Y = 0.02$ mho, all purely resistive. (*a*) Find each short-circuit driving-point admittance and the short-circuit transfer admittance. (*b*) Find each open-circuit driving-point impedance and the open-circuit transfer impedance. (*c*) Show that these satisfy equations 6-3. (Considering the reference arrows of Figure 2a, be specially careful of the signs of transfer functions.) $§6$

12-10. In the T network of Figure 10c, $Z_s = 10 + j\omega3 \cdot 10^{-3}$, $Z_r = 5 + j\omega2 \cdot 10^{-3}$, and the shunt element marked Y has an *impedance* of $10 + j\omega5 \cdot 10^{-3}$ ohm. Find the short-circuit transfer admittance, writing it as a rational function of linear factors, giving the numerical values of poles and zeros, and show the locations of poles and zeros in a sketch of the complex s plane. $§6*$

12-11. Equations 10-7 give the transmission parameters for an unsymmetrical T network. Find similar $ABCD$ parameters for an unsymmetrical Π. Does $A = D$? Does $AD - BC = 1$? Do these parameters reduce to the symmetrical-network parameters of equations 12-4? §10*

12-12. Each branch of a T network is 4 ohms of resistance. Each branch of a Π network is 12 ohms of resistance. Find the transmission parameters, $ABCD$, of these two networks cascaded. Is the combination of the two cascaded an electrically symmetrical network? §12*

12-13. For the Π network shown, (a) find image impedance z_0 at $\omega = 10^6$, at $\omega = 10^5$, and at $\omega = 10^3$; (b) compute input impedance when this Π is terminated in z_0, using $\omega = 10^6$ (only) for this question. §15*

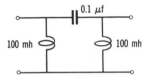

Problems 12-13 and 12-21.

12-14. Find z_{01} and z_{02} for the bridge circuit shown. Redraw the circuit as a lattice. §15*

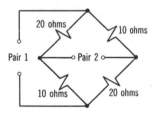

Problems 12-14 and 12-15.

12-15. In the circuit of Problem 12-14 connect image impedance z_{02} at terminal pair 2, and verify the value by computing input impedance to terminal pair 1. §15

12-16. Find the image impedance of the network shown. How does it vary with frequency? What is the name of this type of network? §15

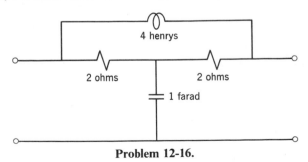

Problem 12-16.

12-17. In an L network (see Figure 1a) the series element is $L = 0.045$ henry; the shunt element is $C = 0.1875 \cdot 10^6$ farad. Compute both image impedances, z_{01} and z_{02}. (a) Let $f = 1000$ Hz, and (b) let $f = 500$ Hz. Note: This L is a matching network for 1000-Hz use.   §16*

12-18. For the network shown, find (a) image impedance at $\omega = 1$, and (b) the transfer function V_{in}/V_{out} with image-impedance termination. Find also, (c) image impedance and (d) image transfer function at $\omega = 1/2$.   §17*

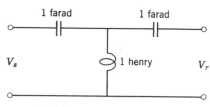

Problems 12-18 and 12-19.

12-19. Verify that your two image impedances of Problem 12-18 are correct, using the ladder method of Chapter 4. Draw a phasor diagram at each frequency, showing all voltages and currents. From these ladder results, compute V_{in}/V_{out} at each frequency.   §17

12-20. Problem 12-12 considers a network of T and Π cascaded. Terminate this network in its image impedance and find its image transfer function V_{in}/V_{out}. Find α and β. Find the ratio of power in to power out in decibels.   §23

12-21. For the network of Problem 12-13, terminated in its image impedance, find α and β at $\omega = 10^6$, at $\omega = 10^5$, and at $\omega = 10^3$. At each frequency, give attenuation in decibels.   §24*

12-22. In view of the discussion of physical meaning in Section 26, how do you account for the number of poles and zeros in the filter network of Figure 25a?
                  §26

CHAPTER
13

THE FOURIER SERIES AND INTEGRAL

1. NONSINUSOIDAL WAVES

There are several reasons for the special attention that we have given to sinusoidal ac waves. Most voltage and current waves *are* in fact approximately sinusoidal. This is a convenient form of voltage to deliver from an electric power generator, and if the voltage is sinusoidal (and the circuit elements are linear) the current is sinusoidal also. It is the natural form to obtain from an electronic oscillator in which the waves are shaped by resonance. Likewise, all types of frequency sensitivity, such as the selectivity of tuned circuits or of filter circuits, depend on waves being sinusoidal. Finally, analysis is better developed and much easier for sinusoidal waves, and we may suspect that an engineer could for that reason prefer a sinusoidal wave.

Now, however, a truly decisive reason for developing circuit analysis in terms of sinusoidal waves appears: any wave—a wave of any form whatever—can be analyzed into the sum of sinusoidal waves of different frequencies. Each of the sinusoidal components is then treated by methods that are now familiar to us. Finally, the components are recombined (by superposition) into a nonsinusoidal wave, giving the answer to the problem. The scope of application of analysis has not, therefore, been restricted by discussing sinusoidal waves. Rather, we have provided a means of attacking problems involving any wave form by means of Fourier analysis.

309

2. ADDITION OF COMPONENTS

The sum of two or more sinusoidal waves of different frequencies is a total wave that is not sinusoidal. In Figure 2a a wave of fundamental frequency and a second-harmonic wave are added, and the total is shown as a solid line. In Figure 2b a fundamental and a third harmonic are added.

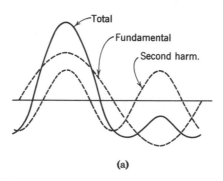

(a)

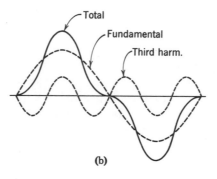

(b)

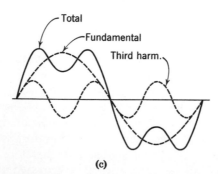

(c)

FIGURES 2a,b,c   Components of complex waves.

Although neither of these total waves is sinusoidal, there is a significant difference between them. In the wave containing a third harmonic, the positive and negative half-cycles are alike; both are peaked. In the wave containing a second harmonic, the positive half-cycle is peaked and the negative half-cycle is flattened. We shall find that something of this sort is always true. When there are only odd-numbered harmonics (third, fifth, and so forth), the positive and negative half-cycles are similar, but if even harmonics are present (second, fourth, and so on) they are not.

The addition of a third harmonic does not necessarily result in peaking the wave; Figure 2c shows a wave that is flattened by the addition of a third harmonic. The difference is in the phase relation between fundamental and third harmonic. In Figure 2b their crests are both positive or both negative at the same time, and so add; in Figure 2c their crests are of opposite polarity and subtract. With even harmonics, as in Figure 2a, if one pair of crests adds, the next subtracts, and that is the reason for the difference of the positive and negative half-wave forms. Thus, by adding harmonic components of many frequencies, with various amplitudes and phase relations, the total wave may be given a tremendous variety of forms.

3. ANALYSIS

Indeed, such a sum of fundamental and harmonic components may give a total wave of *any* form. This fact is usually expressed the other way around by saying that any wave may be analyzed into a Fourier series.† Let us see how such obviously valuable analysis is to be done.

When the above statement is expressed in symbols, it is

$$f(t) = \tfrac{1}{2}a_0 + a_1 \cos \omega_1 t + a_2 \cos 2\omega_1 t + a_3 \cos 3\omega_1 t + \cdots$$
$$+ b_1 \sin \omega_1 t + b_2 \sin 2\omega_1 t + b_3 \sin 3\omega_1 t + \cdots \quad (3\text{-}1)$$

The independent variable is here written t; any letter could be used, but in most of our work the independent variable is *time* so t is a convenient symbol. The function is called $f(t)$; this is quite general and may represent current or voltage or, indeed, any other physical quantity. The first term of the series is a constant, $\tfrac{1}{2}a_0$. If we are talking about current, this is the "dc component" of the wave. The terms with coefficients a_1 and b_1 together constitute the fundamental component. The terms with coefficients a_2 and b_2 give the second-harmonic component, and so on. The constant ω_1 is the radian frequency of the fundamental component.

---

† The mathematician recognizes certain limitations on the function that can be represented by a Fourier series, but no physically possible waves are ruled out. For example, the function must be single-valued and may have only a finite number of finite discontinuities in a finite interval.

The series of equation 3-1 can be made to equal any desired function through any finite length of time as the fundamental component of the series goes through one complete cycle. Let us call the beginning of this period t_1 and the end of the period t_2. The length of the period is $T = t_2 - t_1$. Since T is the period of the fundamental component,

$$\omega_1 T = 2\pi \quad \text{or} \quad T = \frac{2\pi}{\omega_1} \quad \text{or} \quad \omega_1 = \frac{2\pi}{T} \qquad (3\text{-}2)$$

These relations will be convenient in coming pages. Also we shall find it convenient to write x instead of $\omega_1 t$, just to save writing, and in terms of x:

$$x = \omega_1 t = 2\pi \frac{t}{T} \quad \text{or} \quad t = \frac{x}{\omega_1} = \frac{xT}{2\pi} \qquad (3\text{-}3)$$

The Fourier series, to be useful, must converge. That is, the sum of the series, as more and more of the higher harmonics are added, must approach a limit. When we set $f(t)$ equal to the series in equation 3-1, the meaning of the equality is that the limit approached by the series is $f(t)$.†

To find the series that represents any given $f(t)$, we need to evaluate the coefficients a_0, a_1, b_1, a_2, b_2, and so on. We shall now see how this is done. The method is based on the fact that the sine and cosine functions constitute an *orthogonal* system; that is, the average of their cross products is zero as specified in the following integrals. (These definite integrals, which are easily verified by integration, will be used in the next paragraphs.) If m and n are any integers,

$$\int_0^{2\pi} \sin mx \, dx = 0 \qquad \int_0^{2\pi} \cos mx \, dx = 0 \qquad \int_0^{2\pi} \sin mx \cos nx \, dx = 0$$

$$(3\text{-}4)$$

If m and n are *different* integers,

$$\int_0^{2\pi} \sin mx \sin nx \, dx = 0 \qquad \int_0^{2\pi} \cos mx \cos nx \, dx = 0 \qquad (3\text{-}5)$$

but if m and n are *equal* integers equations 3-5 become

$$\int_0^{2\pi} (\sin mx)^2 \, dx = \pi \qquad \int_0^{2\pi} (\cos mx)^2 \, dx = \pi \qquad (3\text{-}6)$$

The limits of integration in equations 3-4, 3-5, and 3-6 are written 0 and 2π. This is correct, but it would have been equally correct and more general

---

† An exception to this, which need not concern us at the moment, is at a point of discontinuity; where the function is continuous the series converges to $f(t)$, and where the function is discontinuous the series converges to the midpoint of the discontinuity.

to have written them C and $C + 2\pi$. The sufficient condition is that the integration be over a full cycle.

By way of explanation, let us study the curves of Figures 3a to c. In 3a the curves of $\sin x$ and $\cos x$ are shown, and the product of these two, which is $\sin x \cos x$. The shaded area is the integral of $\sin x \cos x$, and counting area below the axis as negative it is plain that the integral through a complete cycle is zero. This agrees with equations 3-4.

In 3b there are curves for $\sin x$ and $(\sin x)^2$. The area under $(\sin x)^2$ is all above the axis, so the integral is not zero; rather, the average of the $(\sin x)^2$ function is $\frac{1}{2}$, making the integral through 1 cycle equal to π, as in equations 3-6. The integral of $(\cos x)^2$ is of course the same.

Figure 3c shows two sine waves of different frequency and the wave that is their product. The area under the product curve is zero. This, though a single example, illustrates equations 3-5. Other illustrations can be devised until the reader is satisfied that equations 3-4 through 3-6 are always correct.

Now let us see how these facts of orthogonality can be used to determine the coefficients of the series in equation 3-1. Let us give attention to some one of the coefficients; we shall choose at random a_3. To find a_3, first multiply through equation 3-1 by $\cos 3\omega_1 t$. For convenience, let us write x for $\omega_1 t$ throughout, and write $f(x)$ instead of $f(t)$. Then multiply both sides of the equation by dx and integrate from $x = 0$ to $x = 2\pi$:

$$\int_0^{2\pi} f(x)\cos 3x\, dx = \int_0^{2\pi} \tfrac{1}{2}a_0 \cos 3x\, dx + \int_0^{2\pi} a_1 \cos x \cos 3x\, dx$$

$$+ \int_0^{2\pi} a_2 \cos 2x \cos 3x\, dx + \int_0^{2\pi} a_3(\cos 3x)^2\, dx + \cdots$$

$$+ \int_0^{2\pi} b_1 \sin x \cos 3x\, dx + \int_0^{2\pi} b_2 \sin 2x \cos 3x\, dx + \cdots$$

$$(3\text{-}7)$$

This looks bad, but actually it is not, for the first term of the right-hand member, being the integral of $\cos 3x$ through 1 cycle, is zero; the second and third terms are zero by equations 3-5; the fourth term, by equations 3-6, equals $a_3 \pi$; and the last two terms by equations 3-4 are zero. Also, all other terms of the series that are not explicitly written in the equation are zero. Hence the final result is

$$\int_0^{2\pi} f(x)\cos 3x\, dx = a_3 \pi \qquad (3\text{-}8)$$

or

$$a_3 = \frac{1}{\pi}\int_0^{2\pi} f(x)\cos 3x\, dx \qquad (3\text{-}9)$$

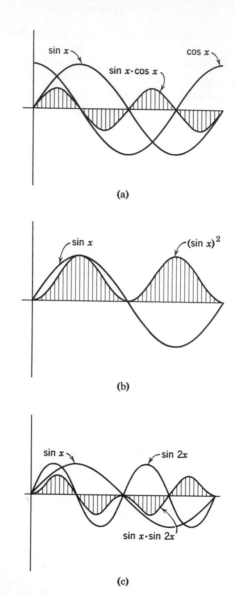

FIGURES 3a,b,c   Products of sinusoidal waves.

Here we have an expression for a_3 in terms of an integral. If the function $f(x)$ is known, the integral can be evaluated and hence the coefficient a_3 can be found. Of course, it may not be easy to integrate, but it is theoretically possible.

Thus we have succeeded in finding the value of one of the coefficients of the Fourier series, and since this particular coefficient is in no way different from the others the same method can be used for all of them. If we generalize from a_3 to a_n, we can write

$$a_n = \frac{1}{\pi} \int_C^{C+2\pi} f(x)\cos nx \, dx \tag{3-10}$$

A similar development after multiplying by a sine function gives

$$b_n = \frac{1}{\pi} \int_C^{C+2\pi} f(x)\sin nx \, dx \tag{3-11}$$

In these equations the limits of integration are written C and $C + 2\pi$, according to the paragraph that follows equation 3-6.

In these equations 3-10 and 3-11 we have the means of evaluating all the coefficients of the series, repeated here with x for $\omega_1 t$:

$$f(x) = \tfrac{1}{2}a_0 + a_1 \cos x + a_2 \cos 2x + a_3 \cos 3x + \cdots$$
$$+ b_1 \sin x + b_2 \sin 2x + b_3 \sin 3x + \cdots \tag{3-12}$$

The coefficient a_0 is found from equation 3-10 with $n = 0$, noting that $\cos 0 = 1$; this is sometimes written as a separate formula but there is no need to do so.

Here we have the usual form of Fourier series. For brevity, the formal symbol for summation can be used; the series is then written

$$f(x) = \tfrac{1}{2}a_0 + \sum_{m=1}^{\infty} (a_m \cos mx + b_m \sin mx) \tag{3-13}$$

This means exactly the same as equation 3-12; it is easier to write and will, with practice, prove to be easier to read also.†

---

† The letter m is used in equation 3-13 to mean any integer; this equation can be substituted into equations 3-10 and 3-11 to prove their validity without confusion between m and n when m and n are different integers.

It will be recognized that the Fourier series converges to the function $f(t)$ through the interval between the limits of integration. Beyond these limits the series may or may not converge to $f(t)$. The series, being cyclic, will repeat periodically beyond the limits of the integration; the function $f(t)$ will be equal to the series outside of the limits of integration only if $f(t)$ also is repetitive, but there is no requirement that $f(t)$ be a repetitive function. The sign $=$ is used with this meaning in all such equations as 3-12 and 3-13, implying convergence through a limited region. Some mathematicians prefer to use a different sign, such as ($=$) to have this meaning.

4. THE SQUARE WAVE

The square wave of Figure 4a can be expressed mathematically as follows (writing x for $\omega_1 t$):

$$f(t) = +1 \quad \text{when} \quad 0 < x < \pi$$
$$f(t) = -1 \quad \text{when} \quad \pi < x < 2\pi \tag{4-1}$$

Two equations are required to describe the wave through the interval 0 to 2π; $f(t) = 1$ when x is between 0 and π, and $f(t) = -1$ when x is between π

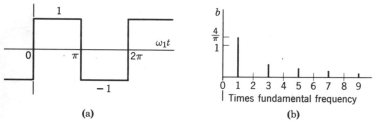

(a)   (b)

FIGURE 4   **(a)**   **A square wave, and**   **(b)**   **its spectrum (equation 4-8).**

and 2π. Consequently two integrals are necessary to find each a or b; we must integrate first from 0 to π and then from π to 2π. Let us find, for example, b_3. By equation 3-11,

$$b_3 = \frac{1}{\pi} \left(\int_0^\pi (+1)\sin 3x \, dx + \int_\pi^{2\pi} (-1)\sin 3x \, dx \right) \tag{4-2}$$

In this simple example the integration is easy; we integrate, then substitute the limits, and simplify the expression:

$$b_3 = \frac{1}{\pi} \left(\int_0^\pi \sin 3x \, dx - \int_\pi^{2\pi} \sin 3x \, dx \right)$$

$$= \frac{1}{\pi} \left\{ \left[-\frac{1}{3}\cos 3x \right]_0^\pi + \left[\frac{1}{3}\cos 3x \right]_\pi^{2\pi} \right\}$$

$$= \frac{1}{\pi} \left(\frac{1+1}{3} + \frac{1+1}{3} \right) = \frac{4}{3\pi} \tag{4-3}$$

  Thus b_3 is evaluated. All the other coefficients could be evaluated in the same way, one by one. This is a tedious process, however, and a good deal of time and labor can be saved by dealing in generalities. For instance, it is possible to show that all the a coefficients are zero for this square wave.

Applying equation 3-10 to the square wave, we write for a_n, where n may have the values 0, 1, 2, 3, ...,

$$a_n = \frac{1}{\pi}\left(\int_0^{\pi}(+1)\cos nx\,dx + \int_{\pi}^{2\pi}(-1)\cos nx\,dx\right)$$

$$= \frac{1}{n\pi}\left\{\left[\sin nx\right]_0^{\pi} - \left[\sin nx\right]_{\pi}^{2\pi}\right\}$$

$$= \frac{1}{n\pi}(2\sin n\pi - \sin n\cdot 2\pi) \tag{4-4}$$

Since $\sin n\pi$ and $\sin n2\pi$ are both zero for any integer value of n, a_n is zero for all n. There is one possible exception: if $n = 0$, equation 4-4 has the indeterminate form 0/0. To find a_0 we go back to the expression for a_n before integration and, letting $n = 0$, write

$$a_0 = \frac{1}{\pi}\left(\int_0^{\pi}dx - \int_{\pi}^{2\pi}dx\right) = \frac{1}{\pi}(\pi - 0 - 2\pi + \pi) = 0 \tag{4-5}$$

The b coefficients are not all zero. Using equation 3-11,

$$b_n = \frac{1}{\pi}\left(\int_0^{\pi}(+1)\sin nx\,dx + \int_{\pi}^{2\pi}(-1)\sin nx\,dx\right)$$

$$= \frac{1}{n\pi}\left\{-\left[\cos nx\right]_0^{\pi} + \left[\cos nx\right]_{\pi}^{2\pi}\right\}$$

$$= \frac{1}{n\pi}(1 - 2\cos n\pi + \cos n\cdot 2\pi) \tag{4-6}$$

Here n may be any integer, but the result depends on whether n is an even or odd integer. For all *even* values of n, equation 4-6 reduces to

$$b_n = \frac{1}{n\pi}(1 - 2 + 1) = 0 \qquad \text{for even } n \tag{4-7}$$

whereas for *odd* values of n,

$$b_n = \frac{1}{n\pi}(1 + 2 + 1) = \frac{4}{n\pi} \qquad \text{for odd } n \tag{4-8}$$

Specifically, $b_1 = 4/\pi$, $b_3 = 4/3\pi$, $b_5 = 4/5\pi$, and so on.

The solution for a series to represent the square wave of Figure 4a is now complete. It is

$$\text{Square wave (Figure 4a)} = \frac{4}{\pi}\left(\sin \omega_1 t + \frac{1}{3}\sin 3\omega_1 t + \frac{1}{5}\sin 5\omega_1 t + \cdots\right)$$

$$\tag{4-9}$$

It will be seen that the third-harmonic component of a square wave is one-third as large as the fundamental, the fifth harmonic is one-fifth as large, and so on. There are no even harmonics, which is consistent with the fact that the negative half-waves are similar to the positive half-waves. There is no constant component.

It is interesting to plot a "spectrum" of the square wave. In Figure 4b the amplitudes of the harmonics are shown. The only frequencies existing in the spectrum are those corresponding to odd-integer harmonics. This is a "line spectrum."

5. THE SAW-TOOTH WAVE

Figure 5a shows another simple yet interesting wave. Letting $x = \omega_1 t$, it can be expressed conveniently from $-\pi$ to $+\pi$ as:

$$f(t) = \frac{x}{\pi} \quad \text{when} \quad -\pi < x < +\pi \tag{5-1}$$

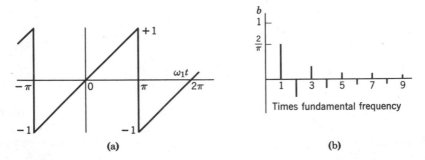

FIGURE 5   (a)   A saw-tooth wave, and   (b)   its spectrum (equation 5-5).

It is not quite so convenient, on the other hand, to express this wave from 0 to 2π; two equations would be required, and the second would not be entirely simple. But by letting $C = -\pi$ in equation 3-10

$$a_n = \frac{1}{\pi} \int_{-\pi}^{\pi} \frac{x}{\pi} \cos nx \, dx$$

$$= \frac{1}{\pi^2} \int_{-\pi}^{\pi} x \cos nx \, dx = \frac{1}{\pi^2} \left[\frac{1}{n^2} \cos nx + \frac{x}{n} \sin nx \right]_{-\pi}^{\pi}$$

$$= \frac{1}{\pi^2} \left(\frac{1}{n^2} (\cos n\pi - \cos n\pi) + \frac{1}{n} (\pi \sin n\pi - \pi \sin n\pi) \right) = 0 \tag{5-2}$$

This expression is zero for any value of n except possibly $n = 0$. If $n = 0$,

$$a_0 = \frac{1}{\pi^2} \int_{-\pi}^{\pi} x\, dx = \frac{1}{2\pi^2} (\pi^2 - \pi^2) = 0 \qquad (5\text{-}3)$$

Hence all the a's are zero. The series contains no cosine terms and no constant term.

Next we find b_n for the saw-tooth wave:

$$b_n = \frac{1}{\pi} \int_{-\pi}^{\pi} \frac{x}{\pi} \sin nx\, dx = \frac{1}{\pi^2} \int_{-\pi}^{\pi} x \sin nx\, dx$$

$$= \frac{1}{\pi^2} \left[\frac{1}{n^2} \sin nx - \frac{x}{n} \cos nx \right]_{-\pi}^{\pi} = \frac{2}{n^2 \pi^2} (\sin n\pi - n\pi \cos n\pi) \quad (5\text{-}4)$$

If n is any even integer, $\sin n\pi = 0$ and $\cos n\pi = 1$, and $b_n = -2/n\pi$. If n is an odd integer, $\sin n\pi = 0$ but $\cos n\pi = -1$, making $b_n = +2/n\pi$. Thus

$$b_1 = \frac{2}{\pi} \qquad b_2 = -\frac{2}{2\pi} \qquad b_3 = \frac{2}{3\pi} \qquad b_4 = -\frac{2}{4\pi} \qquad \cdots \qquad (5\text{-}5)$$

and the final result, the expansion of the saw-tooth wave into a Fourier series, is

Saw-tooth wave (Figure 5a)

$$= \frac{2}{\pi} \left(\sin \omega_1 t - \frac{1}{2} \sin 2\omega_1 t + \frac{1}{3} \sin 3\omega_1 t - \frac{1}{4} \sin 4\omega_1 t + \cdots \right) \qquad (5\text{-}6)$$

In this wave there are no cosine terms and no constant term. All sine terms are present, both even and odd harmonics, the amplitude being inversely proportional to the order of the harmonic but with reversal of sign in alternate terms. These facts are shown graphically in the spectrum of Figure 5b.

6. SYNTHESIS OF WAVES

It is all very well to say that the series of equation 4-9 represents a square wave and that of equation 5-6 a saw-tooth wave, and indeed the mathematics leading to these series is perfectly sound, but we can hardly be expected to be satisfied until we have tried a few examples to see how they work. Let us try to synthesize a wave from its Fourier components, adding them together to see if they really do give us back the wave with which we started.

Equation 4-9 tells us that the terms

$$\frac{4}{\pi} \left(\sin \omega_1 t + \frac{1}{3} \sin 3\omega_1 t + \frac{1}{5} \sin 5\omega_1 t + \cdots \right) \qquad (6\text{-}1)$$

should add to give a square wave of unit height. This may be tried. In Figure 6a we draw a sine wave and a third-harmonic wave of one-third the amplitude and add them. The result is the solid line shown.

This line is redrawn as a dashed line in (b), and a fifth-harmonic wave of one-fifth amplitude is drawn and added to it. The result, shown as a solid line, is beginning to look hopefully like a square wave.

It is transferred to (c), and a seventh harmonic is added. The approximation to a square wave is becoming better.

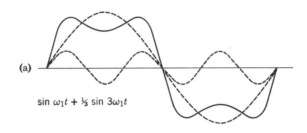

(a)

$\sin \omega_1 t + \frac{1}{3} \sin 3\omega_1 t$

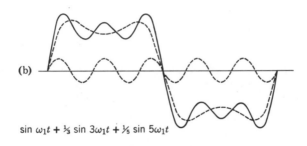

(b)

$\sin \omega_1 t + \frac{1}{3} \sin 3\omega_1 t + \frac{1}{5} \sin 5\omega_1 t$

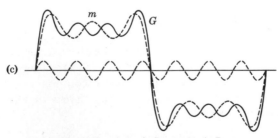

(c)

$\sin \omega_1 t + \frac{1}{3} \sin 3\omega_1 t + \frac{1}{5} \sin 5\omega_1 t + \frac{1}{7} \sin 7\omega_1 t$

FIGURES 6a,b,c   Synthesis of a square wave.

As higher harmonics are added, always decreasing in size, the approximation becomes continually better. A ripple always remains to mar the flat top. Adding another harmonic reduces the ripple and raises its frequency. The general shape becomes nearer and nearer the square wave.†

7. THREE KINDS OF SYMMETRY

We have seen, by evaluating the coefficients, that some waves have no even harmonics. Some waves have no cosine terms, some have no sine terms, some have no constant term. If we could predict without computation that certain waves would surely have no terms in one or another of these categories, it would save a good deal of work. This, in fact, we can do. We shall show that waves with certain types of symmetry are lacking certain harmonics:

1. Odd functions have only sine terms.
2. Even functions have no sine terms.
3. If there is half-wave symmetry, only odd harmonics are present.

When the mathematician speaks of an odd function, he means one like x, x^3, x^5, and so on, for which

$$f(x) = -f(-x) \qquad (7\text{-}1)$$

For instance, consider the function x^3. Note that if $x = 2$, $x^3 = 8$, and if $x = -2$, $x^3 = -8$. Equation 7-1 is therefore satisfied, and x^3 is an odd function. The name " odd " function comes from the number in the exponent.

An even function, on the other hand, is one like x^2, x^4, x^6, and so on (with even exponents) for which

$$f(x) = f(-x) \qquad (7\text{-}2)$$

For instance, if $x = 2$, $x^2 = 4$, and if $x = -2$, $x^2 = 4$ also. This is an example of an even function that satisfies equation 7-2.

We may generalize. The sum of odd functions is itself odd. Thus, if $f(x) = x + x^3$, we know that $f(x)$ is an odd function.

The sum of even functions is itself even. Thus, if $f(x) = K + x^2 + x^4$, we know that $f(x)$ is an even function. Note that the constant K is to be classed with the even functions, for it has the same value whether x is positive or negative.

---

† Note, however, the tendency of the wave to grow " ears " as at G. This is known as the Gibbs phenomenon. On either side of a discontinuity of the original function the partial sum of a Fourier series (found by using a finite number of terms) tends to overshoot. As more terms are used, the " ears " become narrower but not shorter, remaining about 9% of the discontinuity. The *infinite* series, however, converges exactly to the function except at the point of discontinuity, where it converges to the midpoint of the break.

The following conclusions are now drawn:

First, all sine waves are odd. Therefore the sum of sine waves will be odd, and a Fourier series containing only sine components represents an odd function.

Second, all cosine functions are even, and a series of cosine terms must represent an even function.

Third, if an even function and an odd function are added, the sum will be neither even nor odd. Hence a series containing both sine and cosine components represents a function that is not even and not odd.

From these conclusions we deduce that (1) series representing *odd functions have only sine terms* and (2) series representing *even functions have cosine terms* and, possibly, a constant term but do not contain any sine terms.

Thus, the square wave of Figure 4a is odd; this we know from symmetry, and it could have been predicted without computation that its Fourier series of equation 4-8 would contain only sine terms. Also by symmetry, the saw-tooth wave of Figure 5a is likewise odd, and equation 5-6 contains only sine terms.

The third type of symmetry in which we are interested is *half-wave symmetry*. For *any* periodic function with a period T,

$$f(t) = f(t + T) \qquad (7\text{-}3)$$

This says, symbolically, that a function with the value $f(t)$ at any time t will have the same value again at the later time $t + T$. This is obviously true of a sinusoidal wave, T being the time of 1 cycle. The value of the wave at any instant will be repeated 1 cycle later, and again another cycle later, and so on.

In connection with Figures 2b and 2c we spoke of half-wave symmetry, meaning that the shape of the negative half-cycle of a wave is the same as that of the positive half-cycle, though of course inverted. To be more precise, a wave with half-wave symmetry satisfies the equation

$$f(t) = -f\left(t + \frac{T}{2}\right) \qquad (7\text{-}4)$$

Consider a sinusoidal wave of fundamental frequency; it has half-wave symmetry and satisfies equation 7-4. Consider the second harmonic, as shown in Figure 2a. If T is the period of the fundamental frequency, the second harmonic does not satisfy equation 7-4, for $f(t) = +f(t + T/2)$, so it does not have half-wave symmetry. Consider the third harmonic, as shown in Figures 2b or 2c; the third harmonic does have half-wave symmetry.

Generalizing, any wave described by a Fourier series with only fundamental, third harmonic, fifth harmonic, and other *odd* harmonics *will* have half-wave symmetry. A wave containing any second, fourth, or other *even* harmonic *will not* have half-wave symmetry.

Conversely, *a wave, however complex, that has half-wave symmetry will contain no even harmonics.*

As an example, the square wave of Figure 4a does have half-wave symmetry; it passes the test of equation 7-4. Its Fourier series therefore has only odd-harmonic terms, as in equation 4-9 and in the spectrum of Figure 4b.

The saw-tooth wave of Figure 5a gives another example. This saw-tooth wave does not satisfy the test of equation 7-4; it does not have half-wave symmetry. Hence there are even harmonics in its series, as substantiated by equation 5-6 and Figure 5b.

8. LIMITS OF INTEGRATION

If there is symmetry, the limits of integration for finding Fourier coefficients can be reduced. If the wave being analyzed is *either* even *or* odd, *or* has half-wave symmetry, we need not integrate from 0 to 2π; instead we can integrate from 0 to π and multiply the result by 2. Because of such symmetry the result of integration repeats each half-period. For example, in equations 4-6 or 5-4 we might have used 0 and π as the limits of integration and then multiplied the result by 2, and the coefficients would have been the same.

It can be shown from equations 3-10 and 3-11 that this is true in general. Let $C = 0$, and note that the product of two even functions is even, and product of two odd functions is even also.

Further, if the wave is *either* even or odd *and also* has half-wave symmetry, it is enough to integrate from 0 to $\pi/2$ and then multiply by 4. In this case the result of integration repeats four times in a period.

Example. The triangular wave of Figure 8a is odd and has half-wave symmetry. Hence the *a*'s are zero and the *b*'s can be found by integrating from 0 to $\pi/2$ because of repetition of the kind indicated in the diagram.†
Then:

$$b_n = \frac{4}{\pi} \int_0^{\pi/2} f(x)\sin nx\, dx$$

$$= \frac{4}{\pi} \int_0^{\pi/2} \frac{2x}{\pi} \sin nx\, dx = \frac{8}{\pi^2} \int_0^{\pi/2} x \sin nx\, dx$$

$$= \frac{8}{\pi^2} \left[\frac{1}{n^2} \sin nx - \frac{x}{n} \cos nx \right]_0^{\pi/2} = \frac{8}{\pi^2} \left(\frac{1}{n^2} \sin n\frac{\pi}{2} - \frac{\pi}{2n} \cos n\frac{\pi}{2} \right) \quad (8\text{-}1)$$

---

† The curve for sin $3x$ is drawn to a reduced scale to avoid confusing the diagram.

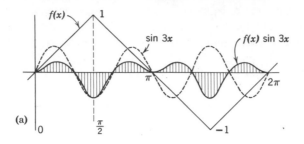

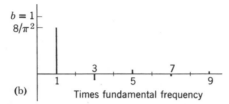

FIGURES 8a,b   A triangular wave and its spectrum (equation 8-2).

We are concerned only† with odd-integer values of n, for all of which $\cos(n\pi/2) = 0$. The value of $\sin(n\pi/2)$ is 1 if $n = 1, 5, 9, \ldots$, so for these values of n, $b_n = 8/\pi^2 n^2$. If $n = 3, 7, 11, \ldots$, $\sin(n\pi/2) = -1$, and $b_n = -8/\pi^2 n^2$.

This completes the analysis, and substitution of these values for the coefficients of equation 3-12 gives:

Triangular wave (Figure 8a)

$$= \frac{8}{\pi^2}\left(\sin \omega_1 t - \frac{1}{9}\sin 3\omega_1 t + \frac{1}{25}\sin 5\omega_1 t - \frac{1}{49}\sin 7\omega_1 t + \cdots\right) \qquad (8\text{-}2)$$

The spectrum of this wave is shown in Figure 8b. It is interesting to notice that this series converges much more rapidly than those previously considered.

9. RECTIFIER WAVES

Both the square wave and the saw-tooth wave that have been described in this chapter find important practical use. The square wave is utilized for testing, and one reason for its usefulness is the fact that it has, as we have

---

† Do *not* use a reduced interval of integration to prove that harmonics are indeed zero if symmetry shows them to be zero; it will not necessarily work. For instance, equation 8-1 is not zero if $n = 2$ although symmetry shows that $b_2 = 0$. Integration through a full period, of course, gives $b_2 = 0$.

seen, so large a harmonic content. Connecting a square-wave generator to a circuit is equivalent to applying voltages of many different frequencies at once, and the response of the circuit to a wide frequency range can be found in a single test. The saw-tooth form is used in television and radar sweep circuits. Figure 5b shows that circuits carrying this wave must transmit frequencies many times the fundamental frequency if the saw-tooth shape is to come through without serious distortion.

Other nonsinusoidal waves of great practical importance are found in the output of rectifiers. In these the dc or constant component is the desirable component and all harmonics are undesirable.

Figure 9a shows the output of a *half-wave rectifier* or simple diode feeding into a resistive load. This wave, like any other, is subject to Fourier analysis. The first step is to locate the vertical axis. There are several possibilities, but the one shown in the figure looks the most hopeful because it makes the wave an even function. Thus there will be no sine terms. There is obviously a constant term, for the average is not zero, and there is not half-wave symmetry. We must expect a series, then, of the form

$$f(t) = \tfrac{1}{2}a_0 + a_1 \cos \omega_1 t + a_2 \cos 2\omega_1 t + \cdots \tag{9-1}$$

The easiest limits of integration will be $-\pi$ and $+\pi$. Since the function is zero from $-\pi$ to $-\pi/2$, and from $+\pi/2$ to $+\pi$, the integrand through these ranges is zero, and the coefficients of the series are found (using equation 3-10) by integrating between the limits $-\pi/2$ and $+\pi/2$. Between these limits the function is $f(x) = \cos x$, so

$$a_n = \frac{1}{\pi} \int_{-\pi/2}^{\pi/2} \cos x \cos nx \, dx \tag{9-2}$$

If $n \neq 1$,

$$a_n = \frac{1}{\pi} \left[\frac{\sin(n-1)x}{2(n-1)} + \frac{\sin(n+1)x}{2(n+1)} \right]_{-\pi/2}^{\pi/2}$$

$$= \frac{1}{\pi} \left(\frac{\sin(n-1)\dfrac{\pi}{2}}{n-1} + \frac{\sin(n+1)\dfrac{\pi}{2}}{n+1} \right) \tag{9-3}$$

If $n = 1$,

$$a_1 = \frac{1}{\pi} \left[\frac{x}{2} + \frac{\sin 2x}{4} \right]_{-\pi/2}^{\pi/2} = \frac{1}{2\pi} \left(\frac{\pi}{2} + \frac{\pi}{2} \right) = \frac{1}{2} \tag{9-4}$$

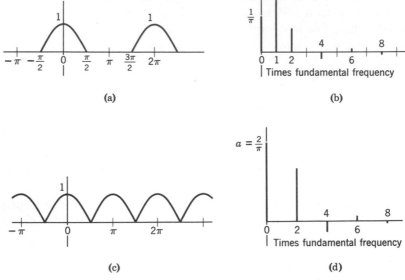

FIGURE 9 (a) A half-wave rectifier output wave, and (b) its spectrum (× 4 on vertical scale) (equation 9-5). (c) A full-wave rectifier output wave, and (d) its spectrum (× 4 on vertical scale) (equation 9-6).

The coefficients a_n are evaluated one at a time by introducing the appropriate values of n. The results are

$$n = 0, \; a_0 = \frac{2}{\pi}$$

$$n = 1, \; a_1 = \tfrac{1}{2} \qquad n = 3, \; a_3 = 0 \qquad\qquad n = 5, \; a_5 = 0$$

$$n = 2, \; a_2 = \frac{2}{3\pi} \qquad n = 4, \; a_4 = -\frac{2}{15\pi} \qquad n = 6, \; a_6 = \frac{2}{35\pi}$$

Placing these in the series gives

Wave of half-wave rectifier (Figure 9a)

$$= \frac{1}{\pi}\left(1 + \frac{\pi}{2}\cos \omega_1 t + \frac{2}{3}\cos 2\omega_1 t - \frac{2}{15}\cos 4\omega_1 t + \frac{2}{35}\cos 6\omega_1 t - \cdots\right) \qquad (9\text{-}5)$$

It is interesting that the only odd harmonic present is the first, the fundamental. All even harmonics are present, but those of higher order are quite small. See Figure 9b.

A *full-wave rectifier*, which is quite commonly a bridge of diodes, has the output wave shown in Figure 9c. Analysis is quite similar to that of the half-wave example just given. The coefficients are easily found, a little forethought in choosing the limits of integration being well repaid. The result is

Wave of full-wave rectifier (Figure 9c)

$$= \frac{2}{\pi}\left(1 + \frac{2}{3}\cos 2\omega_1 t - \frac{2}{15}\cos 4\omega_1 t + \frac{2}{35}\cos 6\omega_1 t - \cdots\right) \qquad (9\text{-}6)$$

It will be seen that the constant term, the useful dc output of the rectifier, is twice as great from a full-wave as from a half-wave rectifier, the crests of the two output waves being the same. This would be expected, since each dc term is the *average* of the output function.

It will also be seen from the equation or from the spectrum of Figure 9d, that this series contains only even harmonics of ω_1. Apparently we might have chosen twice as high a frequency to be the fundamental frequency of the function. In Figure 9c the curve is seen to repeat not once but twice every cycle. Ordinarily the shortest period of repetition is called a cycle, but in this case we prefer to use ω_1, the frequency of the input ac wave, as the fundamental frequency of the rectified output wave even though there is no component of this frequency in the output. Ordinarily a Fourier series is not permitted to consist of even-harmonic terms only.

10. ANOTHER TRIGONOMETRIC FORM OF SERIES

If it is preferred, any Fourier series can be written with sine and cosine terms of the same frequency combined. Thus equation 3-12 or 3-13 can be expressed as

$$f(t) = \tfrac{1}{2}a_0 + c_1 \cos(\omega_1 t - \theta_1) + c_2 \cos(2\omega_1 t - \theta_2)$$
$$+ c_3 \cos(3\omega_1 t - \theta_3) + \cdots + c_n \cos(n\omega_1 t - \theta_n) + \cdots \qquad (10\text{-}1)$$

or as

$$f(t) = \tfrac{1}{2}a_0 + \sum_{n=1}^{\infty} c_n \cos(n\omega_1 t - \theta_n) \qquad (10\text{-}2)$$

where

$$c_n = \sqrt{a_n^2 + b_n^2} \qquad \theta_n = \tan^{-1}\frac{b_n}{a_n} \qquad (10\text{-}3)$$

An example showing the convenience of this form will appear in the next section.

Instead of the cosine series of equations 10-1 and 2, a sine series can be used. Thus:

$$f(t) = \frac{1}{2} a_0 + \sum_{n=1}^{\infty} c_n \sin(n\omega_1 t + \varphi_n) \quad \text{where} \quad \varphi_n = \tan^{-1} \frac{a_n}{b_n} \quad (10\text{-}4)$$

11. USE IN NETWORK ANALYSIS

There are three main reasons why Fourier analysis of a wave into its constant, fundamental, and harmonic components must be thoroughly familiar. First and most significant, all scientists and engineers think and talk in terms of fundamentals and harmonics. To understand what is said or published, it is necessary to be fully aware of the Fourier series concept and all that it implies.

Second, the Fourier series leads by easy stages to the Fourier integral, and that in turn to the Laplace integral, and Laplace transformation is one of our most powerful tools for analysis of linear systems.

Third, we can find the response of a linear network to any applied disturbance of *periodic* form by Fourier analysis, a use that will now be demonstrated. Steps can here be taken to find the current produced by a *periodic* voltage that are helpfully parallel to the steps that will later be taken to find the current produced by a *non-periodic* voltage.

Filter for Rectifier. The circuit of Figure 11a is used to remove ac ripple from the output of a full-wave rectifier. The rectifier supplies the following voltage v_r at its terminals:

$$v_r = \frac{2}{\pi} V_m \left(1 + \frac{2}{3} \cos 2\omega_1 t - \frac{2}{15} \cos 4\omega_1 t + \cdots \right) \quad (11\text{-}1)$$

in which $\omega_1 = 377$. Find the voltage across the 2000-ohm load.

The rectifier will be considered an independent voltage source. Each element of the load and filter is linear; therefore the principle of superposition applies and Fourier-analysis methods can be used. Let us work with the dc component and the two harmonic components shown in equation 11-1.

For each component the circuit consisting of filter and load constitutes a voltage divider. Let the impedance to v_r be called Z_r and the impedance to v_f be called Z_f. The components of these voltages (or more properly of their transforms V_r and V_f) are related by the ratios of impedances, and at each frequency

$$\frac{V_f}{V_r} = \frac{Z_f}{Z_r} \quad (11\text{-}2)$$

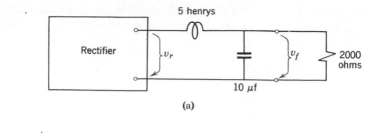

(a)

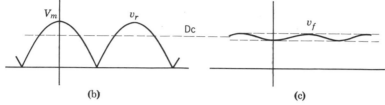

(b)   (c)

**FIGURE 11   (a)   A filter circuit for the output of a full-wave rectifier,
(b)   the rectifier output voltage, and   (c)   the filtered voltage.**

The impedances, of course, have different values for each harmonic frequency.

Z_f is the impedance of the load and shunt capacitor in parallel. Z_r is equal to Z_f plus the impedance of the choke coil. Thus

$$Z_f = \frac{(2000)(-j10^5/n\omega_1)}{2000 - j10^5/n\omega_1} \qquad Z_r = Z_f + jn\omega_1 L \qquad (11\text{-}3)$$

Computing numerical values for the dc, second-harmonic, and fourth-harmonic component frequencies,

$$
\begin{array}{cccc}
n = & 0 & 2 & 4 \\
Z_f = & 2000\underline{/0^\circ} & 132.4\underline{/-86.2^\circ} & 66.3\underline{/-88.1^\circ} \\
Z_r = & 2000\underline{/0^\circ} & 3638\underline{/89.9^\circ} & 7474\underline{/90.0^\circ}
\end{array}
$$

The entire dc component of v_r (equation 11-1) appears across the load and this fact gives the first term of a series for the filter output voltage:

$$v_f = \frac{2}{\pi} V_m(1 + \cdots) \qquad (11\text{-}4)$$

From equation 11-2, the second-harmonic component ($n = 2$) of the load voltage is less in amplitude than the second harmonic of v_r, the rectifier voltage, in the ratio of 132.4 to 3638, and its phase angle is greater by $-86.2 - 89.9 = -176.1$ degrees. Another term of v_f is now computed from these figures:

$$v_f = \frac{2}{\pi} V_m \left[1 + \frac{132.4}{3638} \cdot \frac{2}{3} \cos(2\omega_1 t - 176.1°) + \cdots \right] \tag{11-5}$$

The fourth-harmonic component is reduced in magnitude and delayed in phase in a similar manner, giving another term for the series:

$$v_f = \frac{2}{\pi} V_m \left[1 + \frac{132.4}{3638} \cdot \frac{2}{3} \cos(2\omega_1 t - 176.1°) \right.$$

$$\left. - \frac{66.3}{7474} \cdot \frac{2}{15} \cos(4\omega_1 t - 178.1°) + \cdots \right] \tag{11-6}$$

Multiplication gives the coefficients:

$$v_f = \frac{2}{\pi} V_m [1 + 0.0242 \cos(2\omega_1 t - 176.1°)$$

$$- 0.0012 \cos(4\omega_1 t - 178.1°) + \cdots] \tag{11-7}$$

This is the answer to the problem. It could be put in somewhat more elegant form, but this will serve.

The principal conclusion is illustrated in Figures 11b and c. Whereas the output of the rectifier is a series of sinusoidal half-waves, the output of the filter is so nearly pure direct current that the ripple is greatly exaggerated to be seen in the diagram. It has, according to our results, a second-harmonic ripple that is 2.4% of the dc component, and a fourth-harmonic term so much smaller that it is negligible.

Ripple filters of this type are quite commonly used in practice in the output circuits of power-supply rectifiers. The foregoing analysis has given enough insight into the operation of such a filter to permit drawing the following conclusions regarding *principles of design*:

1. Reactance of the capacitor (at the frequency of the lowest harmonic to be eliminated) should be *substantially smaller* than the load resistance.
2. Reactance of the inductor (at the lowest harmonic frequency) should be *several times greater* than the capacitor reactance. .
3. The amplitude of a voltage harmonic will then be reduced, from filter-section input to filter-section output, approximately in the ratio of capacitor reactance to inductor reactance.†

---

† This ratio may also be written $1/\omega^2 LC$. A slightly better approximation is $1/(\omega^2 LC - 1)$; see Terman (2), Section 11-8. As a numerical check on these approximations, note that in the preceding example the second-harmonic reactances are $X_C = -132.6$, $X_L = 3770$. The ratio (disregarding sign) gives an approximation of second-harmonic ripple output by the filter of 0.0352. A better approximation, using Terman's formula, is 0.0365. The exact ratio, computed above, is 0.0364.

The filter of this example is good enough for many purposes. However, for uses that are highly sensitive to voltage fluctuation, it will not do. When better filtering is needed, another stage of filtering can be added. It is more efficient to use additional inductance and capacitance in another cascaded filter section than to increase the size of the elements in a one-section filter.

12. RMS VALUE AND POWER

When a current of some nonsinusoidal wave form, such as a rectifier wave of Figure 11b or 11c, passes through a resistor, the power loss and consequent heating of the resistor are determined by the rms value of the wave. How can the rms value of odd-shaped waves be found?

The rms value of any wave is found from its Fourier series as follows: *the rms value of the total wave is the square root of the sum of the squares of the rms values of the components.* That is:

$$I_{rms} = \sqrt{I_0{}^2 + I_1{}^2 + I_2{}^2 + \cdots} \tag{12-1}$$

where I_0 is the constant or dc component of current, and $I_1, I_2, \ldots$ are the rms values of the various harmonics.

This formula can be illustrated by computing the rms value of the full-wave rectifier output. Using in equation 12-1 the first four components of equation 9-6,

$$I_{rms} = \frac{2}{\pi} \sqrt{1^2 + \frac{1}{2}\left(\frac{2}{3}\right)^2 + \frac{1}{2}\left(\frac{2}{15}\right)^2 + \frac{1}{2}\left(\frac{2}{35}\right)^2} = 0.707 \tag{12-2}$$

As a matter of fact, we know the true rms value of this rectified wave of unit crest: it is the same as the rms value of the unrectified wave of unit crest and is therefore $\frac{1}{2}\sqrt{2} = 0.707$. The numerical accuracy of equation 12-2 is seen to be good even though we have neglected the seventh and higher harmonics.

The *proof* of equation 12-1 (which is sometimes known as Parseval's theorem) is simple in concept. Since the algebra tends to become rather depressing, it will be best to give only an outline of the proof and anyone who wishes can fill in the details.

By equation 12-4 of Chapter 2, the rms value of any wave is so defined that

$$I_{rms}{}^2 = \frac{1}{T} \int_0^T i^2 \, dt \tag{12-3}$$

T being the time of 1 cycle of the wave. The Fourier series for current, as in equation 10-1, is now substituted for i and the square is found term by term (permitted for a uniformly convergent series); many of the terms of the square integrate to zero, and equation 12-1 remains.

$$\frac{1}{T} \int_0^T [f(t)]^2 \, dt = a_0{}^2 + \frac{1}{2} \sum_{n=1}^{\infty} (a_n{}^2 + b_n{}^2)$$

A theorem closely related to Parseval's theorem may be stated as follows: *total average power* is the sum of the powers of the dc component, the fundamental frequency, and the several harmonics *taken separately*. That is,

$$P = P_0 + P_1 + P_2 + \cdots$$
$$= V_0 I_0 + |V_1| |I_1| \cos \varphi_1 + |V_2| |I_2| \cos \varphi_2 + \cdots \quad (12\text{-}4)$$

The proof appears when Fourier expansions of voltage and current in the form of equation 10-1 are substituted into the expression for average power:

$$P = \frac{1}{T} \int_0^T vi \, dt \quad (12\text{-}5)$$

Equation 12-4 says that the total average power to a network is the sum of the powers of the several Fourier components, dc and ac, as if superposition of quantities of power were possible. This seems in contradiction to Section 1 of Chapter 10. However, the orthogonality of Fourier components leads to zero values for the integrals of products of different harmonic components, and thus gives the simple summation of equation 10-4.

13. AN EXPONENTIAL FORM OF SERIES

We have seen how much easier it is to work with exponential functions than with sinusoidal functions, and this suggests a tremendously useful change in the form of Fourier series. By expressing each cosine and sine term as the sum or difference of exponential functions, the series is changed from this form:

$$f(t) = \tfrac{1}{2}a_0 + a_1 \cos \omega_1 t + a_2 \cos 2\omega_1 t + a_3 \cos 3\omega_1 t + \cdots$$
$$+ b_1 \sin \omega_1 t + b_2 \sin 2\omega_1 t + b_3 \sin 3\omega_1 t + \cdots \quad (13\text{-}1)$$

to this form:

$$f(t) = \cdots + A_{-2} e^{-j2\omega_1 t} + A_{-1} e^{-j\omega_1 t} + A_0 + A_1 e^{j\omega_1 t} + A_2 e^{j2\omega_1 t} + \cdots$$
$$(13\text{-}2)$$

The change is made by letting $\cos x = \tfrac{1}{2}(e^{jx} + e^{-jx})$ and $\sin x = -j\tfrac{1}{2}(e^{jx} - e^{-jx})$. When this is done the coefficients combine as follows:[†]

$$A_0 = \tfrac{1}{2}a_0 \qquad A_1 = \tfrac{1}{2}(a_1 - jb_1) \qquad A_{-1} = \tfrac{1}{2}(a_1 + jb_1) \qquad \cdots (13\text{-}3)$$

or in general

$$A_n = \tfrac{1}{2}(a_n - jb_n) \quad \text{and} \quad A_{-n} = \tfrac{1}{2}(a_n + jb_n) \quad (13\text{-}4)$$

---

† The negative subscripts may be surprising at first glance, but they mean neither more nor less than other subscripts: they merely identify the terms of the series, counting each way from the dc component which is A_0.

As these equations show, the A's are in general complex, and they appear in conjugate pairs: $A_{-n} = A_n{}^*$.

Equation 13-2 may be written more compactly as

$$f(t) = \sum_{m=-\infty}^{+\infty} A_m e^{jm\omega_1 t} = \sum_{m=-\infty}^{+\infty} A_m e^{jmx} \tag{13-5}$$

where m is any integer and $x = \omega_1 t$.

The A coefficients of the exponential series are most conveniently found from the integral

$$A_n = \frac{1}{2\pi} \int_0^{2\pi} f(t) e^{-jn\omega_1 t} \, d(\omega_1 t) = \frac{1}{2\pi} \int_0^{2\pi} f(x) e^{-jnx} \, dx \tag{13-6}$$

where n is any integer, positive, negative, or zero, and $f(t)$ is the function to be expressed as a Fourier series in the form of equation 13-2 or 13-5.

Equation 13-6 is analogous to equations 3-10 and 11 for finding the coefficients of the trigonometric series. It is used in the same way, but with less work—less work, because there is only one integration to be done instead of two, and also because this integration is almost always easier to perform.

14. SYMMETRY IN EXPONENTIAL SERIES

The three kinds of symmetry discussed in connection with trigonometric series are helpful with exponential series also. From equations 13-4 it is seen that if there are only sine terms (so that all the a_n coefficients are zero), A_n is purely imaginary for all values of n. If, on the other hand, there are no sine terms (all b_n terms equal zero), A_n is purely real for all n. Finally, if even harmonics do not exist in the trigonometric form, A_n is zero for even values of n. Hence we may say of the exponential series:

1. For odd functions, all A's are purely imaginary.
2. For even functions, all A's are purely real.
3. If there is half-wave symmetry, $A_n = 0$ for even n.
4. Furthermore, it is always true that $A_{-n} = A_n{}^*$.

The interval of integration given in equation 13-6 is from 0 to 2π, but actually it can be over any complete fundamental cycle, as, for example, from $-\pi$ to $+\pi$. If there is half-wave symmetry the interval of integration can be shortened to half a cycle, just as it can with trigonometric series. *Note particularly*, however, that the interval of integration cannot be shortened because the function $f(x)$ is either an even or an odd function. This is possible with the trigonometric series because $\cos nx$ and $\sin nx$ are themselves even and odd, respectively, but e^{jnx} is neither even nor odd, and hence for the exponential series the limits of integration must be T seconds apart unless there is half-wave symmetry.

15. A SQUARE WAVE

Figure 15a shows a square wave, with a flat top at one volt and a flat bottom at zero. Let us place the vertical axis as in the figure so that the wave is an even function; this makes all the values of A real. There is not half-wave symmetry (though only because of a constant component).

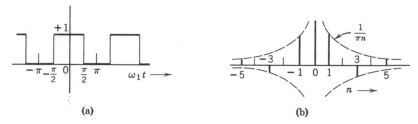

(a)   (b)

FIGURES 15a,b A square wave drawn as an even function, and its spectrum (equation 15-6).

The coefficients are found from equation 13-6, with limits of integration $-\pi$ and $+\pi$ instead of 0 and 2π. Limits of $-\pi$ and $+\pi$ are convenient for, in this interval, the function is non-zero only between $-\pi/2$ and $+\pi/2$ and:

$$A_n = \frac{1}{2\pi} \int_{-\pi/2}^{\pi/2} e^{-jnx}\, dx \tag{15-1}$$

If $n = 0$,

$$A_0 = \frac{1}{2\pi}\left(\frac{\pi}{2} + \frac{\pi}{2}\right) = \frac{1}{2} \tag{15-2}$$

If $n \neq 0$,

$$A_n = \frac{1}{-jn2\pi}\left[e^{-jnx}\right]_{-\pi/2}^{\pi/2} = \frac{1}{n\pi}\frac{e^{jn\pi/2} - e^{-jn\pi/2}}{2j} = \frac{\sin(n\pi/2)}{n\pi} \tag{15-3}$$

If n is even, $A_n = 0$. If n is odd, there are two possibilities:

$$A_n = \frac{1}{\pi n} \quad \text{if } n = \ldots, -11, -7, -3, +1, +5, +9, \ldots \tag{15-4}$$

$$A_n = -\frac{1}{\pi n} \quad \text{if } n = \ldots, -9, -5, -1, +3, +7, +11, \ldots \tag{15-5}$$

The spectrum is shown in Figure 15b. Synthesis gives:

$$\text{Square wave of Figure 15a} = \frac{1}{2} + \sum \left(\pm \frac{1}{\pi n} e^{jn\omega_1 t} \right) \qquad (15\text{-}6)$$

using† the $+$ sign for $n = \dots, -11, -7, -3, +1, +5, +9, \dots$, and $-$ for $n = \dots, -9, -5, -1, +3, +7, +11, \dots$.

It is not easy to visualize the terms of the series when they are complex exponentials, as it was when they were trigonometric waves. This is the disadvantage of the exponential form.

16. A RECURRENT PULSE

Other examples of the expansion of waves into exponential series could now be given: triangular waves, saw-tooth waves, rectifier waves, or waves of other interesting forms. One important and interesting wave is the series of rectangular pulses of Figure 16a; such pulses are used in circuits for television, radar, computers, and many coded communication systems. Fourier analysis shows whether they will be faithfully transmitted through communication or computer networks. Also, and even more important, our Fourier-series analysis of this flat-topped pulse leads into Fourier *integral* analysis and so into *Laplace transformation*.

Let the duration of a pulse be D, the time from pulse to pulse be T, and let the ratio $T/D = k$. One cycle of the fundamental radian frequency ω_1 is the time T, so $\omega_1 T = 2\pi$. Since we find it more convenient to use the dimensionless variable x than to use the time t, where $x = \omega_1 t$, we write $k = \omega_1 T/\omega_1 D$, and in Figure 16a the duration of the pulse on a scale of x is $\omega_1 D = \omega_1 T/k$. Since $\omega_1 T = 2\pi$, the duration of the pulse in terms of x can be written $2\pi/k$, and when the vertical axis is placed as in the diagram, the pulse lasts from $x = -\pi/k$ to $+\pi/k$. These, then, are the limits of integration

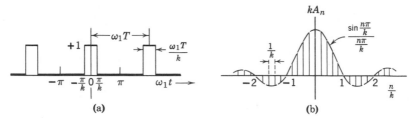

(a) (b)

FIGURES 16a,b A rectangular pulse and its spectrum (drawn for $k = 5$) (equation 16-3).

---

† This alternation of sign can be provided for by writing $(-1)^{(n-1)/2}$ in place of $\pm$ and specifying "for odd n."

in the following analysis in which the coefficients of a Fourier series are obtained:

$$A_n = \frac{1}{2\pi} \int_{-\pi/k}^{\pi/k} e^{-jnx} \, dx \tag{16-1}$$

If $n = 0$,

$$A_0 = \frac{1}{2\pi} \left(\frac{\pi}{k} + \frac{\pi}{k} \right) = \frac{1}{k} \tag{16-2}$$

If $n \neq 0$,

$$A_n = \frac{1}{-jn2\pi} \left[e^{-jnx} \right]_{-\pi/k}^{\pi/k} = \frac{1}{n\pi} \frac{e^{jn\pi/k} - e^{-jn\pi/k}}{2j}$$

$$= \frac{1}{k} \frac{\sin(n\pi/k)}{n\pi/k} \tag{16-3}$$

This can also be written, if preferred,

$$\frac{A_n}{\omega_1} = \frac{D}{2\pi} \frac{\sin(n\pi/k)}{n\pi/k} \tag{16-4}$$

The spectrum is shown in Figure 16b. In both parts of Figure 16 the ratio k is taken to be 5. If k were larger or smaller the vertical lines in Figure 16b, each of which represents a harmonic, would be closer together or farther apart, but they would be contained within the same envelope which is $(\sin n\pi/k)/(n\pi/k)$. (It is interesting to note that Figure 15a shows a special case of this flat-topped pulse for which $k = 2$.)

Equation 16-3 gives the analysis of the rectangular pulse. Formal synthesis to regain the rectangular pulse, which may be called $f(t)$, is accomplished as usual by using equation 13-5. The rectangular pulse is the sum of the components:

$$f(t) = \sum_{n=-\infty}^{\infty} \frac{1}{k} \frac{\sin(n\pi/k)}{n\pi/k} e^{jn\omega_1 t} \tag{16-5}$$

In this example a pulse of duration D is followed in T seconds by another pulse, and so on. Let us now consider that the time between pulses is increased, so that T becomes much greater, though D remains the same. Then the ratio $k = T/D$ becomes greater, and the fundamental frequency $\omega_1 = 2\pi/T$ becomes correspondingly less. The vertical lines that represent harmonics in Figure 16b are now packed closer together, but always within the same envelope. Hence Figure 16b will serve for a pulse with any repetition rate by merely putting in more vertical lines as T is increased and the time between pulses becomes greater.

As the pulses become less frequent, with T becoming greater while D remains the same, the limit is seen to be a *single* pulse of duration D, with other pulses so much earlier or so much later that they do not matter. In this limit, $T = \infty$, $k = \infty$, and ω_1 approaches zero. The frequency difference between successive harmonics is ω_1, and as this becomes smaller the lines of the spectrum are more closely packed together so that the lines merge, in the limit, into a continuous spectrum in which there is a value of amplitude A for *every* frequency.

Since, by equation 16-3, the amplitudes all become smaller, all approaching zero, one might guess that in the limit all usefulness of the concept would be lost. On the contrary, the limit points the way to the valuable Fourier integral.

17. THE FOURIER INTEGRAL

Analysis of a time function into the components of a Fourier series, and synthesis of the components to regain the time function, are given in equations 13-6 and 13-5, respectively. With a few minor changes, these equations can be restated as follows:

$$A_n = \frac{\omega_1}{2\pi} \int_{-T/2}^{T/2} f(t) e^{-j\omega_n t} \, dt \qquad f(t) = \sum_{n=-\infty}^{+\infty} A_n e^{j\omega_n t} \qquad (17\text{-}1)$$

The changes include:

1. Changing the variable of integration from x to t, where $x = \omega_1 t$, and putting the constant ω_1 outside the integral sign.
2. Writing T for the time of one period of the function, so that $T = 2\pi/\omega_1$.
3. Changing the limits of integration from values of x to values of t, and providing for integration through one period of the function by integrating from $t = -T/2$ to $+T/2$.
4. Writing ω_n for $n\omega_1$; that is, $\omega_n = n\omega_1$.
5. In the synthesis, writing n instead of m, a trivial change.

Now let us consider the possibility of extending the limits of integration so greatly that, instead of representing a function through a finite time (however long), we represent it through all time. It is not good enough merely to extend the limits of integration to $-\infty$ and $+\infty$, for increasing T makes ω_1 approach zero, and all values of A_n in equation 17-1 tend to disappear. We lose our function A_n.

But a function that will not be lost as T increases without limit is A_n/ω_1. As A_n becomes small, so does ω_1, and the ratio does *not* approach zero. To have something to call it, let us designate this ratio by the symbol g_n; by definition

$$g_n = \frac{A_n}{\omega_1} \tag{17-2}$$

Rewriting the pair of equations 17-1 in terms of g_n,

$$g_n = \frac{1}{2\pi}\int_{-T/2}^{T/2} f(t)e^{-j\omega_n t}\,dt \qquad f(t) = \sum_{n=-\infty}^{\infty} g_n e^{j\omega_n t}\omega_1 \tag{17-3}$$

We can now find g_n when T increases without limit.

Now let T become infinite in equation 17-3. This makes ω_1 small, approaching zero, and we call it $d\omega$; it is the fundamental frequency, and also the difference between the frequencies of successive harmonics. As ω_1 becomes $d\omega$, the lines of the harmonic spectrum become closer and tend to merge as the separation between them vanishes, and what was for the Fourier series a line spectrum of discrete harmonics becomes for the Fourier integral a continuous spectrum of all frequencies. The discrete frequencies of the harmonics in the series were called ω_n; they now merge into the continuous frequency variable ω. Similarly the amplitudes of the harmonics had discrete values in the series as indicated by g_n in equation 17-3, but as the discrete harmonic frequencies become a continuous frequency variable we write $g(\omega)$ instead of g_n. Thus the following changes are made. Let:

$$T \to \infty \qquad \omega_1 \to d\omega \qquad \omega_n \to \omega \qquad g_n \to g(\omega) \tag{17-4}$$

With these changes, the analysis of equations 17-3 becomes equation 17-5. The synthesis of equations 17-3 becomes the integral† in equation 17-6.

$$g(\omega) = \frac{1}{2\pi}\int_{-\infty}^{\infty} f(t)e^{-j\omega t}\,dt \tag{17-5}$$

$$f(t) = \int_{-\infty}^{\infty} g(\omega)e^{j\omega t}\,d\omega \tag{17-6}$$

These equations are the *Fourier integral pair*. They will analyze a non-repeating time function as the Fourier series analyzes a cyclic time function. As an example, consider a *single* flat-topped pulse.

18. ANALYSIS OF THE RECTANGULAR PULSE

Figure 18a shows a rectangular pulse of duration $D = 2\tau$. Find $g(\omega)$ by means of the Fourier integral.

We use equation 17-5 (sometimes called the prism transformation because it gives the spectrum) and compute:

---

† It will be recognized that *integration* is the limit of summation.

$$g(\omega) = \frac{1}{2\pi} \int_{-\tau}^{\tau} (1)e^{-j\omega t}\, dt = -\frac{1}{2\pi j\omega} \left[e^{-j\omega t} \right]_{-\tau}^{\tau}$$

$$= \frac{1}{\pi\omega} \frac{e^{j\omega\tau} - e^{-j\omega\tau}}{2j} = \frac{1}{\pi\omega} \sin \omega\tau = \frac{\tau}{\pi} \frac{\sin \omega\tau}{\omega\tau} \qquad (18\text{-}1)$$

Although this integration is fundamentally between the limits $-\infty$ and $+\infty$, it is only necessary in this example to integrate from $-\tau$ to $+\tau$, for everywhere else the function being integrated is zero. The result, $g(\omega)$, gives g as a function of frequency, ω in this expression being of course the variable, and τ being a constant, the half-length of the pulse analyzed.

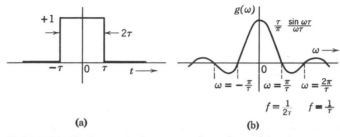

(a)   (b)

FIGURES 18a,b   A single rectangular pulse and its spectrum.

The similarity of this $g(\omega)$ to A_n/ω_1 in equation 16-4 is apparent. The similarity is even more obvious in Figures 16b and 18b. The difference is equally important: the single rectangular pulse contains *all* frequencies from zero to infinity (except those that are integer multiples of $1/2\tau$) in the proportions shown in Figure 18b. Thus Figure 18b shows a *continuous* spectrum, Figure 16b a *line* spectrum.

19. SYNTHESIS OF THE RECTANGULAR PULSE

The spectrum $g(\omega)$ has been obtained in equation 18-1 by direct transformation, substituting the rectangular pulse for $f(t)$ in the analysis equation 17-5. Inversely, it should be possible to substitute the spectrum into the synthesis equation 17-6 and regain the rectangular pulse as $f(t)$. To do so would correspond to adding terms of a series.

This inverse transformation can, indeed, be done. Substituting $g(\omega)$ from equation 18-1 into the synthesis equation 17-6 gives for $f(t)$:

$$f(t) = \int_{-\infty}^{\infty} \frac{\tau}{\pi} \frac{\sin \omega\tau}{\omega\tau} e^{j\omega t}\, d\omega \qquad (19\text{-}1)$$

This integral can be evaluated, though not without some intricacy of algebra, and with its infinite limits it is found to have a value of 1 when t is between $-\tau$ and $+\tau$ and to be zero at other times both earlier and later. But this is exactly the original pulse, the $f(t)$ of Figure 18a, so the synthesis has thus been achieved by integration.

20. CONVERGENCE

There are definite limits to the theoretical applicability of the Fourier integral, but in the analysis of physically possible forms of voltage or current the only important limitation is that the pulse to be analyzed must not go on forever undiminished. That is, a pulse of finite length can be analyzed, but we cannot handle the step function that results when a battery is connected into a circuit and left connected from that time forth.

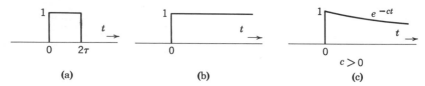

FIGURE 20   A Fourier transform can be found for a pulse   (a),   but not for a step function   (b)   unless a convergence factor is used   (c).

Thus we have found the spectrum of the pulse of Figure 20a, but if a similar analysis is attempted for the step function of Figure 20b there is trouble with the upper limit of integration which, for the step function, must be infinity:

$$g(\omega) = \frac{1}{2\pi} \int_{-\infty}^{\infty} f(t)e^{-j\omega t}\, dt = \frac{1}{2\pi} \int_{0}^{\infty} 1e^{-j\omega t}\, dt \tag{20-1}$$

Then

$$g(\omega) = \frac{1}{2\pi} \left[\frac{1}{-j\omega} e^{-j\omega t} \right]_{0}^{\infty} \tag{20-2}$$

This integration gives trouble with the upper limit, for no value can be assigned to $e^{-j\infty}$. If the exponent were real the exponential would be zero, but with a purely imaginary exponent the function is periodic and does not approach a limit. Hence the Fourier integral cannot be evaluated; it is said not to *converge*.

Although the function of Figure 20b, with a continuous value of 1 through positive time, does not give a Fourier transform, the somewhat similar function of Figure 20c that is zero for negative time and equal to e^{-ct} through positive time (c is real and positive) does give a Fourier transform:

$$g(\omega) = \frac{1}{2\pi} \int_0^\infty e^{-ct} e^{-j\omega t}\, dt = \frac{1}{2\pi} \left[\frac{1}{-(c+j\omega)} e^{-ct} e^{-j\omega t} \right]_0^\infty$$

$$= \frac{1}{2\pi} \frac{1}{c+j\omega} \tag{20-3}$$

Here, in the upper limit, e^{-ct} becomes zero for infinite t.

But, by letting c approach zero, Figure 20c can be made as much like Figure 20b as is desired. The limit of the exponential function as $c \to 0$ is the constant 1; at the same time, in equation 20-3, the limit of $g(\omega)$ as $c \to 0$ is

$$\lim_{c \to 0} g(\omega) = \frac{1}{2\pi} \frac{1}{j\omega} \tag{20-4}$$

Thus a spectrum is found as a limit.

This suggests the possibility of treating other functions in the same way. Suppose $f(t)$ is a time function that is zero for negative time but has, during positive time, a value such that the Fourier integral does not converge. We multiply this function $f(t)$ by e^{-ct} (specifying that c is real and positive) and call the product $f_1(t)$:

$$f_1(t) = f(t)e^{-ct} \tag{20-5}$$

We now use this $f_1(t)$ in the Fourier integral:

$$g_1(\omega) = \frac{1}{2\pi} \int_0^\infty f(t)e^{-ct} e^{-j\omega t}\, dt \tag{20-6}$$

The integral of equation 20-6 converges, giving a value for $g_1(\omega)$, if the function $f(t)$ is a steady battery voltage, a steadily alternating voltage, or some other steady voltage for which, alone, the Fourier integral does not converge. By this use of e^{-ct} as a *convergence factor* we are able to find a limit for $g_1(\omega)$, and in this sense

$$g(\omega) = \lim_{c \to 0} g_1(\omega) \tag{20-7}$$

Equation 20-6 can be written

$$g_1(\omega) = \frac{1}{2\pi} \int_0^\infty f(t)e^{-(c+j\omega)t}\, dt \tag{20-8}$$

We now introduce a new symbol s, which is defined as

$$s = \sigma + j\omega \tag{20-9}$$

and let c be a particular value of σ, a positive real constant value. Equation 20-8, now a function of s, is

$$g_1(s) = \frac{1}{2\pi} \int_0^\infty f(t)e^{-st}\, dt \qquad (20\text{-}10)$$

or

$$2\pi g_1(s) = \int_0^\infty f(t)e^{-st}\, dt \qquad (20\text{-}11)$$

21. THE LAPLACE TRANSFORM

Equation 20-11 has the form of the Laplace transform, for which the customary symbol is $F(s)$. Let us therefore write

$$F(s) = \int_0^\infty f(t)e^{-st}\, dt \qquad (21\text{-}1)$$

Clearly $F(s) = 2\pi g_1(s)$, and the Laplace transform is the same as the Fourier transform with a convergence factor, except for the trivial change that the constant 2π is contained within $F(s)$. We may again note, however, the change that was made when a convergence factor was introduced: the lower limit of integration in equation 21-1 is zero, whereas in equation 17-5 the lower limit was $-\infty$.†

Equation 21-1 is the direct Laplace transformation. The inverse Laplace transformation, analogous to the inverse Fourier transformation of equation 17-6, is:

$$f(t) = \frac{1}{2\pi j} \int_{s=c-j\infty}^{c+j\infty} F(s)e^{st}\, ds \qquad (21\text{-}2)$$

The same steps that change the direct Fourier transformation into the direct Laplace transformation also relate the inverse transformations. That is, the variable of integration is changed from ω to s, where $s = \sigma + j\omega$. The Fourier path of integration is from $\omega = -\infty$ to $+\infty$, and this would correspond to a Laplace path of integration from $s = -j\infty$ to $+j\infty$. However, the Laplace integration is taken along the line $\sigma = c$, so the integration is from $s = c - j\infty$ to $c + j\infty$; this use of c is comparable to a convergence factor. The constant 2π, of course, must be taken out of $F(s)$. Equation 21-2 is the result.

---

† For the Fourier transform with a convergence factor, or for the one-sided or unilateral Laplace transform of equation 21-1 (the bilateral form will not be discussed here), we have the choice of specifying that $f(t) = 0$ for $t < 0$, or alternatively specifying that $f(t)$ is not necessarily zero for $t < 0$ and restricting both direct and inverse transformations to $t \geq 0$. The latter alternative is less restrictive on $f(t)$ and is the more useful, but the former is easier to visualize.

22. TRANSFORMATION AND THE COMPLEX PLANE

It is interesting to consider the Laplace transformation in its relation to the complex s plane. We are given a time function $f(t)$ and wish to find its transform $F(s)$. Equation 21-1 defines $F(s)$ (within limits), and from it we compute a formula for $F(s)$. This formula for $F(s)$ has a value corresponding to each point in the s plane, and it is convenient to visualize a surface to represent the magnitude of $F(s)$, lying at a height $|F(s)|$ above the s plane. At certain values of s, called poles, the formula for $F(s)$ becomes infinite.

Clearly, the location of poles of $F(s)$ depends in some way on the given $f(t)$. Let us suppose the given $f(t)$ is the sum of components of exponential form, and one of these components is $e^{s_1 t}$ (through positive time). The transform for this component is

$$F(s) = \int_0^\infty e^{s_1 t} e^{-st}\, dt \qquad (22\text{-}1)$$

Note that s is a variable, whereas s_1 is a specific value. Writing $s = \sigma + j\omega$, and letting $s_1 = \sigma_1 + j\omega_1$, equation 22-1 can be written:

$$F(s) = \int_0^\infty e^{(\sigma_1 - \sigma)t + j(\omega_1 - \omega)t}\, dt \qquad (22\text{-}2)$$

If σ is greater than σ_1 integration is straightforward, and

$$F(s) = \frac{1}{s - s_1} \qquad (22\text{-}3)$$

This is the desired formula† for $F(s)$, and we see that it has a pole at $s = s_1$. This pole can be marked by a cross on the s plane.

Depending on the form of the given time function $f(t)$, its transform $F(s)$ may have poles anywhere in the s plane. Several poles are shown in Figure 22a.

Inverse transformation is accomplished by the integral of equation 21-2. This integration is along the dashed line of Figure 22a which is a vertical line to the right of all poles of $F(s)$, at a constant distance c from the vertical axis. One way to evaluate this integral is to use the principles of integration in the complex plane, from which it can be shown that the required integral is determined by the *residues* at each pole of $F(s)$ (see Section 19 of Chapter 14). The value of the integral, and hence the value determined for $f(t)$, is not

---

† If σ is equal to or less than σ_1, the integral cannot be evaluated. $F(s)$ is defined by equation 21-1 for all values of s that lie to the right of any poles of $F(s)$ in the s plane, but elsewhere in the s plane (except at poles), $F(s)$ is defined by *analytic continuation*, and this is accomplished by letting the same formula describe $F(s)$ in the left-hand region as in the right-hand region. Such "extension by preservation of form" gives a function, analytic except at poles, that is unique.

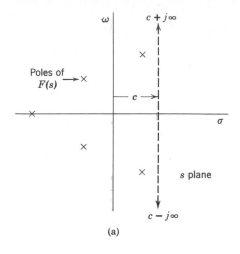

(a)

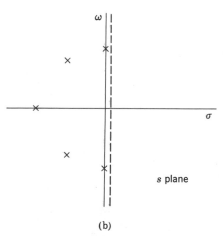

(b)

FIGURE 22   (a)   The path of integration in equation 21-2 must be to the right of all poles of $F(s)$.   (b)   The constant c can approach zero if there are no poles in the right half-plane.

dependent on the value of c provided only that the line of integration lies to the right of all poles of $F(s)$. This restriction requires that the time function e^{st} must grow (with time) more rapidly than any component of the time function $f(t)$.

The magnitude of c does not need to be specified in the Laplace transformation. It may be 2, or 10, or any other number, provided it is greater than σ at any pole of $F(s)$. If $F(s)$ has poles only in the left half-plane, c can be zero and integration can be along the vertical axis; this corresponds to using the *Fourier* integral. If $F(s)$ has poles on the vertical axis and perhaps in the left half-plane, but not in the right half-plane, as in Figure 22b, c can *approach* zero (from the right). This corresponds to using the Fourier integral with a convergence factor.

It is interesting that the Fourier integral synthesizes a time function by summing an infinity of *steady* sinusoidal components (each of infinitesimal size), whereas the Laplace integral synthesizes the time function by summing an infinity of *exponentially growing* sinusoidal components (each of infinitesimal size). These sinusoidal components result from considering the exponential components in conjugate pairs; for Fourier analysis the exponents are imaginary, and for the Laplace synthesis they are complex.

It is recognized that this chapter does not give a *proof* of Fourier or Laplace transformation. Rather, a plausibility argument is offered, and rigorous proof is left to a mathematical study. It is the *use* of the transform method that is so valuable to us, and the next two chapters are devoted to such applications. It is suggested that these last two sections be reviewed from time to time during the study of the next chapters, for each will clarify the other.

23. SUMMARY

A nonsinusoidal periodic wave can be approximated as the sum of a *constant* value plus a *fundamental*-component wave plus a number of *harmonic* components. If the series is extended to higher harmonics without limit, the resulting infinite series approaches the given periodic wave as a limit. The series of trigonometric terms is a *Fourier* series.

Analysis of the given wave into a Fourier series is achieved by integration, using the *orthogonality* of trigonometric functions. The *square* wave and the *saw-tooth* wave give examples.

Addition of the components of a square wave illustrates the way in which their sum approaches the given square wave.

Analysis is aided by three possible types of *symmetry* in a given wave. The given wave may be: an *even* function, an *odd* function, or a function with *half-wave* symmetry. If there is symmetry, certain components of the series *vanish*.

The Fourier series is *cyclic* at fundamental frequency. Its sum approaches the given function through this period. The *limits of integration* for purposes of analysis must in general be one fundamental cycle apart. However, if the given function has one of the foregoing types of *symmetry*, the limits of integration may be only one-half cycle apart, or if there are two types of symmetry, one-quarter cycle apart.

With rectifier waves as an example, *network analysis* by Fourier series is illustrated. A nonsinusoidal voltage is applied to a linear network. This voltage is *analyzed* into its sinusoidal Fourier components. Each component of voltage is divided by the *corresponding impedance* to obtain the components of current. Current components are then *summed* to find the total current.

Certain principles of *rectifier design* are adduced.

The *effective rms* value of a voltage or current is found from its Fourier components. The *power* to an element or network is found from the Fourier components of applied *voltage* and entering *current*.

Another *trigonometric* form of Fourier series is given.

An *exponential* form of series is deduced. This is important because it leads to the Fourier integral.

Symmetry causes certain components of the exponential series to vanish.

A *recurrent pulse* of rectangular form is used to illustrate the exponential series. The separation between pulses is allowed to become greater. In the limit, a single pulse remains alone. In the limit, the Fourier series becomes the *Fourier integral*.

The *Fourier integral pair* is given. As the Fourier *series* analyzes a given function through a *limited* time, so the Fourier *integral* analyzes a function through *all* time.

The Fourier integral must *converge*. If, for a given function, the Fourier integral does not converge, it is possible that the integral will converge when the given function is multiplied by a *convergence factor*.

A *Laplace transformation* is a Fourier transformation with a built-in convergence factor. It considers a given time function as the sum of exponentially *expanding* components. It is correspondingly less restricted than the Fourier transformation.

Laplace analysis and Fourier analysis are considered in relation to the *complex s plane*. In particular, inverse transformation is an integral in the s plane along the line $s = c$, where c is a positive real constant for Laplace transformation, or c is zero (or approaches zero) for Fourier transformation.

PROBLEMS

13-1. Substitute $f(t)$ from equation 3-12 or 3-13 into equation 3-10 to prove that it is thereby reduced to an identity.   §3

13-2. Current is 1 ampere for 1 millisecond, from $t = -0.5$ to $+0.5$ millisecond. Current is then zero for 3 milliseconds, from $t = 0.5$ to 3.5 milliseconds. The current pulse is repetitive and continues to be repeated with a 4-millisecond period. (*a*) Sketch the wave of current, and (*b*) find its analysis in the form of equation 3-1. Give terms through the fifth harmonic.   §5*

13-3. An analysis of the triangular wave of Figure 8a is given as equation 8-2. Plot the four components given in that equation and add them. Does it seem likely that the infinite series will converge to the triangular wave? §6

13-4. The first five terms of the output of a half-wave rectifier (Figure 9a) are given in equation 9-5. Plot these five component terms and add them. Does it seem likely that the infinite series will converge to the rectified half-wave? §6

13-5. (a) The diagram shows a function from $x = 0$ to $x = \pi/2$. Complete the function over the full period from $x = -\pi$ to $x = +\pi$ with the requirement that the Fourier series of the function must contain only odd-harmonic cosine terms. (b) Find the first nonvanishing coefficient of the Fourier series of the function, as completed. §8*

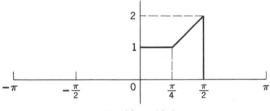

Problem 13-5.

13-6. Show that any single-valued function that is neither an even function nor an odd function is the sum of an even function and an odd function. Give formulas for finding the even part of a function $f(t)$, and the odd part of $f(t)$. Prove that one of these parts is indeed an even function, the other an odd function, and that their sum is $f(t)$. §8*

13-7. It is stated in Section 8 that "if the wave being analyzed is even ... we can integrate from 0 to π and multiply the result by 2." Prove that this statement is correct. §8

13-8. It is stated in Section 8 that "if the wave being analyzed is odd ... we can integrate from 0 to π and multiply the result by 2." Prove that this statement is correct. §8

13-9. Voltage is 10 volts for 3 milliseconds and then zero for 1 millisecond, repeated periodically. Sketch the wave and analyze it in the form of equation 3-1, selecting zero time at such a point that the series will contain no sine terms. Give the coefficients of all terms through the fifth harmonic. The series may be obtained by any means, but the method must be shown. §8*

13-10. Considering symmetry, what terms will not appear in the Fourier series (form of equation 3-1) for the function shown? Determine (by any means) the remaining terms, and give the series through the third harmonic. §8*

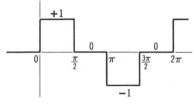

Problem 13-10.

13-11. Given $f(t) = (\omega t/\pi)^2$, find the Fourier series that converges to this $f(t)$ between $\omega t = -\pi$ and $\omega t = +\pi$, and that repeats with a period of 2π. Find $a_0, a_1, a_2, a_3, b_1, b_2, b_3$. Sketch the component waves to show that the series can reasonably be expected to approach $f(t)$.   §8*

13-12. Given $f(t) = e^{t/\pi}$, find the Fourier series that converges to this $f(t)$ between $t = -\pi$ and $+\pi$. Find the coefficients through the third harmonic; sketch the component waves and see whether they approach $f(t)$.   §8

13-13. Half of a cycle of a wave, from $x = 0$ to $x = \pi$, is described by the following equation:

$$y = 100\left(\frac{4x}{\pi} - \frac{4x^2}{\pi^2}\right)$$

The wave has half-wave symmetry. The appearance of the wave is somewhat sinusoidal, but not precisely so. Write a Fourier series for the wave, giving all terms through the seventh harmonic.   §8*

13-14. Find the Fourier series (through the fifth harmonic) for the clipped sine wave shown.   §8

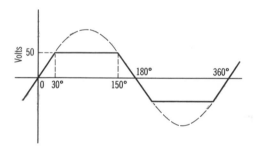

Problem 13-14.

13-15. Derive equation 9-6 for the rectified full wave.   §9

13-16. A triangular voltage wave (Figure 8a) is applied to the terminals of a pure inductance. Find a series for current. Sketch the resulting wave.   §11*

13-17. The ripple filter of Figure 11a is used in the output of a 60-Hz *half*-wave rectifier (Figure 9a). What are the frequencies of the two largest ripple components in the filter output? What percentage of the dc component is the amplitude of each of these? Sketch the sum of the dc and the largest ripple component.   §11*

13-18. The filter of Problem 13-17 is considered unsatisfactory for use with a half-wave rectifier. Compare the improvement resulting from the following two changes: (*a*) double both L and C in the present filter, or (*b*) put another identical filter section in tandem between the present filter and the load.

§11*

13-19. A Hay impedance bridge is shown; its purpose is to measure L and R_L. With the values shown (ohms, farads, henrys) the bridge is balanced for the fundamental frequency, which is 1000 Hz. Unfortunately the applied voltage wave is practically triangular (Figures 8a and b), with a crest value of 1 volt.

Assuming infinite galvanometer impedance, find the rms value of third harmonic voltage across the galvanometer G.   §11*

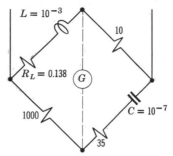

$L = 10^{-3}$

$R_L = 0.138$ G

10

1000

$C = 10^{-7}$

35

Problem 13-19.

13-20. From the series, find the rms values of the waves of Figures 4a (square), 5a (saw-tooth), 8a (triangular), 9a (rectified half-wave), and 9c (rectified full wave).   §12*

13-21. Voltage $v = 30(\cos \omega t - \frac{1}{3}\cos 3\omega t)$ with $\omega = 377$ is applied to $R = 1.0$ ohm and $L = 2.65$ millihenrys in series. (a) Compute steady-state current. (b) Sketch the voltage wave and its components, and also the current wave and its components, giving particular attention to phase relations. (c) Compute power to the circuit.   §12*

13-22. Current $i = a + b\cos \omega t$ flows through resistance R. Compute the average power consumed in the resistor from $P = $ Average (vi). Compute P also from equation 12-4. Do the results agree?   §12

13-23. Substitute equation 13-5 into equation 13-6 to prove that the exponential expansion is correct.   §13

13-24. A square wave of voltage described by an exponential series with $V_n = 2V_{max}/jn\pi$ (for odd n) is applied to a coil having resistance R and inductance L. $R = 1$ ohm; $\omega_1 L = 1$ ohm. Find (a) the impedance (written for the nth harmonic) and (b) the transform of current. (c) Express the current as an exponential series, giving terms from A_{-4} to A_4. Do you recognize the series as describing any known wave? From other information, what would you expect to be the form of the current wave?   §15*

13-25. A function $f(t)$ is zero from $t = -10$ to $t = -2$. It is then equal to 1 from $t = -2$ to $t = +2$. It is then zero from $t = +2$ to $t = +10$. It then repeats with a period of 20. Expand in an exponential Fourier series, finding terms through the third harmonic.   §16*

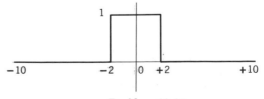

1

-10 -2 0 $+2$ $+10$

Problem 13-25.

13-26. Draw a rectangular pulse (as in Figure 16a) for which $k = 10$, and (as in 16b) draw its spectrum.   §16

13-27. By means of the Fourier integral, find $g(\omega)$ for a pulse that has a value of -1 from time $-\tau$ to 0, and a value of $+1$ from time 0 to $+\tau$, and is zero at all other times. Draw sketches to show this given pulse, which is $f(t)$, and its spectrum $g(\omega)$.   §18*

13-28. A single triangular pulse is shown in the figure. Use the Fourier integral to find its spectrum. Sketch $g(\omega)$, the spectrum.   §18*

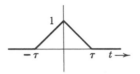

Problem 13-28.

13-29. Find the Fourier transform $g(\omega)$ of the time function $f(t)$. For $t < 0$, $f(t) = e^{\alpha t}$; for $t > 0, f(t) = 0; \alpha > 0$.   §18

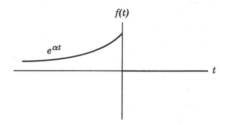

Problem 13-29.

13-30. A pulse of current, as shown in the diagram, is $i = e^{\alpha t}$ for all negative time, and is $i = e^{-\alpha t}$ for all positive time. By means of the Fourier integral, find the spectrum of this pulse.   §18*

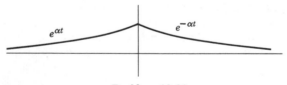

Problem 13-30.

13-31. If $f(t) = te^{-t}$ for $t > 0$, and $f(t) = 0$ for $t < 0$, find $g(\omega)$. Sketch f as a function of t, and g as a function of ω.   §18*

13-32. Compute and sketch the Fourier transform $g(\omega)$ for the following time function $f(t)$: between $t = -1$ and $t = +1, f(t) = \cos(\pi t/2)$; before $t = -1$ and after $t = +1, f(t) = 0$.   §18

13-33. In the figure, $f(t) = \cos \omega_0 t$ for $-\tau < t < +\tau$. This is the radio-frequency current during a *dot* of telegraph code, or a television spot of uniform brightness. (*a*) Show that the Fourier-integral spectrum of this signal is

$$g(\omega) = \frac{\tau}{2\pi} \left[\frac{\sin(\omega - \omega_0)\tau}{(\omega - \omega_0)\tau} + \frac{\sin(\omega + \omega_0)\tau}{(\omega + \omega_0)\tau} \right]$$

[*Hint:* Let $\cos \omega_0 t = \frac{1}{2}(e^{j\omega_0 t} + e^{-j\omega_0 t})$ before integrating.] (*b*) Letting $\omega_0 \tau = 8\pi$, sketch $g(\omega)$ and compare with Figure 18b. §18

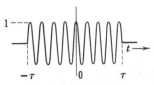

Problem 13-33.

13-34. By means of the Fourier integral, find $g(\omega)$ for a pulse that has a value of $+1$ from time 0 to $+2\tau$ and is zero at all other times. Show that $|g(\omega)|$ equals the $g(\omega)$ of equation 18-1. §18*

13-35. A circuit consists of resistance R and inductance L in series. Voltage $v(t)$ is applied to this circuit; $v(t) = 0$ for all negative time, and $v(t) = V_m e^{-\alpha t}$ for all positive time (α is a positive real constant). Find the resulting current $i(t)$ by the following process: transform $v(t)$ by Fourier transformation, and divide the transform of voltage by the impedance. Sketch the voltage and the current. §18*

13-36. Can you show that if a function $f_1(t)$ is even, its Fourier transform $gF_1(\omega)$ is even or odd or real or imaginary or has any other simple symmetry? Can you show that if a function $f_2(t)$ is odd, its Fourier transform $g_2(t)$ is even or odd or real or imaginary or has any other simple symmetry? §17

$$a_n = \frac{2}{T} \int_{t_1}^{t_2} f(t) \cos n\omega_0 t \, dt$$

$$\frac{2\pi}{T} = \omega_0$$

$$b_n = \frac{2}{T} \int_{t_1}^{t_2} f(t) \sin n\omega_0 t \, dt$$

$$f(t) = -f(-t) \quad odd; \; a_n^0$$

$$f(t) = f(-t) \quad even; \; b_n^0$$

$$a_0 = \frac{1}{T} \int_{t_1}^{t_2} f(t) \, dt$$

$\bullet$ HWS = only odd harmonics

CHAPTER
14

THE LAPLACE TRANSFORMATION

1. THE LAPLACE TRANSFORM PAIR

The usual form of the direct Laplace transformation is (as in equation 21-1 of Chapter 13):

$$F(s) = \int_0^\infty f(t)e^{-st}\, dt \tag{1-1}$$

The inverse transformation (as in equation 21-2 of Chapter 13) is:

$$f(t) = \frac{1}{2\pi j} \int_{c-j\infty}^{c+j\infty} F(s)e^{st}\, ds \tag{1-2}$$

The principal use of the Laplace transformation is in the solution of differential equations. A differential equation, or a set of simultaneous differential equations, or integro-differential equations, can be transformed into algebraic equations which are much easier to handle; the algebraic equations are solved and then inverse transformation gives the desired result. The transformation process is highly useful with linear differential equations in general, as will be discussed in Section 10. First, however, let us consider transformation for solving electric network problems, for this is a familiar use of differential equations and we already know a number of special techniques (such as the representation of voltage and current by complex transforms, and the concept of impedance and admittance) which we use with Laplace solutions.

Suppose a voltage is to be applied to a network, and current is to be found

352

by Laplace transformation; the method is not unfamiliar. First, the voltage, whatever its form, is expressed as the sum of components; this is the direct Laplace transformation. Then this voltage transform is divided by the network impedance; the result is the current transform. Finally the actual current is determined from the current transform by the inverse transformation in what amounts to adding the components of current. That is all. In our earlier examples we shall assume that initial current is zero, but the restriction will later be removed.

2. AN EXAMPLE: PAIR 1

Let a constant battery voltage of 1 volt be applied to R and L in series, as in Figure 2a, by closing a switch at $t = 0$. Find the current.

First, we analyze the voltage function into exponential components by equation 1-1. The voltage, a unit step function, is often represented by the symbol $u(t)$, and as shown in Figure 2b it has two values:

$$u(t) = \begin{cases} 0 & \text{if } t < 0 \\ 1 & \text{if } t > 0 \end{cases} \tag{2-1}$$

This unit function is $f(t)$ in equation 1-1, and since its value is 1 between the limits of integration, equation 1-1 for this particular function is merely

$$F(s) = \int_0^\infty (1)e^{-st}\, dt = -\frac{1}{s}(0 - 1) = \frac{1}{s} \tag{2-2}$$

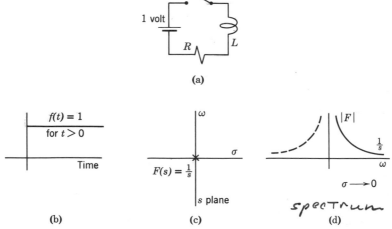

(a)

(b)   (c)   (d)

FIGURE 2   (a)   A unit step of voltage.   (b)   A step function,   (c)   the pole of the transform, and   (d)   the profile of the transform along the ω axis. Pair 1.

TABLE 14-2 Laplace Transform Pairs

Pair No.	$f(t)$, the Time Function, for $t > 0$ (the Voltage or Current)	$F(s)$, the Laplace Transform (the Spectrum)	See Equation	See Figure
1	1	$\dfrac{1}{s}$	Chap. 14 2-2	Chap. 14 2b
2	$e^{-\alpha t}$	$\dfrac{1}{s + \alpha}$	3-1	3a
3	$\dfrac{1}{\alpha}(1 - e^{-\alpha t})$	$\dfrac{1}{s(s + \alpha)}$	3-2	3d
4	$\dfrac{1}{s_1 - s_2}(e^{s_1 t} - e^{s_2 t})$	$\dfrac{1}{(s - s_1)(s - s_2)}$	4-2	4a
5	$te^{-\alpha t}$	$\dfrac{1}{(s + \alpha)^2}$	4-4	4d
6	$\dfrac{1}{\omega_1}(e^{-\alpha t}\sin \omega_1 t)$	$\dfrac{1}{(s - s_1)(s - s_2)}$ $s_1, s_2 = -\alpha \pm j\omega_1$	4-7	4g
7	$\dfrac{1}{\omega_1}\sin \omega_1 t$	$\dfrac{1}{(s - j\omega_1)(s + j\omega_1)} = \dfrac{1}{s^2 + \omega_1^2}$	4-10	4j

	$f(t)$		$F(s)$	Pole-zero plot		
8	$\cos \omega_1 t$		$\dfrac{s}{(s-j\omega_1)(s+j\omega_1)} = \dfrac{s}{s^2+\omega_1^2}$		Chap. 15 14-2	Chap. 15
9	$\dfrac{\sin(\omega_1 t - \theta) + e^{-\alpha t}\sin\theta}{\omega_1\sqrt{\alpha^2+\omega_1^2}}$ where $\theta = \tan^{-1}\dfrac{\omega_1}{\alpha}$		$\dfrac{1}{(s+\alpha)(s^2+\omega_1^2)}$		14-13	14b
10	$\dfrac{1}{s_1 - s_2}\left(s_1 e^{s_1 t} - s_2 e^{s_2 t}\right)$		$\dfrac{s}{(s-s_1)(s-s_2)}$		Chap. 14 13-6	
11	$\dfrac{e^{s_1 t}}{(s_1 - s_2)(s_1 - s_3)}$ $+ \dfrac{e^{s_2 t}}{(s_2 - s_1)(s_2 - s_3)}$ $+ \dfrac{e^{s_3 t}}{(s_3 - s_1)(s_3 - s_2)}$		$\dfrac{1}{(s-s_1)(s-s_2)(s-s_3)}$			
12	t		$\dfrac{1}{s^2}$		Chap. 15 13-2	
13	δ		1		Chap. 16 2-4	Chap. 16 1a

Thus $1/s$ is the transform of the step function. This is **pair 1** of Table 14-2 on page 354 (and inside the back cover).

The spectrum is plotted in Figure 2d. The step function has components of all frequencies, with the low frequencies relatively large and the high frequencies small, the amplitude being inversely proportional to the frequency. This interpretation results from letting s in equation 2-2 equal $c + j\omega$, and then letting c approach zero.

The impedance function of the circuit is:

$$Z(j\omega) = R + j\omega L \quad \text{or} \quad Z(s) = R + Ls \tag{2-3}$$

Now we divide the voltage transform by the impedance function to get the current transform. To avoid confusion we shall hereafter use the symbol $V(s)$ for the transform of voltage [instead of $F(s)$], and $I(s)$ for the transform of current. Then

$$V(s) = \frac{1}{s}$$

$$I(s) = \frac{V(s)}{Z(s)} = \frac{1/s}{R + Ls} = \frac{1}{L}\frac{1}{s(s + R/L)} \tag{2-4}$$

Thus we have found the spectrum of current, the transform of the time function of current. As in circuit problems worked by means of Fourier series (and even as in circuit problems using complex numbers to represent a single steady-state frequency), the final step of solution is to obtain the actual time function of current. We know the spectrum of current; how can we find the actual total current itself?

The obvious suggestion is to use equation 1-2 to find the inverse Laplace transform. There is, however, a practical difficulty: the integral of that equation is not easy to evaluate.

Since equation 1-2 is difficult, perhaps equation 1-1 can be used. This, the direct transformation, might be worked backwards. Can we find some time function that can be inserted as $f(t)$ in equation 1-1 to yield an $F(s)$ of the form $1/s(s + \alpha)$, as needed in equation 2-4?

This is indeed the correct question to ask, but to it we have no present answer. We have had no experience with the Laplace transformation that might provide results upon which to draw. That lack, however, can be remedied. Let us transform a number of useful-looking time functions by equation 1-1, tabulating the results as we go, so that we shall be provided with a stock of functions of s that may serve as solutions for circuit problems.

3. TRANSFORMATION OF FUNCTIONS

Pair 2. Examples of Chapter 6 suggest that the exponential time function is likely to appear in the solutions of circuit equations. Let us apply the Laplace transformation to analysis of a function that is 0 for $t < 0$ and $e^{-\alpha t}$ for $t > 0$. This exponential time function is substituted for $f(t)$ in equation 1-1, giving

$$F(s) = \int_0^\infty e^{-\alpha t} e^{-st}\, dt = \int_0^\infty e^{-(s+\alpha)t}\, dt = \frac{1}{s + \alpha} \qquad (3\text{-}1)$$

Figure 3a shows the exponential function of time. The spectrum shown in Figure 3c is obtained from the transform $1/(s + \alpha)$ by letting $s = j\omega$. Figure 3b is a plot of poles and zeros of $F(s)$ in the s plane.

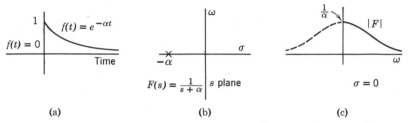

(a)   (b)   (c)

FIGURES 3a,b,c   **An exponential function and its transform. Pair 2.**

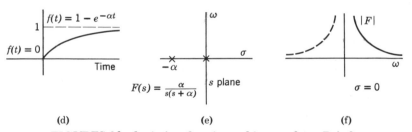

(d)   (e)   (f)

FIGURES 3d,e,f   **A time function and its transform. Pair 3.**

In equation 3-1 the transform of $e^{-\alpha t}$ is found. It is not the solution to the problem of Section 2; it does not turn out to match the current transform of equation 2-4. It may, however, be useful in some later problem, so it should not be forgotten. Indeed, it is not too early to start collecting transform pairs in a table. Though we have only two pairs, let us start Table 14-2, and enter these as pairs 1 and 2 of the table. This table will be extended as more pairs are found, and can then serve in much the same way as a table of integrals; when we have a time function and want its spectrum, or when we have a spectrum and want its time function, we shall look in the table to try to find a similar form.

Pair 3. So far we have pretended that we do not know the answer to the problem of Section 2. As a matter of fact, we know from Chapter 6 that the answer is a current of the general form $(1 - e^{-\alpha t})$. Guided by our knowledge, let us insert this time function as $f(t)$ in the Laplace transformation equation 1-1:

$$F(s) = \int_0^\infty (1 - e^{-\alpha t})e^{-st}\, dt = \int_0^\infty [e^{-st} - e^{-(s+\alpha)t}]\, dt$$

$$= \frac{1}{s} - \frac{1}{s+\alpha} = \frac{\alpha}{s(s+\alpha)}$$

$$(3\text{-}2)$$

The result is entered as pair 3 in Table 14-2, with α transposed for convenience.

Now pair 3 is of the form of the current transform in equation 2-4, · provided $\alpha = R/L$. Hence we finally arrive at a solution of the problem of Section 2. Equation 2-4 gives the spectrum or transform of an unknown current as

$$I(s) = \frac{1}{L}\frac{1}{s(s+R/L)} \qquad (3\text{-}3)$$

and we find from pair 3 that the corresponding time function of current is

$$i(t) = \frac{1}{L}\frac{1}{\alpha}(1 - e^{-\alpha t}) = \frac{1}{R}(1 - e^{-(R/L)t}) \qquad (3\text{-}4)$$

Figure 3d shows the time function. Its transform has two poles in the s plane, one at the origin and one at $-\alpha$. The spectrum looks much like that of the step function (Figure 3c); they are indistinguishable at the low-frequency end, but the high-frequency content of Figure 3f is somewhat less.

4. MORE TRANSFORM PAIRS

Pair 4. Let us continue deriving transform pairs. The natural current in a circuit of R, L, and C in series may be described by three different time functions, and for the overdamped case (large R) current is of the form (for $t > 0$):

$$f(t) = e^{s_1 t} - e^{s_2 t} \qquad (4\text{-}1)$$

To analyze this with the Laplace transformation, equation 1-1:

$$F(s) = \int_0^\infty (e^{s_1 t} - e^{s_2 t})e^{-st}\, dt$$

$$= \frac{s_1 - s_2}{(s - s_1)(s - s_2)}$$

$$(4\text{-}2)$$

If s_1 and s_2 are negative real quantities, as they are for the circuit in question, the integration gives no trouble and we add this analysis to our table as pair 4.

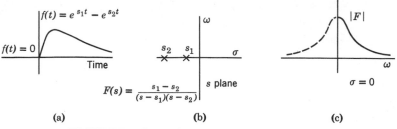

$$F(s) = \frac{s_1 - s_2}{(s - s_1)(s - s_2)}$$

(a)   (b)   (c)

FIGURES 4a,b,c   The overdamped surge. Pair 4.

Figure 4a shows the time function, 4b the two poles in the complex plane, and 4c the spectrum.

Pair 5.   A time function that may arise from a similar circuit is (for $t > 0$):

$$f(t) = te^{-\alpha t} \tag{4-3}$$

Analyzing by equation 1-1,

$$F(s) = \int_0^\infty (te^{-\alpha t})e^{-st}\, dt = \frac{1}{(s + \alpha)^2} \tag{4-4}$$

This time function is shown in Figure 4d. Because the denominator of equation 4-4 is a square, the pole is a second-order pole.

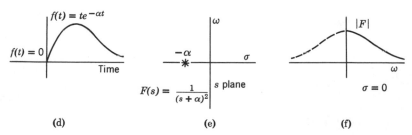

$$F(s) = \frac{1}{(s + \alpha)^2}$$

(d)   (e)   (f)

FIGURES 4d,e,f   The critically damped surge. Pair 5.

Pair 6.   The damped sine wave (Figure 4g) that describes oscillatory current in the RLC circuit is (for $t > 0$):

$$f(t) = e^{-\alpha t} \sin \omega_1 t \tag{4-5}$$

Hence

$$F(s) = \int_0^\infty (e^{-\alpha t} \sin \omega_1 t)e^{-st}\, dt = \frac{\omega_1}{(s + \alpha)^2 + \omega_1^2} \tag{4-6}$$

This is the form that results directly from integration, but the poles of $F(s)$ do not readily appear in equation 4-6. We therefore expand the squared term and factor the resulting quadratic to obtain

$$F(s) = \frac{\omega_1}{(s - s_1)(s - s_2)} \tag{4-7}$$

where s_1 and s_2 represent

$$s_1 = -\alpha + j\omega_1 \qquad s_2 = -\alpha - j\omega_1 \tag{4-8}$$

This shows clearly the poles of $F(s)$. They are a conjugate complex pair, as in Figure 4h. It is interesting, although hardly unexpected, to find that this time function, so different in appearance from the two preceding, is nevertheless analyzed into a closely related transform.

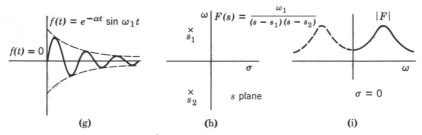

FIGURES 4g,h,i   The damped sine wave. Pair 6.

The spectrum, Figure 4i, can be visualized from the positions of the poles (possibly thinking in terms of a rubber sheet above the s plane). It will clearly have a symmetrical pair of humps in the positive and negative frequency ranges, where the ω axis comes closest to the poles; it will droop, but not to zero, at the origin.

Pair 7.   The limiting case, the discontinuous undamped sine wave, is (for $t > 0$)

$$f(t) = \sin \omega_1 t \tag{4-9}$$

as shown in Figure 4j.

The transform may be found by integration:

$$F(s) = \int_0^\infty (\sin \omega_1 t) e^{-st} \, dt = \frac{\omega_1}{s^2 + \omega_1^2} = \frac{\omega_1}{(s - j\omega_1)(s + j\omega_1)} \tag{4-10}$$

(It is not, in fact, necessary to integrate, for this result is more easily obtained from equation 4-6 by letting α approach zero.)

The poles of $F(s)$ are this time an imaginary pair. The spectrum therefore is as shown in Figure 4l, for the ω axis of the s plane passes directly through the poles.

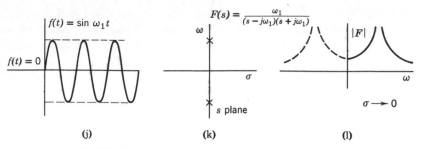

$$F(s) = \frac{\omega_1}{(s - j\omega_1)(s + j\omega_1)}$$

$f(t) = \sin \omega_1 t$

$f(t) = 0$

s plane

$\sigma \longrightarrow 0$

(j) (k) (l)

FIGURES 4j,k,l   The undamped sine wave. Pair 7.

5. EXAMPLES

A Circuit Example.   In Figure 5a, the switch is closed at $t = 0$, connecting a battery of independent voltage v_0 to R and C. Find current, the capacitor being initially uncharged.

Step 1.   The transform of the voltage is found, using pair 1:

$$V(s) = \frac{v_0}{s} \tag{5-1}$$

Step 2.   The voltage transform is divided by the impedance:

$$Z(s) = R + \frac{1}{sC} = \frac{sCR + 1}{sC} \tag{5-2}$$

$$I(s) = \frac{V(s)}{Z(s)} = \frac{v_0 sC}{s(sCR + 1)} = \frac{v_0}{R} \frac{1}{(s + 1/RC)} \tag{5-3}$$

Step 3.   Letting $1/RC = \alpha$, pair 2 applies, and

$$i(t) = \frac{v_0}{R} e^{-\alpha t} \tag{5-4}$$

as in Figure 5b.

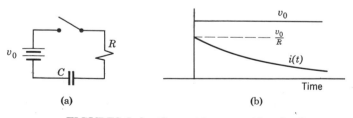

v_0

R

v_0

C

v_0

$\frac{v_0}{R}$

$i(t)$

Time

(a) (b)

FIGURES 5a,b   Current in a capacitive circuit.

A Network Example.   In Figure 5c the switch is closed at $t = 0$, connecting a 12-volt battery to a network of two loops with values of resistance and inductance shown. Find i_1 and i_2 thereafter, both currents being initially zero.

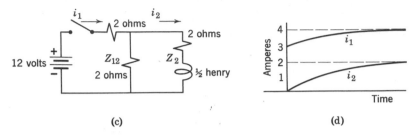

(c)   (d)

FIGURES 5c,d   A network and its two loop currents.

Step 1.   The transform of voltage, by pair 1, is:

$$V(s) = \frac{12}{s} \qquad (5\text{-}5)$$

Step 2.   For i_1 the input impedance is needed, and this is:

$$Z(s) = 2 + \frac{2(\frac{1}{2}s + 2)}{2 + \frac{1}{2}s + 2} = \frac{4s + 24}{s + 8} \qquad (5\text{-}6)$$

Hence:

$$I_1(s) = \frac{V(s)}{Z(s)} = \frac{12}{s}\frac{s + 8}{4s + 24} = 3\frac{s + 8}{s(s + 6)} \qquad (5\text{-}7)$$

The current i_2 is a fraction of i_1, and using the methods of Chapter 4 the transforms are related as follows:

$$I_2(s) = I_1(s)\frac{Z_{12}(s)}{Z_{12}(s) + Z_2(s)} = 3\frac{s + 8}{s(s + 6)}\frac{2}{\frac{1}{2}s + 4} = \frac{12}{s(s + 6)} \qquad (5\text{-}8)$$

Step 3.   The time function for current i_2 is found from the transform I_2 by pair 3 of the table, noting that the number 12 is merely a constant coefficient:

$$i_2(t) = \frac{12}{6}(1 - e^{-6t}) = 2(1 - e^{-6t}) \qquad (5\text{-}9)$$

The time function for current i_1 would correspondingly be found from the transform I_1 of equation 5-7, but unfortunately this transform does not correspond to any pair yet computed, or listed in our table. However, we may recognize that

$$I_1(s) = 3\frac{1}{s+6} + 24\frac{1}{s(s+6)} \tag{5-10}$$

If it can now be assumed that the transform of the sum of two functions of s is the sum of the transforms, we can recognize each term of equation 5-10 and write, from pairs 2 and 3:

$$i_1(t) = 3e^{-6t} + \tfrac{24}{6}(1 - e^{-6t}) = 4 - e^{-6t} \tag{5-11}$$

The two currents, which have now been found, are plotted in Figure 5d.

In computing these currents we made a number of assumptions. One of these, relating to the sum of two transforms, seems reasonable, but will be considered together with a number of other *transformations of operations* in the next few sections.

6. TRANSFORMATION OF OPERATIONS

A number of time functions, the step, the exponential, the damped sine, and others, have been transformed in preceding paragraphs by means of the transformation equation, 1-1. The resulting pairs have been listed in Table 14-2 (and also inside the back cover of the book) along with a number of other useful transformations of functions. This transformation of functions can be carried on indefinitely, but it is found that certain generalizations can be discovered that often save time and trouble. Perhaps a transform can be recognized as the sum of two known transforms, or as the product of two known transforms, or as the integral or the derivative of a known transform. A number of theorems relating to the transformation of *operations* will now be given.[†] These are listed in Table 14-7 (and inside the front cover).

As a matter of notation, we shall hereafter follow the customary practice of writing the script letter $\mathscr{L}$ to represent the direct Laplace transformation. That is, $\mathscr{L}f(t) = \int_0^\infty f(t)e^{-st}\,dt$. If, for instance, we know that $F(s)$ is the transform of $f(t)$, we can write $F(s) = \mathscr{L}f(t)$, and this means the same as equation 1-1. In a similar way, $\mathscr{L}^{-1}$ is the inverse operation, and it is easier to write $f(t) = \mathscr{L}^{-1}F(s)$ than to write equation 1-2.

The transform of a constant times a function is the constant times the transform of the function. That is, if

$$\mathscr{L}f(t) = F(s) \quad \text{then} \quad \mathscr{L}Af(t) = A\mathscr{L}f(t) = AF(s) \tag{6-1}$$

(where A is not a function of either t or s).

[†] Credit for all theorems in this chapter goes to Gardner and Barnes. Their classic book is recommended to the reader for more extensive study, more precise mathematics, more comprehensive tables, and many other aspects of Laplace analysis.

The validity of this theorem follows from equation 1-1, for when the integral relation is fully written out it is evident from our knowledge of calculus that

$$\int_0^\infty Af(t)e^{-st}\, dt = A \int_0^\infty f(t)e^{-st}\, dt \qquad (6\text{-}2)$$

as in equation 6-1.

A special case that may be emphasized to avoid possible later confusion appears when $f(t)$ is a time function that is constant through positive time. For instance, if a time function is $Bu(t)$, its transform is $\mathscr{L}Bu(t) = B/s$. Since the Laplace transform of equation 1-1 is concerned only with positive time, it is easy to write a time function that is constant through positive time as B. It is essential to notice in every transformation that is made whether a constant is merely a coefficient or whether it is indeed a time function of constant value.

7. LINEARITY

The transform of the sum of functions is the sum of the transforms. If

$$\mathscr{L}f_1(t) = F_1(s) \quad \text{and} \quad \mathscr{L}f_2(t) = F_2(s)$$

then

$$\mathscr{L}[f_1(t) + f_2(t)] = \mathscr{L}f_1(t) + \mathscr{L}f_2(t) = F_1(s) + F_2(s) \qquad (7\text{-}1)$$

This also is evident when the integrals are written out.

These transformations of *operations*, and others, are collected in Table 14-7. The relations of equations 6-1 and 7-1 were used in the foregoing examples without proof.

8. REAL DIFFERENTIATION

Differentiation of a function transforms to multiplication by s, with an added constant. That is, if

$$\mathscr{L}f(t) = F(s) \qquad (8\text{-}1)$$

then

$$\mathscr{L}\left[\frac{d}{dt}f(t)\right] = sF(s) - f(0_+) \qquad (8\text{-}2)$$

Here $f(0_+)$ means the value of $f(t)$ when $t = 0$, and the $+$ indicates that if there is a discontinuity in $f(t)$ at $t = 0$, then $f(0_+)$ is the limit of $f(t)$ as t approaches 0 from the positive side.

TABLE 14-7 Transformation of Operations

Pair	Operation on Time Function	Operation on Laplace Transform	See Equation
			Chap. 14
I	$Af(t)$	$AF(s)$	6-1
II	$f_1(t) + f_2(t)$	$F_1(s) + F_2(s)$	7-1
III	$\dfrac{d}{dt} f(t)$	$sF(s) - f(0_+)$	8-2
IV	$\dfrac{d^2}{dt^2} f(t)$	$s^2 F(s) - sf(0_+) - f'(0_+)$	8-15
V	$\int f(t)\,dt$	$\dfrac{F(s)}{s} + \dfrac{1}{s} \int f(t)\,dt \Big|_{t=0_+}$	9-11
VI	$\int_0^t f(t)\,dt$	$\dfrac{F(s)}{s}$	9-15
VII	$e^{-\alpha t} f(t)$	$F(s + \alpha)$	
VIII	$e^{-\alpha t} \int_0^t e^{\alpha t} f(t)\,dt$	$\dfrac{F(s)}{s + \alpha}$	
			Chap. 16
IX	$f(t - \alpha) \cdot u(t - \alpha)$	$e^{-s\alpha} F(s)$	4-6
X	$f_1(t) * f_2(t)$	$F_1(s) \cdot F_2(s)$	5-4
	Given: $\mathscr{L}f(t) = F(s)$	$\mathscr{L}f_1(t) = F_1(s) \qquad \mathscr{L}f_2(t) = F_2(s)$	

The integral that is indicated by equation 8-1 is

$$\int_0^\infty f(t) e^{-st}\,dt = F(s) \tag{8-3}$$

It is convenient to let $f'(t)$ indicate the derivative of $f(t)$:

$$\frac{d}{dt} f(t) \triangleq f'(t) \tag{8-4}$$

(The symbol $\triangleq$ means "equals by definition.") The proof now uses the method of integration by parts, for which the general theorem is

$$\int u\,dv = uv - \int v\,du \tag{8-5}$$

We let

$$u = f(t) \qquad\qquad dv = e^{-st}\,dt \tag{8-6}$$

$$du = df(t) = \frac{df(t)}{dt}\,dt = f'(t)\,dt \qquad v = -\frac{1}{s} e^{-st} \tag{8-7}$$

Equation 8-3 is expanded by equation 8-5, using equations 8-6 and 7:

$$F(s) = -\left[f(t)\frac{1}{s}e^{-st}\right]_{t=0}^{\infty} + \int_{0}^{\infty}\frac{1}{s}e^{-st}f'(t)\,dt \qquad (8\text{-}8)$$

Since the real component of s is positive, the first term is zero when $t = \infty$. When $t = 0$, the exponential is 1 and $f(t)$ is $f(0_+)$. Hence

$$F(s) = \frac{1}{s}f(0_+) + \frac{1}{s}\int_{0}^{\infty}f'(t)e^{-st}\,dt \qquad (8\text{-}9)$$

Rearranging, we obtain

$$\int_{0}^{\infty}f'(t)e^{-st}\,dt = sF(s) - f(0_+) \qquad (8\text{-}10)$$

But the left-hand member is the Laplace transform of the derivative of $f(t)$, and a mere change of notation gives equation 8-2.

Also we may consider how the *second derivative* and other higher derivatives can be transformed by an extension of the preceding theorem. We know for the first derivative that if

$$\mathscr{L}f(t) = F(s) \quad \text{then} \quad \mathscr{L}f'(t) = sF(s) - f(0_+) \qquad (8\text{-}11)$$

Similarly, if

$$\mathscr{L}f'(t) = F'(s) \quad \text{then} \quad \mathscr{L}f''(t) = sF'(s) - f'(0_+) \qquad (8\text{-}12)$$

where

$$f''(t) \triangleq \frac{d}{dt}f'(t) = \frac{d^2}{dt^2}f(t) \qquad (8\text{-}13)$$

In equation 8-11, $F'(s) = \mathscr{L}f'(t) = sF(s) - f(0_+)$, so equation 8-12 becomes

$$\mathscr{L}f''(t) = s[sF(s) - f(0_+)] - f'(0_+) \qquad (8\text{-}14)$$

which means that

$$\mathscr{L}\frac{d^2}{dt^2}f(t) = s^2F(s) - sf(0_+) - f'(0_+) \qquad (8\text{-}15)$$

where $f(0_+)$ is the value of $f(t)$ at $t = 0$ (as approached from the positive side), and $f'(0_+)$ is the value of the derivative, $f'(t)$, at $t = 0$ (as approached from the positive side).

9. REAL INTEGRATION

Integration of a Function Transforms to Division by s.   More precisely,

$$\mathscr{L}\left[\int_{0}^{t}f(t)\,dt\right] = \frac{F(s)}{s} \qquad (9\text{-}1)$$

Proof is much like that of equation 8-2. Let it be given that

$$F(s) = \mathscr{L}f(t) = \int_0^\infty f(t)e^{-st}\, dt \qquad (9\text{-}2)$$

The indefinite integral of $f(t)$ will be called $g(t)$:

$$\int f(t)\, dt \triangleq g(t) \qquad (9\text{-}3)$$

and its transform $G(s)$:

$$\mathscr{L}g(t) \triangleq G(s) = \int_0^\infty g(t)e^{-st}\, dt \qquad (9\text{-}4)$$

We shall again use the general formula for integration by parts:

$$\int u\, dv = uv - \int v\, du \qquad (9\text{-}5)$$

this time letting

$$u = e^{-st} \qquad\qquad dv = f(t)\, dt \qquad (9\text{-}6)$$

$$du = -se^{-st}\, dt \qquad v = \int f(t)\, dt = g(t) \qquad (9\text{-}7)$$

Equation 9-2 is expanded by equation 9-5, using equations 9-6 and 7:

$$F(s) = \left[e^{-st}g(t) \right]_{t=0}^{\infty} + \int_0^\infty g(t)se^{-st}\, dt \qquad (9\text{-}8)$$

The exponential with exponent $-\infty$ is zero, and if $g(t) = g(0_+)$ when $t = 0$ (or approaches zero from the positive side), then

$$F(s) = -g(0_+) + s\int_0^\infty g(t)e^{-st}\, dt$$

$$= -g(0_+) + sG(s) \qquad (9\text{-}9)$$

$$= -g(0_+) + s\mathscr{L}g(t) \qquad (9\text{-}10)$$

Rearranging,

$$\mathscr{L}g(t) = \mathscr{L}\left[\int f(t)\, dt \right] = \frac{F(s)}{s} + \frac{g(0_+)}{s} \qquad (9\text{-}11)$$

Here we have the Laplace transform of the indefinite integral.

Since it is often more convenient to have the transform of the definite integral, we may put limits on equation 9-3 to get

$$\int_0^t f(t)\, dt = \left[g(t) \right]_0^t = g(t) - g(0_+) \qquad (9\text{-}12)$$

Transforming,

$$\mathscr{L}\left[\int_0^t f(t)\,dt\right] = \mathscr{L}g(t) - \mathscr{L}g(0_+) \tag{9-13}$$

Since $\mathscr{L}g(t)$ is given by equation 9-11, and, by pair 1,

$$\mathscr{L}g(0_+) = \frac{g(0_+)}{s} \tag{9-14}$$

equation 9-13 becomes

$$\mathscr{L}\left[\int_0^t f(t)\,dt\right] = \frac{F(s)}{s} \tag{9-15}$$

This is equation 9-1, the transform of the definite (or semi-indefinite) integral.

Repeated integration can clearly be considered an extension of this single integration. Repeated *indefinite* integration transforms with additional terms, as in equation 9-11. The transform of the nth *definite* integral (as in equation 9-15) is $F(s)/s^n$.

10. SOLUTION OF DIFFERENTIAL EQUATIONS

In Section 5 there are examples of Laplace solutions of electrical network problems. Our use of impedance tends to disguise the fact that these are the solutions of differential equations; we have used familiar techniques to make the work easier, and differential equations were not written. Now, however, having seen the Laplace method at its easiest, it is time to consider a more general application.

A differential equation relates an unknown function $y(t)$ and its derivatives to a known function $x(t)$. As an example

$$A\frac{d}{dt}y(t) + By(t) = x(t) \tag{10-1}$$

It is not necessary that t be time; t is a convenient letter. If the equation also includes terms that are *integrals* of $y(t)$, it may be called an *integro-differential* equation.

Each term of the equation is a function of t. Thus, whether $x(t)$ is sinusoidal or exponential or constant, it is to be considered a function of t. The problem is to find the unknown $y(t)$.

The solution of a differential equation, by whatever means, requires a known set of conditions from which to start. Let us consider that, at $t = 0$,

$y(t)$ has a value that will be called $y(0)$, its derivative has a value $y'(0)$, its second derivative is $y''(0)$, and so on, and as many of these are known as are needed.

The first step in Laplace analysis is to transform the differential equation. This is done by letting first one side and then the other be $f(t)$ in the Laplace transformation equation:

$$F(s) = \int_0^\infty f(t)e^{-st}\, dt \qquad (10\text{-}2)$$

Thus, as an example, transforming equation 10-1 in this manner gives

$$\int_0^\infty \left[A\frac{dy(t)}{dt} + By(t) \right] e^{-st}\, dt = \int_0^\infty x(t)e^{-st}\, dt \qquad (10\text{-}3)$$

or

$$\mathcal{L}\left[A\frac{dy(t)}{dt} + By(t) \right] = \mathcal{L}x(t) \qquad (10\text{-}4)$$

Note that equation 10-3 is not concerned with either $x(t)$ or $y(t)$ for $t < 0$, but only for $t > 0$. This means that if t is time, as in our solutions it usually is, the equation relates $x(t)$ and $y(t)$ for positive time, after the reference instant $t = 0$ at which initial conditions are known, but not for negative time. It is possible that $x(t) = 0$ and $y(t) = 0$ through all negative time (in circuit terms, that there was no voltage and no current until $t = 0$), but for equation 10-3 this is not necessary. The essential fact is that we know the values for y and its derivatives (the initial conditions) at the instant $t = 0$, and our solution for $y(t)$ proceeds from that time on.

Continuing our example with equation 10-1, let us find $y(t)$, given that $x(t) = C$, a constant, for all $t > 0$, and that initially, at $t = 0$, $y(0_+) = 0$. First we use pairs from Table 14-7 to make the transformations indicated in equation 10-4. The transform of $dy(t)/dt$ is $sY(s) - y(0_+)$, but we are given that $y(0_+) = 0$. The transform of $y(t)$ is $Y(s)$. We are given that $x(t) = C$, and the transform of C is C/s. Therefore the transformed equation is:

$$AsY(s) + BY(s) = \frac{C}{s} \qquad (10\text{-}5)$$

The next step is to solve for $Y(s)$:

$$Y(s) = \frac{C}{s(As + B)} = \frac{C}{A}\frac{1}{s(s + B/A)} \qquad (10\text{-}6)$$

We now find the inverse transform of each side of this equation; that is, we return to the "time domain" by going from right to left in Table 14-2. Using pair 3, inverse transformation of equation 10-6 gives

$$y(t) = \frac{C}{A}\frac{A}{B}[1 - e^{-(B/A)t}] = \frac{C}{B}[1 - e^{-(B/A)t}] \qquad (10\text{-}7)$$

and this is the required solution.

Let us test this solution. Does it satisfy the initial condition that $y(0) = 0$? When $t = 0$ in equation 10-7 the quantity in brackets is zero, so indeed $y(0) = 0$. Is equation 10-7 a solution of equation 10-1 for $t > 0$? Let us try substitution:

$$\frac{AC}{B}\left[\frac{B}{A}e^{-(B/A)t}\right] + \frac{BC}{B}[1 - e^{-(B/A)t}] = C \qquad (10\text{-}8)$$

so the given equation is satisfied. Our two checks are successful.

11. NON-ZERO INITIAL CONDITIONS

But suppose that when $t = 0$, y is not zero, but is given to be 2. That is, $y(0) = 2$. For this new initial condition we have, instead of equation 10-5:

$$A[sY(s) - 2] + BY(s) = \frac{C}{s} \qquad (11\text{-}1)$$

Then

$$Y(s) = \frac{C/s + 2A}{As + B} = \frac{C}{A}\frac{1}{s(s + B/A)} + \frac{2}{s + B/A} \qquad (11\text{-}2)$$

Inverse transformation uses pairs 3 and 2 to give

$$y(t) = \frac{C}{A}\frac{A}{B}[1 - e^{-(B/A)t}] + 2e^{-(B/A)t}$$

$$= \frac{C}{B} + \left(2 - \frac{C}{B}\right)e^{-(B/A)t} \qquad (11\text{-}3)$$

This is the required solution.

To check, is the initial y equal to 2? If $t = 0$ the exponential in equation 11-3 is 1, and $y(0) = 2$, as given. Is equation 11-3 a solution of the differential equation? Substituting into equation 10-1,

$$-A\frac{B}{A}\left(2-\frac{C}{B}\right)e^{-(B/A)t} + B\left|\frac{C}{B} + \left(2-\frac{C}{B}\right)e^{-(B/A)t}\right|$$

(11-4)

$$= (-2B + C)e^{-(B/A)t} + C + (2B - C)e^{-(B/A)t} = C$$

Thus the solution checks in the equation.

12. INVERSE TRANSFORMATION

Equations have been solved in the preceding sections by *implicit* inverse transformation. That is, in order to change back from the s domain to the time domain we have looked in our table of Laplace transform pairs to find a pair of the required form. In doing so we have been guided by our knowledge of the transformation of operations, such as recognizing that the transform of a sum is the sum of the transforms, and we have often made helpful algebraic changes in $F(s)$ before attempting its inverse transformation.

An algebraic change that frequently helps is to expand $F(s)$ by the method of *partial fractions*, for $F(s)$ can then be written as the sum of easily recognized functions. We shall often find, indeed, that the only transform pair then needed is the familiar:

$$\mathscr{L}e^{-\alpha t} = \frac{1}{s + \alpha}$$

(12-1)

It was seen in Chapter 5 (Section 13) that network functions (including admittances, impedances, and transfer functions) for lumped, linear networks can be put in the form:

$$F(s) = H\frac{(s - s_a)(s - s_b) \cdots (s - s_m)}{(s - s_1)(s - s_2) \cdots (s - s_n)}$$

(12-2)

In this form the function can be seen as the product of n terms, each having a denominator of the form $(s - s_k)$. This is not directly helpful because the inverse transform of a product is not easily known (although convolution will be considered in Chapter 16). However, the method of *partial fractions* permits us to change the product of fractions to the sum of fractions, and that is just what we need.

How can we expand equation 12-2 into the sum of fractions? First, it is necessary that equation 12-2 be a *proper* fraction; that is, n must be greater than m. If this is not so, the numerator can be divided by the denominator, giving one or more terms of impulsive nature (see Chapter 16) and a remainder that *is* a proper fraction.† Let us therefore consider a function in which m is less than n.

If the poles of $F(s)$ are all simple, meaning that s_1, s_2, and so on, are all different, so that the denominator does not have multiple roots, a partial-fraction expansion takes the form:

$$F(s) = \frac{K_1}{s - s_1} + \frac{K_2}{s - s_2} + \cdots + \frac{K_n}{s - s_n} \tag{12-3}$$

The values of K are constant and they are called the *residues* at their respective poles. These residues must, of course, be computed.

With the network function in the form of equation 12-3, it is evident that $f(t)$, the inverse transform of $F(s)$, is the sum of the transforms of the individual terms and with the aid of equaton 12-1 it can be written:

$$f(t) = K_1 e^{s_1 t} + K_2 e^{s_2 t} + \cdots + K_n e^{s_n t} \tag{12-4}$$

Note that a pole can be at the origin, making s_1 equal zero, in which case $K_1 e^{s_1 t} = K_1$, a constant. In equation 12-4 we have the time function, the inverse transform of $F(s)$, that is the answer to our problem. The residues must be computed.

13. PARTIAL FRACTIONS

Instead of continuing with the general form of network function shown in equation 12-2, it will perhaps be clearer to discuss an example. Suppose we are given the following network function, with the specification that there are only simple poles $(\alpha \neq \beta)$:

$$F(s) = \frac{Ms + N}{(s + \alpha)(s + \beta)} \tag{13-1}$$

---

† For instance, consider the improper function (in which $m = n$):

$$F(s) = \frac{As^2 + Bs + C}{s^2 + Ds + E} \tag{12-2a}$$

Division of numerator by denominator gives:

$$F(s) = A + \frac{Fs + G}{s^2 + Ds + E} \tag{12-2b}$$

Inverse transformation to obtain $f(t)$ from $F(s)$ gives:

$$f(t) = \mathscr{L}^{-1}F(s) = A\delta(t) + \mathscr{L}^{-1}\frac{Fs + G}{s^2 + Ds + E} \tag{12-2c}$$

where $A\delta(t)$ is the constant A times the impulse function $\delta(t)$ which is discussed in Chapter 16 (see pair 13). The remainder, with the numerator $Fs + G$, is a proper fraction and can be expanded into partial fractions.

and we are required to find its inverse transform $f(t)$. We first write this function as a sum:

$$\frac{Ms + N}{(s + \alpha)(s + \beta)} = \frac{A}{s + \alpha} + \frac{B}{s + \beta} \tag{13-2}$$

The general method is to multiply both sides of the equation by the left-hand denominator, giving

$$Ms + N = A(s + \beta) + B(s + \alpha)$$
$$= (A + B)s + (A\beta + B\alpha) \tag{13-3}$$

This is more than an equation, it is an identity: we require it to be true for *all* values of s. Therefore the coefficients of equal powers of s on the two sides of equation 13-3 must be equal, and

$$A + B = M \qquad A\beta + B\alpha = N \tag{13-4}$$

Simultaneous solution yields:

$$A = \frac{M\alpha - N}{\alpha - \beta} \qquad B = \frac{N - M\beta}{\alpha - \beta} \tag{13-5}$$

These residues are the coefficients of the terms in equation 13-2, and inverse transformation gives, in the form of equation 12-4:

$$f(t) = Ae^{-\alpha t} + Be^{-\beta t} \tag{13-6}$$

for which A and B as well as α and β are now known.

This section shows inverse transformation by means of partial fractions, and indeed it gives a new pair in which $F(s)$ has one finite zero and two finite poles, a pair that might be added to Table 14-2; indeed, if $M = 1$ and $N = 0$ it reduces to pair 10.

By a variation of this method, the simultaneous solution can sometimes be avoided.

Let us again start with the function $F(s)$ of equation 13-1. The function is to be expanded into a sum as in equation 13-2. Multiply each side by the denominator of A, which is $s + \alpha$:

$$\frac{Ms + N}{s + \beta} = A + B\frac{s + \alpha}{s + \beta} \tag{13-7}$$

Since this equation must be true for *all* values of s, it must be true for $s = -\alpha$. Substitution of $-\alpha$ for s gives

$$\frac{-M\alpha + N}{-\alpha + \beta} = A + 0 \tag{13-8}$$

from which A is known. Similarly, each side of equation 13-2 is multiplied by the denominator of B, which is $s + \beta$, and s is set equal to $-\beta$ to give B:

$$B = \frac{-M\beta + N}{-\beta + \alpha} \tag{13-9}$$

Thus A and B are found. If there are more than two factors in the denominator of equation 13-2, the technique remains essentially the same.

A and B of equations 13-8 and 13-9 are of course the same as in equation 13-5; two methods of computing residues are offered, and the second is usually the easier. The method is generalized in the following formula for finding residues at simple poles:

$$\text{Residue at } s_1 \text{ of } F(s) = \lim_{s \to s_1} (s - s_1)F(s) \tag{13-10}$$

14. RESIDUES FROM THE s PLANE

The foregoing computation of residues at simple poles can be visualized in the s plane in a helpful way.† A network used as an example in Chapter 5, Section 10, gave the network function:

$$Y(s) = \frac{2(s + 8)}{(s + 12)(s + 4)} \tag{14-1}$$

The response of this network to an applied battery of 12 volts (for which $V(s) = 12/s$) is to be found. First,

$$I(s) = \frac{24(s + 8)}{s(s + 12)(s + 4)} \tag{14-2}$$

Entering current $i(t)$ is the inverse transform of this $I(s)$, and the next step is to divide $I(s)$ into partial fractions.

This function has one zero and three poles as shown in Figure 14a, and the coefficient $H = 24$. We need the residues at the poles:

$$I(s) = \frac{K_0}{s} + \frac{K_{12}}{s + 12} + \frac{K_4}{s + 4} \tag{14-3}$$

Let us first find K_4. Multiplying by $(s + 4)$ and then letting $s = -4$ (the method of equation 13-8) we get:

$$K_4 = \frac{24(-4 + 8)}{-4(-4 + 12)} = \frac{24(4)}{(-4)(8)} = -3 \tag{14-4}$$

---

† See R. N. Clark, *Automatic Control Systems*, for discussion. (See Appendix 2.)

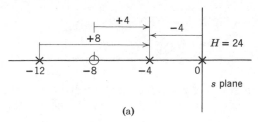

(a)

FIGURE 14a   To find K_4.

Now let us look at Figure 14a, and at equation 14-4. The numerator of the equation is H (we know that $H = 24$) times the vector in the s plane drawn from the zero (at $s = -8$) to the pole (at $s = -4$) at which the residue is being determined; this vector is $+4$. The denominator of the equation is the vector drawn from the pole at $s = -12$ to the pole (at $s = -4$) at which the residue is being obtained (this vector is 8), times the vector drawn from the pole at the origin to the pole in question (this vector is -4). The residue is the ratio.

Interpretation of the residue as the ratio of vectors can be generalized, and proof of the generalization comes from using the method on a function with the general form of equation 12-2:

$$\begin{bmatrix} \text{The residue} \\ \text{at a pole} \end{bmatrix} = \frac{H \cdot \begin{bmatrix} \text{Product of vectors from each} \\ \text{zero to the pole in question} \end{bmatrix}}{\begin{bmatrix} \text{Product of vectors from each other} \\ \text{pole to the pole in question} \end{bmatrix}} \quad (14\text{-}5)$$

Note that if there are no zeros the bracket in the numerator is 1.

It is now easy to find the residues at the other two poles of this function. Referring to Figure 14a, but ignoring the vectors drawn in that figure and visualizing other vectors to the pole at the origin where $s = 0$, mere mental calculation gives $K_0 = 24(8)/(4)(12) = 4$. Similarly, for the pole at $s = -12$, $K_{12} = 24(-4)/(-12)(-8) = -1$. With these values inserted, equation 14-3 gives the breakdown of equation 14-2 into partial fractions.

Our purpose in this section is to show the analysis of a function into partial fractions. It is of course obvious that the inverse transforms of these partial fractions are familiar exponential time functions, and the physical network problem with which we started can be concluded by writing the entering current as the inverse transform of equation 14-3:

$$i(t) = K_0 + K_4 e^{-4t} + K_{12} e^{-12t} = 4 - 3e^{-4t} - e^{-12t} \quad (14\text{-}6)$$

We have thus completed the problem of Section 11 of Chapter 5.

15. INTERPRETATION AT REAL POLES

When the inverse transform of a function is to be found, a number of important conclusions can be drawn from little more than a glance at the pole-zero map.

1. There are as many exponential terms in the time function as there are real (non-zero) poles on the map. If the map has a pole at the origin, there is a constant term. For each pair of conjugate complex poles on the map there is an oscillation.

2. Time constants are inversely proportional to distance of poles to the left; if far to the left, the term decays quickly.

3. The sign of the residue at a real pole, and hence the sign of the corresponding exponential term, depends on the number of poles *and* zeros to the right of the pole in question. The sign is $+$ if the number is even; and $-$ if odd. Thus in Figure 14a there are no poles or zeros to the right of the origin, so (since 0 is an even number) the sign of the constant term is $+$. There is 1 pole to the right of -4, so the term containing e^{-4t} in equation 14-6 has a negative sign, and so on. The reason for this rule is obvious from equation 14-5, remembering that the vector *to* s_1 *from* s_2 is $(s_1 - s_2)$ in sign as well as magnitude (or, if complex, in angle).

4. Residues at complex poles are considered in Section 16.

5. If a pole and a zero *coincide* on the map they should both be ignored, for their factors (in such an equation as 14-2) cancel each other. If a pole and a zero are close (compared to the distances to other poles), the residue at the pole is small and both pole and zero may *perhaps* be ignored as an approximation. Thus in Figure 14a the zero at -8 is not far from the pole at -12, and this is one reason for the term containing e^{-12t} being smaller than the other terms in equation 14-6. If the zero were at -11 instead of -8, the effect would be much more marked.

6. The residue at a *remote* pole is small. A remote pole is one that is at a considerable distance from a cluster of other poles, and the number of poles in the cluster must be at least 2 greater than the number of zeros in that cluster. The reason is seen from equation 14-5: for each pole in the cluster, the denominator contains one large factor because of the remote pole, but for the remote pole itself, the denominator contains *two* or more large factors. In Figure 14a, the pole at -12 is not very remote from the poles at 0 and at -4, but its remoteness is enough to contribute to the small size of the term containing e^{-12t}. The effect would be much greater if the pole were at -1 instead of -4, or if there were an additional pole at -1 as well. Hence it may be possible, as an approximation, to neglect a pole that is *remote to the left*, for the corresponding exponential term is both small and quickly over. When a remote pole at s_n is ignored, the coefficient H must be divided by

$-s_n$ to keep the residue at the origin and hence the steady final current correct. Thus, we have already computed that K_0 in Figure 14a is

$$K_0 = 24\frac{8}{(4)(12)} = 4 \tag{15-1}$$

If the pole at $s = -12$ is neglected, the (-12) is lost from the denominator, but the coefficient must be divided by $-s_1 = 12$, and the result is the same:

$$K_0 = \frac{24}{12}\cdot\frac{8}{4} = 4 \tag{15-2}$$

16. INTERPRETATION AT COMPLEX POLES

If there are complex poles they appear in conjugate pairs; the denominator of the network function has factors such as $(s - s_1)$ and $(s - s_1{}^*)$. To each conjugate pair of poles on the map there corresponds an oscillation in the time function, a term of the form

$$f(t) = Ae^{-t/\tau}\sin(\omega_1 t - \varphi) \tag{16-1}$$

Values of A, τ, φ, and ω_1 can be determined from the residues. The residues are most easily found by multiplying and dividing vectors (as in previous sections) but since the vectors are now complex it is probably easier to measure than to compute them. First, map the pair of complex poles and all other poles and zeros of a function $F(s)$ in the s plane (Figure 16c is an example). Let us call the upper pole of the complex pair s_1; draw (or imagine) vectors to s_1 from all other poles and from zeros, and measure their lengths and angles on the map. Then, to find the oscillation corresponding to this particular pair of poles, use the following rules. (These rules are readily obtained from principles that we already know.) Referring to Figures 16a and b:

1. Let the *upper* pole of the pair be at s_1, a value known from the network function. Since s_1 is complex,

$$s_1 = \sigma_1 + j\omega_1 \tag{16-2}$$

The residue at this upper pole we call K_1. K_1 is complex, and its magnitude and angle can be written:

$$K_1 = |K_1|\underline{/K_1} \tag{16-3}$$

Compute the *magnitude* $|K_1|$ from vector *lengths* as in equation 14-5.

2. Sketch the exponential envelope within which the oscillation is to be contained. The initial value A of the envelope is

$$A = 2|K_1| \tag{16-4}$$

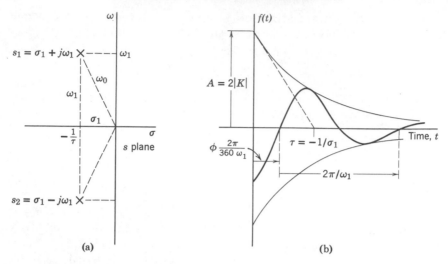

FIGURE 16 (a) **A pair of poles, perhaps among other poles and zeros, and** (b) **the resulting oscillation.**

The time constant τ of the envelope is

$$\tau = -\frac{1}{\sigma_1} \text{ seconds.} \tag{16-5}$$

3. The oscillation passes (upward) through zero at φ degrees, for which time (on the time axis) is $2\pi\varphi/(360\,\omega_1)$ seconds. (This angle φ is the angle by which the oscillation would lag behind a pure sine wave.) Since:

$$\varphi = -\underline{/K_1} - 90 \text{ degrees} \tag{16-6}$$

it follows from equation 14-5 that:

$$\varphi = \sum \text{(Angles of vectors to } s_1 \text{ from other poles)}$$
$$-\sum \text{(Angles of vectors from zeros)} - 90 \text{ degrees} \tag{16-7}$$

Mark this point on the time scale.

4. The period of the oscillation is $2\pi/\omega_1$ seconds. Use this value to mark each crossing of the time axis, and perhaps each point of tangency to the envelope, and draw the desired curve.

These four rules are intended to make it easy to draw an oscillation. They can also be used with equation 16-1 to obtain an analytical result; merely fill in ω_1 from equation 16-2, A from 16-4, τ from 16-5, and φ from 16-7. An example of finding residues at complex poles will now be given.

Example. Given the following $F(s)$, find its inverse transform $f(t)$. (If a physical meaning helps visualization, this problem could be to find voltage across the capacitor of an *RLC* series circuit, each element being 1 unit, when unit step voltage is applied, or, alternatively, to find entering current when unit ramp voltage is applied to the same *RLC* circuit.) Let:

$$F(s) = \frac{1}{s(s^2 + s + 1)} \tag{16-8}$$

There are three poles which can be distinguished by the subscripts 0, 1, and 1*. These are shown in Figure 16c:

$$s_0 = 0 \qquad s_1 = -\frac{1}{2} + j\frac{\sqrt{3}}{2} \qquad s_{1*} = -\frac{1}{2} - j\frac{\sqrt{3}}{2} \tag{16-9}$$

The residue at the pole at the origin is 1 divided by the vectors from the other poles:

$$K_0 = \frac{1}{(1\underline{/60°})(1\underline{/-60°})} = 1 \tag{16-10}$$

This pole gives a constant 1 in the time function.

From the complex poles, using rule 1, $\sigma_1 = -\frac{1}{2}$ and $\omega_1 = \sqrt{3}/2$. The residue at s_1 is complex; its magnitude is 1 divided by the distances from the other two poles which are 1 and $\sqrt{3}$. (These distances could be measured on the *s* plane, but in this example they are obvious.)

$$|K_1| = \frac{1}{(1)(\sqrt{3})} = \frac{1}{\sqrt{3}} \tag{16-11}$$

Hence, by rule 2:

$$A = \frac{2}{\sqrt{3}} \quad \text{and} \quad \tau = 2 \tag{16-12}$$

By rule 3, the angle φ is the angle of the vector to s_1 from s_2, which is 90 degrees, plus the angle of the vector to s_1 from the origin, which is 120 degrees, less 90 degrees, so:

$$\varphi = 90 + 120 - 90 = 120 \text{ degrees} \tag{16-13}$$

The time function can now be written. The pole at the origin yields a constant term in the time function, and the pair of complex poles yields an oscillation of the form given in equation 16-1. Hence:

$$f(t) = 1 + \frac{2}{\sqrt{3}} e^{-t/2} \sin\left(\frac{1}{2}\sqrt{3}t - \varphi\right) \tag{16-14}$$

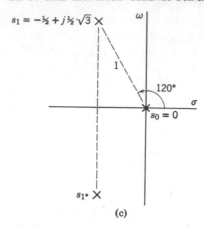

(c)

FIGURE 16c   Equation 16-14 results from these poles.

The angle φ is 120 degrees. Note that both terms of the argument of the sine function must be expressed in the same units (radians or degrees) before they can be combined.

17. A PARTIAL-FRACTION FORM FOR COMPLEX POLES

The residue at a complex pole is, as we have seen, a complex quantity, but the resulting expression for current is real. It is possible to rearrange the partial-fraction terms to avoid complex quantities entirely. This method looks simpler, but probably requires more actual work.

Two of the terms of a partial-fraction expansion, such as $K_1/(s - s_1)$ and $K_{1*}/(s - s_{1*})$ can, if desired, be left together in the form $(As + B)/(s^2 + Ms + N)$. This form with a real quadratic denominator then takes the place of the two fractions with complex denominators. The procedure of Section 13 is then followed, and there are no complex quantities in the computation. An example will perhaps be helpful.

Consider again the function from Section 16:

$$F(s) = \frac{1}{s(s^2 + s + 1)} \tag{17-1}$$

Let us expand, this time, into partial fractions of the form

$$F(s) = \frac{K_0}{s} + \frac{As + B}{s^2 + s + 1} \tag{17-2}$$

Multiply each of these equations by the denominator of equation 17-1 to obtain:

$$\begin{aligned} 1 &= K_0(s^2 + s + 1) + (As + B)s \\ &= s^2(K_0 + A) + s(K_0 + B) + K_0 \end{aligned} \tag{17-3}$$

Equating like powers of s,

$$K_0 = 1 \qquad K_0 + A = 0 \qquad K_0 + B = 0 \tag{17-4}$$

from which $A = -1$ and $B = -1$, so the expansion is:

$$F(s) = \frac{1}{s} - \frac{s+1}{s^2 + s + 1} \tag{17-5}$$

This is the desired expansion.

The advantage of this form is that no complex quantities are involved. The disadvantage is that the inverse transformation that is now required is not trivial. Let us continue with this example to obtain the inverse transform $f(t)$.

The inverse transform of $1/s$ is 1, by pair 1. The rest of $F(s)$ can be divided into two parts:

$$\frac{1}{s^2 + s + 1} + s\frac{1}{s^2 + s + 1} \tag{17-6}$$

The first of these, by pair 6, transforms to $(2/\sqrt{3})e^{-t/2} \sin \frac{1}{2}\sqrt{3}t$. The remaining term is s times the first term, and it will be remembered that multiplication by s in the frequency domain corresponds to differentiation in the time domain. (This is true, by pair III, if $f(0_+) = 0$, which it does in this case because $\sin 0 = 0$.) Therefore the inverse transform of the first term is differentiated to give, for the inverse transform of the second term:

$$\frac{d}{dt}\left(\frac{2}{\sqrt{3}}e^{-t/2} \sin \frac{1}{2}\sqrt{3}t\right) = e^{-t/2}\left(\cos \frac{1}{2}\sqrt{3}t - \frac{1}{\sqrt{3}}\sin \frac{1}{2}\sqrt{3}t\right) \tag{17-7}$$

We now add the transforms that have been obtained for the fractions in equation 17-5, and the result is:

$$f(t) = 1 - \frac{2}{\sqrt{3}}e^{-t/2}\left(\frac{1}{2} \sin \frac{1}{2}\sqrt{3}t + \frac{\sqrt{3}}{2} \cos \frac{1}{2}\sqrt{3}t\right) \tag{17-8}$$

This $f(t)$ is the inverse transform of $F(s)$ in equation 17-5 and hence of $F(s)$ in equation 17-1. With the aid of a trigonometric identity it is seen to be exactly the same as $f(t)$ in equation 16-14, which of course is the same transform obtained by another means. Either method can be used, but the former, using vectors in the s plane, is recommended as being usually the easier.

18. A FORM FOR SECOND-ORDER POLES

We have now considered several ways to find residues at simple poles, either real or complex, but none of our formulas for expansion of a function into partial fractions will work if the function has poles of the second or higher order. If there is a second-order pole, resulting from a factor such as $(s - s_1)^2$ in the denominator of $F(s)$, partial-fraction expansion is still possible but the form must be changed to:

$$F(s) = \frac{A}{s - s_1} + \frac{B}{(s - s_1)^2} + \frac{C}{s - s_2} + \cdots \qquad (18\text{-}1)$$

An example will show the method.

Given

$$F(s) = \frac{4s^2 + 17s + 16}{(s + 2)^2(s + 3)} \qquad (18\text{-}2)$$

Expand $F(s)$ into partial fractions, and write the inverse transform $f(t)$.

First, follow equation 18-1 in writing the form of the expansion as:

$$F(s) = \frac{A}{s + 2} + \frac{B}{(s + 2)^2} + \frac{C}{s + 3} \qquad (18\text{-}3)$$

C can be evaluated as we have done before, in Section 13 (or, if preferred, as in Section 14). Multiply by $(s + 3)$ the right-hand members of equations 18-2 and 18-3 and set them equal to each other; then let $s = -3$. The terms containing A and B both vanish, for each is multiplied by $(-3 + 3) = 0$, and there remains:

$$C = \frac{4s^2 + 17s + 16}{(s + 2)^2}\bigg|_{s = -3} = \frac{36 - 51 + 16}{(-1)^2} = 1 \qquad (18\text{-}4)$$

B can be evaluated in the same way, multiplying equations 18-2 and 18-3 by $(s + 2)^2$ and setting them equal to obtain:

$$B = \frac{4s^2 + 17s + 16}{s + 3}\bigg|_{s = -2} = \frac{16 - 34 + 16}{1} = -2 \qquad (18\text{-}5)$$

The value of A, however, cannot be obtained in this way, and we must fall back on the identity that was used in equation 13-3. Multiply equations 18-2 and 18-3 by the denominator of equation 18-2 to obtain:

$$4s^2 + 17s + 16 = A(s + 2)(s + 3) + B(s + 3) + C(s + 2)^2 \qquad (18\text{-}6)$$

Now equate the coefficients of s^2 on the left and the coefficients of s^2 on the right to give the equation:

$$4 = A + C \qquad (18\text{-}7)$$

(Two other equations could be written, one from the coefficients of the terms in s, and the other by equating the terms that are constant, but these we do not need.) Since C was found in equation 18-4 to be 1, equation 18-7 gives

$$A = 3 \qquad (18\text{-}8)$$

This completes the partial-fraction expansion, giving†

$$F(s) = \frac{3}{s+2} - \frac{2}{(s+2)^2} + \frac{1}{s+3} \qquad (18\text{-}9)$$

The final requirement is to write from pairs 2 and 5 the inverse transform of equation 18-9, and hence of equation 18-2:

$$f(t) = 3e^{-2t} - 2te^{-2t} + e^{-3t} \qquad (18\text{-}10)$$

Thus our inverse transformation of $F(s)$ by means of partial fractions is concluded.

19. EXPLICIT INVERSE TRANSFORMATION

Inverse transformation that is aided by expansion into partial fractions is still, of course, implicit, for we seek to use pairs that were derived from the direct-transformation integral of equation 1-1. *Explicit* inverse transformation would use equation 1-2, which is:

$$f(t) = \frac{1}{2\pi j} \int_{c-j\infty}^{c+j\infty} F(s)e^{st}\, ds \qquad (19\text{-}1)$$

This integral requires integration in the complex s plane along the line $s = c$, from infinitely far down to infinitely far up, c being a real constant that is large enough to place the line of integration to the right of all poles of $F(s)$. If $F(s)$ is at all complicated, simple means of integration are inadequate, and contour integration in the complex plane is usually resorted to (see Bush, Gardner and Barnes, Skilling (4), etc.). The integral along the line $s = c$ is evaluated from residues; the value of the integral is $2\pi j$ times the sum of the residues at the poles of the integrand where (for simple poles) the residue at such a pole as s_1 is defined as in equation 13-10:

$$\text{Residue of } F(s)e^{st} \text{ at } s_1 = \lim_{s \to s_1} (s - s_1)F(s)e^{st} \qquad (19\text{-}2)$$

---

†A partly graphical method that would be convenient if some of the poles were complex is given by R. N. Clark (see bibliography) in his Section 4.3.

If $F(s)$ has a pole at the origin, $s_1 = 0$ and $e^{s_1 t} = 1$, and $f(t)$ includes a constant term.

Note that equation 19-2 gives the residue at a pole of $F(s)e^{st}$. In Section 13 we found residues at poles of $F(s)$. Since e^{st} has no finite poles, the poles of $F(s)e^{st}$ are the poles of $F(s)$. If residues of $F(s)$ are called K_1, K_2, and so on, then residues of $F(s)e^{st}$ are $K_1 e^{s_1 t}$, $K_2 e^{s_2 t}$, and so on. Therefore, multiplying each residue of $F(s)$ by the appropriate $e^{s_n t}$ and adding, we obtain the line integral in the complex plane:

$$f(t) = K_1 e^{s_1 t} + K_2 e^{s_2 t} + \cdots + K_n e^{s_n t} \tag{19-3}$$

This is the result of explicit transformation, but it is the same as equation 12-4 which was obtained by implicit transformation. It is not surprising that explicit transformation and implicit transformation result in the same time function, but it is interesting that the technique of residues is so similar in the two cases. Equation 12-4, which is also equation 19-3, was called by V. Bush the Heaviside expansion theorem.

20. SUMMARY

Linear differential equations can be *transformed* into algebraic equations. *Manipulation* of these algebraic equations to solve for the transform of an unknown quantity is then relatively easy. *Inverse transformation* returns the solution to the original domain and gives the desired answer.

In electrical work, the *time domain* is usually transformed into the complex *frequency* domain and then, after manipulation, the result is returned to the *time* domain to obtain a real current or voltage.

The chapter begins by stating the *direct* and *inverse Laplace transformations*. Direct transformation is then used to obtain a number of transform *pairs*, and these are listed in a table for ready reference (page 354 and inside back cover).

Such transformation of *functions* can be generalized to give transformation of *operations*, including differentiation and integration. These transformations of operations are also collected in a table (page 365 and inside front cover).

Use of transformation in solving *differential equations* is then considered. Various *initial conditions*, both zero and non-zero, are illustrated. These examples are quite general, not necessarily related to networks.

The final step of inverse transformation is greatly aided by expansion into *partial fractions*. A proper fraction $F(s)$ can be expanded into partial fractions; the form to be used depends on whether $F(s)$ has poles that are *simple* or multiple, *real* or complex.

The *technique* is to let s equal s_k at each pole in turn in the product $(s - s_k)F(s)$; the coefficients of the partial fractions result. This method is easy to consider *graphically* as the ratio of distances on the pole-zero map if the poles are *real*; in terms of both distances and angles if the poles are *complex*. This method, whether done analytically or graphically, is for functions with *simple* poles only.

Six rules help formulate $f(t)$ from the poles of $F(s)$, particularly poles on the *real* axis. *Four more* rules relate to *complex* pairs of poles and the resulting oscillations.

A partial-fraction form is given for an $F(s)$ that has poles of the *second order*.

Explicit inverse transformation is possible by integration in the complex s plane, the usual means of integration being the determination of *residues* at poles. Although the reasoning is quite different, the technique is the same as for implicit inverse transformation by means of partial fractions.

PROBLEMS

14-1. A circuit has a 0.01-microfarad capacitor, a 25-ohm resistor, a 7.5-volt battery, and a switch in series. The switch is initially open and the capacitor is uncharged; at $t = 0$ the switch is closed. Find current in the circuit thereafter, using the Laplace transform method. §3*

14-2. A circuit consists of a 250-ohm resistor, a 0.10-microfarad capacitor, and a coil having 200 millihenrys of inductance and negligible resistance, connected in series with a 10-volt battery and a switch. The switch is closed at $t = 0$; find current by Laplace transformation. §5*

14-3. Find the Laplace transform $F(s)$ of the time function $f(t) = e^{-\alpha t} \cos \omega t$. Add the result to Table 14-2. §4*

14-4. Referring to the example of Section 5 and to Figure 5c, find the transform of current in the element Z_{12}, and from this transform find the current through Z_{12} after the switch is closed. Redraw Figure 5d, adding this current. §5*

14-5. Find a transfer function to relate the voltage across the inductance in the circuit of Figure 5c to the applied voltage. Use this transfer function to find the voltage across the inductance when the applied voltage is the 12-volt battery as given. §8

14-6. Repeat Problem 14-5 to find the voltage across the inductance when the suddenly applied voltage is a ramp form with $v(t) = 75t$ for positive time. (You may use pair 12 from Table 14-2 if you wish.) §9

14-7. A famous equation of mechanics (a pendulum with initial displacement) is $(d^2x/dt^2) + m^2x = 0$ with $x(0) = 1$ and $x'(0) = 0$. Find $x(t)$ by Laplace transform method, and check your result. §11*

14-8. Consider the set of equations

$$\frac{dy}{dt} - \frac{x}{2} = 1 \quad \text{and} \quad \frac{dx}{dt} + 2y = 2t$$

Initial conditions are $y(0) = 1$ and $x(0) = 1$. Solve for $y(t)$ and $x(t)$, using Laplace transforms. Check your results in the given equations. §11

14-9. (a) Solve the following equation for $x(t)$ by Laplace transformation:

$$\frac{dx}{dt} + 2x = 2 \qquad x(0) = 1$$

(b) Solve the following simultaneous equations for $x(t)$ and $y(t)$ by Laplace transformation:

$$\frac{dx}{dt} + y = 2 \qquad x(0) = 0$$

$$2\frac{dy}{dt} - x = 0 \qquad y(0) = 0$$

Check your results in the differential equations. §11*

14-10. Differentiate pair 7 to obtain pair 8 of Table 14-2. Then differentiate pair 8 to obtain, if possible, a new pair. §8

14-11. Solve the following by Laplace transformation. Verify all answers by checking in the given differential equation and with the given initial values. §11

(a) Given $3\frac{dy}{dt} + 5y = x, \ x = \begin{cases} 0 \text{ for } t < 0 \\ 6 \text{ for } t > 0 \end{cases}$, and $y(0) = 0$. Find $y(t)$.

(b) Same equation, but let $y'(0) = 0$. Find $y(t)$. Note:

$$y'(0) = \frac{dy}{dt}\bigg|_{t=0}$$

(c) Given $4\frac{dx}{dt} - x = 2$ and $x(0) = -1$. Find $x(t)$.

(d) Given $2\frac{d^2y}{dt^2} + 3\frac{dy}{dt} + 6y = 8$ and $y(0) = 0, \ y'(0) = 0$. Find $y(t)$.

(e) Same equation but let $y(0) = 0$ and $y'(0) = 1$. Find $y(t)$.

(f) Same equation but let $y(0) = 2$ and $y'(0) = 2$. Find $y(t)$.

(g) Given $\frac{dx}{dt} + x = e^{-t}$ and $x(0) = 0$. Find $x(t)$.

14-12. Find $i(t)$ by partial fractions, knowing that $I(s) = 42/(s + 4)(s + 7)(s + 2)$. Check your partial fractions by addition to regain $I(s)$. §15

14-13. Derive the time function of pair 11 from the transform

$$1/(s - s_1)(s - s_2)(s - s_3)$$

by the method of partial fractions. §15

14-14. Expand the following into partial fractions, each of the partial fractions having a denominator that is linear in s (no squares or higher powers of s in the denominators):

(a) $\dfrac{6s^2 + 22s + 18}{s^3 + 6s^2 + 11s + 6}$

(b) $\dfrac{16(s + 1)}{s^4 + 4s^3 + 8s^2 + 8s}$

(c) $\dfrac{2s + 30}{s^2 + 10s + 50}$ §15*

14-15. A 25-volt battery is suddenly applied to a circuit containing 500 millihenrys of inductance, a 0.40-microfarad capacitance, and 100 ohms of resistance in series. Plot poles and zeros of $I(s)$ and find the current that results. Find Q of the circuit.   §16*

14-16. A source of 5 volts (constant, dc) and a switch are in series with $L = 10$ millihenrys, $C = 3.5$ microfarads, and $R = 75$ ohms. Until $t = 0$ the switch is open and the capacitor is uncharged. Then the switch is closed. Plot the poles and zeros of $I(s)$, and find $i(t)$.   §16

14-17. Find the time function corresponding to the transform

$$F(s) = s/(s + \alpha)^2$$

Show the poles and zeros of $F(s)$ in the s plane, and sketch $f(t)$ as a function of time. Add this pair to Table 14-2.   §18*

14-18. In the circuit shown, $R/L = \alpha$. The switch is closed at $t = 0$. The applied voltage is $v = te^{-\alpha t}$. Find $i(t)$. Check $i(t)$ in the differential equation of the circuit. Check that $i(0) = 0$.   §18*

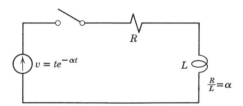

Problem 14-18.

14-19. If

$$\mathscr{L}f(t) = F(s), \quad \text{and} \quad F(s) = \frac{1}{s^2(s + \alpha)}$$

find $f(t)$: (a) by partial fractions, and (b) by real integration. (c) Check the resulting $f(t)$ by direct transformation, using $\mathscr{L}f(t) = F(s)$.   §18

14-20. A transform is $F(s) = 1/s(s - s_1)(s - s_2)$; find the corresponding $f(t)$ by whatever valid method seems simplest, making use of any information available in this chapter.   §15*

14-21. Voltage $v = V_m \cos \omega t$ is suddenly applied at $t = 0$ to a circuit of R and L in series (as in Section 6 of Chapter 6). Write $I(s)$, and use the methods of Sections 14 and 16 to find the current $i(t)$. Show that this current is the same as that of equation 6-11 of Chapter 6. (Note that φ of Chapter 6 is not the same angle as φ of Chapter 14.)   §16

14-22. An alternating voltage $v = 12 \cos 377t$ is suddenly applied at time $t = 0$ to a coil with 600 millihenrys of inductance and negligible resistance. By Laplace transform methods, find the resulting current. Sketch $v(t)$ and $i(t)$.   §16*

14-23. Referring to Table 14-2, derive pair 4 by the method of Section 19, explicit inverse transformation.   §19

CHAPTER
15

LAPLACE APPLICATIONS AND THEOREMS

1. NETWORK PROBLEMS

The Laplace method is now ready for use. We shall find that it saves time and trouble whenever linear differential equations are to be solved.

Problems involving changes of networks are, of course, transient problems. Discrete signals, pulses, any kind of currents following switching, are transient. Communication systems, computers, pulse-code systems, television, switching of power systems, lightning surges and faults, automatic control systems, guidance and tracking systems; these all have transient problems. If the systems are linear, or at least partly linear, or nearly linear, transformation may well be helpful.

2. A STATE-VARIABLE EXAMPLE

In Chapter 7 we derived a pair of differential equations to describe the response of a network. We used a state-variable approach to write these equations and, with computers in mind, we used a numerical solution to find the currents. No formal solution was given in Chapter 7, although the formal answers were stated without proof.

Now we are ready to obtain these formal answers by the easy means of Laplace transformation. This is permissible because the equations are linear;

389

the network is composed of constant elements. The problem serves as a good example of Laplace transformation.

The equations, 5-10 and 5-11 of Chapter 7, are:

$$\frac{di_1}{dt} = -2i_1 + 2i_2 + 120 \tag{2-1}$$

$$\frac{di_2}{dt} = 2i_1 - 5i_2 \tag{2-2}$$

and the initial conditions are $i_1(0) = i_2(0) = 0$.

To solve, the equations are first transformed. I_1 and I_2 are the transforms of i_1 and i_2, and sI_1 is the transform of the derivative because $i_1(0) = 0$ (pair III). The transform of 120, a function that is constant with respect to time, is $120/s$. Hence:

$$sI_1 = -2I_1 + 2I_2 + \frac{120}{s} \tag{2-3}$$

Similarly:

$$sI_2 = 2I_1 - 5I_2 \tag{2-4}$$

Simultaneous solution, which is merely algebraic in the frequency domain, gives:

$$I_1 = \frac{120(s + 5)}{s(s^2 + 7s + 6)} = \frac{120(s + 5)}{s(s + 6)(s + 1)} \tag{2-5}$$

$$I_2 = \frac{240}{s(s + 6)(s + 1)} \tag{2-6}$$

There are several ways to find the inverse transform of equation 2-5. Perhaps we look first through the table of transforms but, not finding this particular function, we decide to use partial fractions. It may not be necessary but it is surely helpful to spot the poles and zeros of I_1 in Figure 2a; from these we see that the time function i_1 consists of a constant term and two real exponentials. The constant will be positive, and both exponentials will be negative.

The residue at the origin is 120 times the vector from the zero, which is 5, divided by the product of 1 and 6, the vectors from the other poles, which is 6. That is:

$$K_0 = 120 \cdot \tfrac{5}{6} = 100 \tag{2-7}$$

Similarly, the residue at the pole at -1 is

$$K_1 = 120 \cdot \frac{4}{(5)(-1)} = -96 \tag{2-8}$$

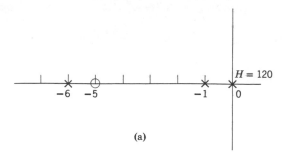

(a)

FIGURE 2a   Poles and a zero in the _s_ plane, equation 2-5.

for the vector from the zero is 4, and the vectors from the other two poles are 5 and -1. Finally, the residue at the pole at -6, which may be called K_6, is:

$$K_6 = 120 \cdot \frac{(-1)}{(-5)(-6)} = -4 \tag{2-9}$$

The current i_1 is therefore:

$$i_1 = K_0 + K_1 e^{-t} + K_6 e^{-6t} = 100 - 96e^{-t} - 4e^{-6t} \text{ amperes} \tag{2-10}$$

and this is equation 5-12 of Chapter 7.

Equation 2-6 gives I_2. This has the same denominator as I_1, and hence the same poles as those we have mapped in Figure 2a, but there is no zero in this function, and the coefficient is 240. Residues at the poles are now to be obtained; Figure 2a can be used again if we ignore the zero at -5 and change H to 240. The time function for i_2 will consist of a constant term and two exponentials, one of which will be negative and the other positive. The residues are to be found, and if the letter C is used,

$$C_0 = 240 \cdot \frac{1}{(1)(6)} = 40 \tag{2-11}$$

Since the function has no zeros, the numerator of this fraction is 1. Next, the residue at the pole at -1 is:

$$C_1 = 240 \cdot \frac{1}{(5)(-1)} = -48 \tag{2-12}$$

Finally, at the pole at -6, the residue is:

$$C_6 = 240 \cdot \frac{1}{(-5)(-6)} = 8 \tag{2-13}$$

The time function for i_2 can now be written:

$$i_2 = 40 - 48e^{-t} + 8e^{-6t} \text{ amperes} \qquad (2\text{-}14)$$

This is the same as equation 5-13 of Chapter 7, and both currents are plotted in Figure 5c of that chapter.

3. EXAMPLE USING ADMITTANCE

In the foregoing example the equations 2-1 and 2 were given, having been formulated in a previous chapter, so it was only the solution of the equations and not their derivation that was illustrated. If our problem actually began with the network, and the equations had not yet been formulated, we might well have chosen a different attack. We might have decided, if starting from the network, to make use of what we already know about admittances and impedances, proceeding as follows. Figure 3a shows the network diagram;

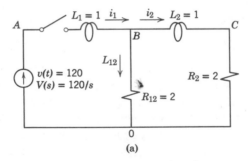

(a)

FIGURE 3a   Units are volts, ohms, and henrys.

this is a repetition of Figure 5a of Chapter 7 which was used to obtain the state equations that were solved in the preceding section. Let us now look at another way to obtain the network equations.

The impedance into which the source feeds is

$$Z_1 = s + \frac{(s + 3)2}{s + 3 + 2} \qquad (3\text{-}1)$$

and its reciprocal is

$$Y_1 = \frac{s + 5}{s^2 + 7s + 6} = \frac{s + 5}{(s + 6)(s + 1)} \qquad (3\text{-}2)$$

There is no initial current.

The applied voltage is a step function that is zero until zero time and 120 volts thereafter, so its Laplace transform is $V(s) = 120/s$. The transform of entering current is therefore:

$$I_1(s) = V(s)Y(s) = \frac{120(s + 5)}{s(s + 6)(s + 1)} \tag{3-3}$$

and this is equation 2-5.

We also wish to find current i_2. Note that the entering current i_1 divides in inverse proportion to branch impedances:

$$I_2 = \frac{2}{(s + 3) + 2}I_1 = \frac{2}{(s + 5)}\frac{120(s + 5)}{s(s + 6)(s + 1)}$$

$$= \frac{240}{s(s + 6)(s + 1)} \tag{3-4}$$

which is equation 2-6. Equations 2-5 and 6 were solved for $i_1(t)$ and $i_2(t)$ in the previous section, and this part of the solution need not be repeated. The purpose of this section is to show that the same answers are obtained by the two methods although the original statements and the intermediate equations are so different.

4. EXAMPLE USING LOOP EQUATIONS

The loop equations of Chapter 8 are also well suited to a problem of this kind. Let us apply them to the same network of Figure 3a. From Kirchhoff's voltage law we write for the voltages around loop 1 of this network:

$$L_1 \frac{di_1}{dt} + R_{12} i_1 - R_{12} i_2 = v(t) \tag{4-1}$$

This is in accord with Chapter 8 on loop equations; see particularly the note to Section 5 of that chapter.

Transforming, i_1 and i_2 become I_1 and I_2. Since $i_1(0)$ is given to be zero, di_1/dt transforms to sI_1. The voltage $v(t)$ is 120 volts, unchanging with time, so its transform is $120/s$. Hence the transformed equation, with numerical values inserted for the parameters, is

$$1sI_1 + 2I_1 - 2I_2 = \frac{120}{s} \tag{4-2}$$

or

$$(s + 2)I_1 - 2I_2 = \frac{120}{s} \tag{4-3}$$

Similarly for the transforms of voltages around loop 2:

$$-2I_1 + (s + 5)I_2 = 0 \tag{4-4}$$

Simultaneous solution for I_1 and I_2 is now easy, and gives:

$$I_1 = \frac{120(s + 5)}{s(s + 6)(s + 1)} \tag{4-5}$$

$$I_2 = \frac{240}{s(s + 6)(s + 1)} \tag{4-6}$$

Here again we have equations 3-3 and 3-4, or equations 2-5 and 2-6, ready for the inverse transformation that is shown in Section 2. In using the loop equations we are, in fact, well able to write the transformed equations 4-3 and 4-4 immediately, by inspection of the network diagram without writing the differential equations first.

5. EXAMPLE USING NODE EQUATIONS

The node equations of Chapter 8 are not entirely suitable for the solution of the problem that has now been solved by several other methods, for the direct result of the node method is a node voltage, whereas our problem is to determine the input current. However, when the voltages are known the currents can be found without much trouble.

Referring to Figure 3a, we may identify nodes A, B, C, and O. Using O as the reference node, V_A is forced; $v_A(t)$ is a step function rising to 120 volts at $t = 0$, so $V_A(s) = 120/s$. We shall find it helpful to obtain $V_B(s)$, so we begin by writing current equations at nodes B and C. Current away from node B is:

$$\left(\frac{1}{s} + \frac{1}{s} + \frac{1}{2}\right)V_B - \frac{1}{s}V_A - \frac{1}{s}V_C = 0 \tag{5-1}$$

Current away from node C is:

$$\left(\frac{1}{s} + \frac{1}{3}\right)V_C - \frac{1}{s}V_B = 0 \tag{5-2}$$

Inserting the forced value of $V_A = 120/s$ and solving simultaneously, we find that

$$V_C = \frac{3V_B}{s + 3} \tag{5-3}$$

$$V_B = \frac{240(s + 3)}{s(s^2 + 7s + 6)} \tag{5-4}$$

Now the entering current, called i_1, is to be found. Perhaps the easiest way is to find its transform as the difference $V_A - V_B$ multiplied by the admittance $1/s$ of the branch through which the current flows:

$$I_1 = (V_A - V_B)\frac{1}{s} = \left[\frac{120}{s} - \frac{240(s + 3)}{s(s^2 + 7s + 6)}\right]\frac{1}{s}$$

$$= \frac{120(s + 5)}{s(s + 6)(s + 1)}$$

(5-5)

Thus we find again the transform of the entering current, and the current itself can be found, as from equation 2-5, by inverse transformation. Note that all possible manipulation is done in the frequency domain, before inverse transformation to the time domain.

If i_2 is desired, it can be found in a similar manner. V_C is known from equations 5-3 and 5-4, and this is multiplied by $\frac{1}{3}$ to give I_2 which is then the same as equation 2-6 and can be similarly transformed to give i_2.

The same two network currents, as plotted in Figure 5c of Chapter 7, have now been computed by four different methods. In practice, any one of these four methods can be used, and they give a good check on each other.

6. INFERENCE

In practical work it is often not necessary to find an actual time function of output by inverse transformation. It may be that all necessary information can be obtained from the transform, leaving the final step of inverse transformation unnecessary. Is it the *shape* of an output current that is wanted? The composition of the current as the sum of a constant term, exponential components, or oscillations can be deduced from the poles of the transform. Is it the *speed* of decay? Time constants can be inferred from the poles, too. Is it the *sign*? Whether an exponential component or a steady term is positive or negative is known, also, from the poles and zeros of the transform. Even the relative *magnitudes* of the components can be inferred from the remoteness of poles from other poles, or the proximity of poles to zeros. Is the *initial value* of the transient current needed? This can be told from the initial value theorem, which will now be given. Or the steady value that will be approached after a great time? The *final value* theorem will give this.

These modes of inference were considered and explained in Chapter 14 except the initial and final value theorems which now follow.

7. THE INITIAL AND FINAL VALUE THEOREMS

The initial value theorem tells us that

$$f(0_+) = \lim_{s \to \infty} sF(s) \tag{7-1}$$

The final value theorem says that if the time function approaches a finite or zero value as time becomes great without limit, that value is

$$f(\infty) = \lim_{s \to 0} sF(s) \tag{7-2}$$

Before giving proofs, let us try an example.

Given the following transforms of two currents, find the initial values and the final values of the currents.

$$I_1(s) = \frac{12}{s} \cdot \frac{(s + 8)}{4(s + 6)} \qquad I_2(s) = \frac{12}{s(s + 6)} \tag{7-3}$$

(Let us suppose that these are unfamiliar transforms, although in fact they come from the network of Section 9 of Chapter 6.)

The initial value of $i_1(t)$ is found from equation 7-1 to be:

$$i_1(0_+) = \lim_{s \to \infty} s \frac{12}{s} \cdot \frac{(s + 8)}{4(s + 6)}$$

$$= \lim_{s \to \infty} \frac{12(1 + 8/s)}{4(1 + 6/s)} = 3 \text{ amperes} \tag{7-4}$$

The initial value of $i_2(t)$ is:

$$i_2(0_+) = \lim_{s \to \infty} s \frac{12}{s(s + 6)} = 0 \tag{7-5}$$

The final values are found from equation 7-2:

$$i_1(\infty) = \lim_{s \to 0} s \frac{12}{s} \cdot \frac{(s + 8)}{4(s + 6)} = 4 \text{ amperes} \tag{7-6}$$

$$i_2(\infty) = \lim_{s \to 0} s \frac{12}{s(s + 6)} = 2 \text{ amperes} \tag{7-7}$$

Initial and final values are found as different limits of the same function; it may help the memory to note that the final value is found by letting s approach zero, this being a long-term, dc, zero-frequency condition, whereas the initial value, reached in infinitesimal time, is the high-frequency limit as s approaches infinity.

The initial value thus obtained (for either current) is clearly to be joined to the final current, by an exponential function with a time constant of $\frac{1}{6}$ second. (The results, obtained by actual computation of the time functions, are drawn in Figure 9c of Chapter 6.)

The initial value theorem can also be used to find the initial *derivative* of a time function. We call a function $f(t)$; let us call its time derivative $f'(t)$. Then the initial value of this derivative (or, more precisely, the limit approached by $f'(t)$ as t approaches zero from the positive side) is:

$$f'(0_+) = \lim_{s \to \infty} s[sF(s) - f(0_+)] \tag{7-8}$$

If the initial value of the time function is already known from the initial value theorem to be zero, this is particularly simple, for then:

$$f'(0_+) = \lim_{s \to \infty} s^2 F(s) \tag{7-9}$$

For instance, the initial derivative of i_2 in the foregoing example is found to be:

$$i_2'(0_+) = \lim_{s \to \infty} s^2 \frac{12}{s(s + 6)} = \frac{12}{1(1 + 6/s)} = 12 \tag{7-10}$$

The initial derivative of i_1, however, requires equation 7-8 rather than 7-9 because the initial value of the current is not zero:

$$i_1'(0_+) = \lim_{s \to \infty} s\left[s\frac{12}{s} \cdot \frac{(s + 8)}{4(s + 6)} - 3 \right]$$

$$= \lim_{s \to \infty} s\left[\frac{3s + 24 - 3s - 18}{s + 6} \right] = \lim_{s \to \infty} \frac{6}{1 + 6/s} = 6 \tag{7-11}$$

(These two initial derivatives, thus so easily determined by the theorems, can be checked by computation from the time functions given in Section 9 of Chapter 7, and they can be seen as slopes of the curves for i_1 and i_2 in Figure 9c of that chapter.)

It is entirely possible to obtain initial values of higher derivatives, or the final values of derivatives, from these theorems, but none of these are likely to be very useful.

8. PROOF OF THE THEOREMS

The initial value theorem is proved from equation 8-10 of Chapter 14, which is:

$$\int_0^\infty f'(t)e^{-st}\, dt = sF(s) - f(0_+) \tag{8-1}$$

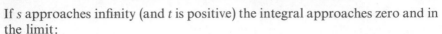

If s approaches infinity (and t is positive) the integral approaches zero and in the limit:

$$0 = \lim_{s \to \infty} sF(s) - f(0_+) \tag{8-2}$$

This, giving equation 7-1, is the required proof of the initial value theorem.

The final value theorem also follows from equation 8-1. Let s approach zero:

$$\lim_{s \to 0} \int_0^\infty f'(t)e^{-st}\, dt = \lim_{s \to 0} \left[sF(s) - f(0_+) \right] \tag{8-3}$$

The limit is with respect to s and the integration is with respect to t, so since s and t are independent, the integral in equation 8-3 can be written:

$$\int_0^\infty f'(t) \lim_{s \to 0} e^{-st}\, dt = \int_0^\infty f'(t)\, dt \tag{8-4}$$

This infinite integral has a finite value if $f'(t)$ decreases exponentially with increasing t, and this is guaranteed if the transform of $f'(t)$ which is $sF(s) - f(0_+)$ has no poles or other singularities in the right half s plane or on the axis of imaginaries. Hence the restriction on the final value theorem is that $sF(s)$ can have poles only in the *left* half plane.

The integral of equation 8-4 can be stated:

$$\lim_{t \to \infty} \int_0^t f'(t)\, dt \tag{8-5}$$

and since the integral from 0 to t of the derivative with respect to t is the function itself, this is equal to:

$$\lim_{t \to \infty} \left[f(t) \right]_0^t = \lim_{t \to \infty} \left[f(t) - f(0_+) \right] \tag{8-6}$$

This expression now becomes the left-hand member in equation 8-3:

$$\lim_{t \to \infty} f(t) - f(0_+) = \lim_{s \to 0} sF(s) - f(0_+) \tag{8-7}$$

From this:

$$\lim_{t \to \infty} f(t) = \lim_{s \to 0} sF(s) \tag{8-8}$$

which is equation 7-2, the final value theorem.

Proof of the initial derivative theorem, equation 7-8, is fairly obvious. If the transform of $f(t)$ is $F(s)$, and the transform of its derivative $f'(t)$ is called $F'(s)$, we know from pair III that

$$F'(s) = sF(s) - f(0_+) \qquad (8\text{-}9)$$

Then, by equation 7-1, the result is equation 7-8:

$$f'(0_+) = \lim_{s \to \infty} sF'(s) = \lim_{s \to \infty} s[sF(s) - f(0_+)] \qquad (8\text{-}10)$$

9. THE EXCITATION FUNCTION

When current is found by dividing the transform of voltage by impedance (or multiplying by admittance), it is implicit that the network is initially at rest. If, as in Figure 9a, a voltage $v(t)$ is connected at $t = 0$ to a circuit of L, R,

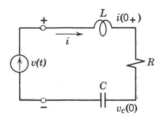

FIGURE 9a   A circuit that is not initially at rest.

and C in series, and if there were no initial current and no initial charge, we might consider that $I(s) = V(s)/Z(s)$ and write

$$I(s) = \frac{V(s)}{sL + R + 1/sC} \qquad (9\text{-}1)$$

If, however, the circuit were not initially at rest, this equation would be incorrect. If there were initial current, or initial charge, we might instead write a differential equation based on Kirchhoff's voltage law:

$$L\frac{di}{dt} + Ri + \left(v_C(0) + \frac{1}{C}\int_0^t i \, dt\right) = v(t) \qquad (9\text{-}2)$$

When this equation is transformed it becomes:

$$L[sI(s) - i(0_+)] + RI(s) + \left[\frac{v_C(0)}{s} + \frac{I(s)}{sC}\right] = V(s) \qquad (9\text{-}3)$$

Solution for the current transform then gives:

$$I(s) = \frac{V(s) + Li(0_+) - v_C(0)/s}{sL + R + 1/sC} \tag{9-4}$$

This suggests that an *excitation function* can be used.

$$\text{Excitation function} = V(s) + Li(0_+) - v_C(0)/s \tag{9-5}$$

and

$$I(s) = \frac{(\text{Excitation function})}{Z(s)} \tag{9-6}$$

Using the previous example of an *LRC* series circuit, equation 9-1 becomes identical with equation 9-4 when the excitation function of equation 9-5 replaces $V(s)$.

Thus we can find entering current when there is excitation in only one loop of a network. If we meet a problem with voltage applied in more than one loop, or with initial stored energy in other loops, super-position can be used. The responses to the various energy sources can be determined one at a time, and then added. Alternatively, the loop equations of the network based on Kirchhoff's voltage law can be used, as in equation 9-2.

It is evident from physical reasoning that initial current or initial charge will alter the input current when a voltage source is inserted in a network, and this is confirmed by equation 9-4, or equation 9-6. It is interesting to consider how initial energy is reflected in the transform $I(s)$. In general, initial energy can alter the coefficient H of the transform of current, or it can give the function an additional zero. Specifically, if a battery is introduced into the network (so that $v(t)$ in equation 9-2, for instance, equals V_m, and $V(s) = V_m/s$) the initial capacitor voltage merely subtracts from the battery voltage,[†] while initial current, by providing electrokinetic momentum, puts a new zero in the $I(s)$ function, thereby altering the residues at all the poles. The time function of current, $i(t)$, is altered correspondingly.

10. AN EXAMPLE OF THE EXCITATION FUNCTION

Figure 10a shows a circuit containing a coil and a resistor. Initially the switch between them is closed; it is opened at $t = 0$. Find current thereafter.

Assuming steady current is established before $t = 0$, the initial current is $i(0) = 12$. The voltage transform is $60/s$. The excitation function is therefore

---

[†] The negative sign of $v_C(0)/s$ results from the convention that a positive current to a capacitor carries a positive charge which produces a positive voltage across the capacitor.

$$\text{Excitation function} = \frac{60}{s} + (0.5)(12) = \frac{60}{s} + 6 \qquad (10\text{-}1)$$

Impedance of the circuit after opening the switch is $Z(s) = 12 + 0.5s$. The current transform is obtained by dividing:

$$I(s) = \frac{60/s + 6}{12 + 0.5s} = \frac{60}{s(12 + 0.5s)} + \frac{6}{12 + 0.5s}$$

$$= \frac{120}{s(s + 24)} + \frac{12}{s + 24} \qquad (10\text{-}2)$$

Current is now found by using pairs 2 and 3 of Table 14-2:

$$i(t) = \frac{120}{24}(1 - e^{-24t}) + 12e^{-24t} = 5 + 7e^{-24t} \qquad (10\text{-}3)$$

This current starts at 12 when $t = 0$ and decreases exponentially to 5; Figure 10b shows the current. It will be seen that initial conditions are satisfied automatically when the excitation function is used.

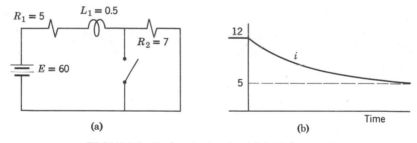

(a)   (b)

FIGURES 10a,b   A circuit with initial current.

This foregoing solution may be looked upon as an application of superposition. The first term of the excitation function gives the result of applying the battery voltage to the circuit after $t = 0$; the second term gives the result of the impulse owing to initial energy in the coil.

11. APPLICATION OF A STEP FUNCTION

Although the transform method of handling network problems can be used with input voltages or currents of any form, some kinds of input are used more frequently than others. One of the most helpful is the step function. The step function corresponds to closing or opening a switch.

More generally, a step function can be used in analysis as an approximation when the actual disturbing function is unknown. In designing an automatic control system, for instance, it is helpful to know how the system will respond to a step-function disturbance although one cannot tell in advance what the actual disturbance will be. The step gives a typical but fairly severe test.

Since the response to a step input is so often interesting, curves have been computed showing the output as a function of time when the input is a step function and the network function is of a specified form. Such precalculated curves are available† if the network function has one pole or two poles or three poles; if there are two poles there may be one zero, or none. Many families of curves are necessary to show these various possibilities. One example, for a network function with two finite poles and no finite zero, is here reproduced from Clark (by permission). Each curve of Figure 11a shows $x(t)/x_{ss}$ as a function of $\omega_0 t$ for a given value of ζ when

$$X(s) = \frac{H}{s(s^2 + 2\zeta\omega_0 s + \omega_0{}^2)} \tag{11-1}$$

Figure 11a can be used as a plot of the time function corresponding to any function of s, however obtained, that can be put in the form of equation 11-1.

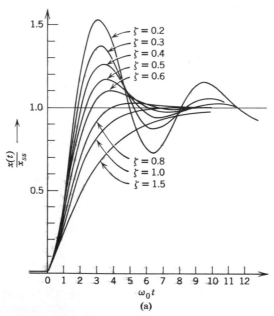

FIGURE 11a   Curves for $x(t)$ when $X(s) = H/s(s^2 + 2\zeta\omega_0 s + \omega_0{}^2)$ (courtesy of R. N. Clark).

---

† See for instance R. N. Clark in bibliography.

Clark's $X(s)$ can be our $F(s)$ or $I(s)$ or $V(s)$, and its inverse transform is $x(t)$. The curves are normalized in terms of x_{ss}, the steady state value, which might also be written $x(\infty)$. To use the curves it is necessary to know ζ, the damping ratio, which may be determined (if it is not already known) by comparison of the transform function with equation 11-1. It is also necessary to know ω_0, also found by comparison with equation 11-1, and to know x_{ss}, found by applying the final value theorem to equation 11-1:

$$x_{ss} = x(\infty) = \frac{H}{\omega_0^2} \qquad (11\text{-}2)$$

The curves of Figure 11a are derived by inverse transformation leading to

$$x(t) = \frac{H}{\omega_0^2}\left[1 + \frac{\omega_0}{\omega_1}e^{-t/\tau}\sin(\omega_1 t - \varphi)\right] \qquad (11\text{-}3)$$

where

$$\omega_1 = \omega_0\sqrt{1 - \zeta^2} \quad \text{and} \quad \tau = \frac{1}{\zeta\omega_0} \qquad (11\text{-}4)$$

and φ is the angle of the vector in the s plane from the origin to the upper pole s_1 of the function $X(s)$. This inverse transformation can be done by partial fractions, using the rules of Section 16 of Chapter 14.

It is interesting, and important in knowing the shape of the curves near the origin, to determine the initial value and the initial derivative of $x(t)$. This can be done by applying the initial value theorem to $X(s)$ in equation 11-1. The initial value is:

$$x(0) = \lim_{s \to \infty} s\left[\frac{H}{s(s^2 + 2\zeta\omega_0 s + \omega_0^2)}\right] = 0 \qquad (11\text{-}5)$$

Since the initial value (for any ζ) is zero, the initial derivative is:

$$x'(0) = \lim_{s \to \infty} s\left[\frac{sH}{s(s^2 + 2\zeta\omega_0 s + \omega_0^2)}\right] = 0 \qquad (11\text{-}6)$$

Thus all the curves of Figure 11a are tangent to the horizontal axis at the origin.

The initial *second* derivative of $x(t)$ can be found also. It is not zero; it is, in fact, H. It is not very likely to be wanted for any practical purpose, however.

Since, in this example, we have $x(t)$ in equation 11-3, the initial value and initial derivatives can be found directly by letting $t = 0$. In this case, however, and perhaps in most, it is a good deal easier to use the initial value theorem, particularly for initial derivatives.

12. EXAMPLE OF PRECALCULATED CURVES

A step of 24 volts is the input to a circuit with L, R, and C in series, and the output of the circuit is the voltage across C. This output voltage, which is amplified and used to operate a control system, must be found as a function of time. The circuit is that shown in Figure 9a. We know that initial current and initial capacitor voltage are both zero. Numerical values are: $L = 0.25$ henrys, $R = 400$ ohms, and $C = 10^{-6}$ farads.

First write the input impedance, $Z = sL + R + 1/sC$. Then Y, the reciprocal of Z, is:

$$Y = \frac{1}{L} \cdot \frac{s}{s^2 + (R/L)s + 1/LC} \tag{12-1}$$

Now $I = YV_{in}$ and $V_{out} = (1/sC)I = (1/sC)YV_{in}$, so:

$$\frac{V_{out}}{V_{in}} = \frac{1}{sC}Y = \frac{1}{LC} \cdot \frac{1}{[s^2 + (R/L)s + 1/LC]} \tag{12-2}$$

This transfer function has two poles and no zero. We now apply the input step function for which the transform is $V_{in} = 24/s$, obtaining:

$$\begin{aligned} V_{out} &= \frac{24}{LC} \cdot \frac{1}{s[s^2 + (R/L)s + 1/LC]} \\ &= \frac{96 \cdot 10^6}{s(s^2 + 1600s + 4 \cdot 10^6)} \end{aligned} \tag{12-3}$$

Since this equation has the form of equation 11-1, we know that the curves of Figure 11a give the answer to our problem. Which curve of the diagram applies? It is necessary to know ζ. Comparing equations 12-3 and 11-1:

$$\omega_0 = 2 \cdot 10^3 \quad \text{and} \quad 2\zeta\omega_0 = 1600 \tag{12-4}$$

so

$$\zeta = \frac{1600}{4 \cdot 10^3} = 0.4 \tag{12-5}$$

Hence the curve in Figure 11a marked $\zeta = 0.4$ is the answer to our problem. Regarding scales in the figure, the steady-state value of output is 24 volts; this is evident from the network diagram, and also results very obviously by using the final value theorem with equation 12-3. Also, it is necessary to know ω_0 in order to evaluate the time scale, and this is found in equation 12-4 to be 2000 rn/sec.

It is helpful to emphasize that ζ is the ratio of actual damping to critical damping, and therefore, in this example:

$$\zeta = \frac{R}{R_{\text{crit}}} = \frac{R}{2\sqrt{L/C}} = \frac{400}{1000} = 0.4 \tag{12-7}$$

This is the same value that was already obtained in equation 12-5 by comparing the actual numbers with equation 11-1.

13. APPLICATION OF A RAMP FUNCTION

Another applied voltage of interest, although perhaps less interest than the step function, is the ramp function kt. A ramp voltage starts at zero and rises in a straight line, proportional to time (pair 12). Obviously this voltage cannot continue forever, but it may last for any finite time. The ramp function can be Laplace transformed, although it could not be Fourier transformed.

Before applying such a voltage to a circuit, let us find the transform of the unit ramp function $f(t) = t$. The unit ramp is the integral of the unit step; that is,

$$t = \int_0^t 1 \, dt \tag{13-1}$$

and since integration transforms to $1/s$, it follows that

$$\mathscr{L}t = \frac{1}{s}\mathscr{L}1 = \frac{1}{s}\left(\frac{1}{s}\right) = \frac{1}{s^2} \tag{13-2}$$

This is pair 12.

Let us apply a ramp voltage

$$v(t) = 50t \quad \text{so} \quad V(s) = \frac{50}{s^2} \tag{13-3}$$

to resistance and capacitance in series, for which

$$Z = R + \frac{1}{sC} \quad \text{or} \quad Y = \frac{1}{R}\frac{s}{(s + 1/RC)} \tag{13-4}$$

It is given that $R = 10^6$ ohms and $C = 10^{-7}$ farads. Let there be no initial charge on the capacitance. Then

$$I(s) = V(s)Y(s) = \frac{V(s)}{R}\frac{s}{(s + 1/RC)} = \frac{50}{R}\frac{1}{s(s + 1/RC)} \tag{13-5}$$

$$i(t) = 50C(1 - e^{-t/RC}) = 5 \cdot 10^{-6}(1 - e^{-10t}) \text{ amperes} \tag{13-6}$$

That is, the current rises in a few tenths of a second to approach 5 microamperes which continues as long as the applied voltage continues to rise. When the applied voltage ceases to rise at 50 volts/second, this original problem ends and a new problem begins, and the initial condition of the new problem is the capacitor voltage at the end of the original problem.

14. APPLICATION OF A SINUSOIDAL FUNCTION

An alternating voltage is often applied to a network by closing a switch. The source voltage is modeled by a discontinuous sine wave if the switch is closed at an instant when the voltage is zero, or by a discontinuous cosine wave if the switch is closed when the voltage is at its maximum. If the switch is closed at some other time an appropriate phase angle must be introduced.

The transform of the discontinuous sine function, a function equal to zero for all negative time and through positive time proportional to $\sin \omega_1 t$, is found in the table as pair 7:

$$\mathscr{L} \frac{1}{\omega_1} \sin \omega_1 t = \frac{1}{s^2 + \omega_1{}^2} \tag{14-1}$$

If we are interested in the discontinuous cosine rather than the discontinuous sine, we note that the cosine function is the derivative of the sine, and by pair III (since $\sin 0 = 0$):

$$\mathscr{L} \cos \omega_1 t = \mathscr{L} \frac{d}{dt} \frac{1}{\omega_1} \sin \omega_1 t = \frac{s}{s^2 + \omega_1{}^2} \tag{14-2}$$

and here we have obtained pair 8.

Should the applied voltage not be either the sine function or the cosine function, but something between, it can be written in terms of a phase angle φ, and by addition of equations 14-1 and 14-2 with appropriate coefficients A and B, a more general pair could be obtained in the form:

$$\mathscr{L} \sin(\omega_1 t - \varphi) = \frac{As + B}{s^2 + \omega_1{}^2} \tag{14-3}$$

As an example, the ac voltage of Figure 14a is suddenly applied to the inductive circuit of L and R (perhaps a transformer) by closing the switch at $t = 0$. Find and plot the entering current.

Admittance is:

$$Y = \frac{1}{L} \cdot \frac{1}{s + R/L} \tag{14-4}$$

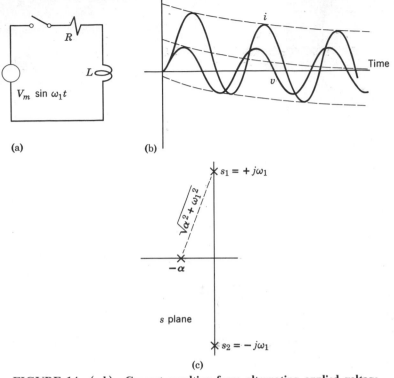

FIGURE 14 (a,b) Current resulting from alternating applied voltage.
(c)   Poles of equation 14-7.

The transform of voltage comes from equation 14-1:

$$V(s) = \mathcal{L} V_m \sin \omega_1 t = \frac{V_m \omega_1}{s^2 + \omega_1{}^2} \tag{14-5}$$

Hence

$$I(s) = V(s)Y = \frac{V_m \omega_1/L}{(s^2 + \omega_1{}^2)(s + R/L)} \tag{14-6}$$

The solution is now to be completed by inverse transformation. Since no transform pair with this $F(s)$ has yet been derived, let us try inverse transformation by the method of partial fractions. The function of equation 14-6 has a general form that can be expanded as follows:

$$\begin{aligned}
F(s) &= \frac{1}{(s + \alpha)(s^2 + \omega_1{}^2)} \\
&= \frac{K_0}{s + \alpha} + \frac{K_1}{s + j\omega_1} + \frac{K_2}{s - j\omega_1}
\end{aligned} \tag{14-7}$$

From this, $f(t)$ is to be obtained in the form:

$$f(t) = K_0 e^{-\alpha t} + A e^{-t/\tau} \sin(\omega_1 t - \varphi) \tag{14-8}$$

with coefficients yet to be determined.

Equation 14-7 has three poles in the s plane as shown in Figure 14c. Using equation 14-5 of Chapter 14, the product of the vectors from the other two poles is the denominator of the residue at the real pole:

$$K_0 = \frac{1}{\alpha^2 + \omega_1{}^2} \tag{14-9}$$

Using the rules of Section 16 of Chapter 14, relating to an oscillation, we find from rule 1 that the magnitude of the residue at the upper imaginary pole is:

$$|K_1| = \frac{1}{(2\omega_1)\sqrt{\alpha^2 + \omega_1{}^2}} \tag{14-10}$$

Then from rule 2:

$$A = 2|K_1| = \frac{1}{\omega_1 \sqrt{\alpha^2 + \omega_1{}^2}} \quad \text{and} \quad \frac{1}{\tau} = \sigma_1 = 0 \tag{14-11}$$

By rule 3, using the sum of angles of vectors from other poles less 90 degrees:

$$\varphi = 90° + \tan^{-1}\frac{\omega_1}{\alpha} - 90° = \tan^{-1}\frac{\omega_1}{\alpha} = \sin^{-1}\frac{\omega_1}{\sqrt{\alpha^2 + \omega_1{}^2}} \tag{14-12}$$

By rule 4, the frequency of the resulting oscillation is ω_1. When these coefficients are entered into equation 14-8 the result is:

$$f(t) = \frac{e^{-\alpha t}}{\alpha^2 + \omega_1{}^2} + \frac{\sin(\omega_1 t - \varphi)}{\omega_1 \sqrt{\alpha^2 + \omega_1{}^2}}$$
$$= \frac{e^{-\alpha t} \sin \varphi + \sin(\omega_1 t - \varphi)}{\omega_1 \sqrt{\alpha^2 + \omega_1{}^2}} \tag{14-13}$$

where φ is the angle defined in equation 14-12. This result gives pair 9 of our table.

Now we use this transform pair to find the current entering the circuit of Figure 14a when the switch is closed. From equation 14-6:

$$i(t) = \frac{V_m}{\sqrt{R^2 + (\omega_1 L)^2}} [e^{-(R/L)t} \sin \varphi + \sin(\omega_1 t - \varphi)] \tag{14-14}$$

where $\sin \varphi = \omega_1 L / \sqrt{R^2 + \omega_1{}^2 L^2}$. This current is plotted in Figure 14b. A number of aspects are worth emphasis. The total current consists of two parts: the natural component of exponential form that dies away with a time constant determined by the circuit alone, and a forced component of sinusoidal form that continues undiminished with an amplitude equal to the applied voltage divided by the impedance of the circuit. The total current can be described as a sinusoidal wave offset by an exponential of equal initial magnitude. The initial current is zero; this is apparent from equation 14-14, and it can be verified by applying the initial value theorem to equation 14-6. The initial derivative, the slope at zero time, is also zero; this is easily seen from the initial value theorem, and somewhat less easily by differentiating equation 14-14.

Note particularly that the steady state is a sinusoidal function and the current cannot be said to have a final value. This corresponds to the fact that the final value theorem is not valid for this function. The final value theorem cannot be applied if the given $F(s)$ has poles on the axis of imaginaries or in the right half plane, and Figure 14c shows the two poles of our function on the axis. Obviously the final value theorem must not be used when alternating voltage is applied to any ordinary network.

It is not part of this particular problem, but it is fairly easy to see what the total current would be if the switch were closed when the sinusoidal source voltage was not instantaneously zero. The applied voltage might then be a discontinuous cosine, or some other sinusoidal function between the sine and the cosine. With any such applied voltage (assuming zero initial current in the circuit) the total current would have to start at zero; it would consist of a forced sinusoidal component of easily computed amount and a sinusoidal component of known time constant with initial value just equal in magnitude to the initial value of the forced component but opposite in sign. It is an interesting sidelight that for any given ratio of R to L in the circuit, there is one instant in each half-cycle when the alternating voltage can be applied with the production of no natural component at all, and with only the forced component flowing from the very beginning. A natural component flows, it will be remembered, to be a bridge between the initial current and the forced component of current, and if no bridge is needed there is no natural or transient component of current.

15. SUMMARY

The *Laplace transformation* method, introduced in Chapter 14, is used in this chapter to solve electric network problems. To show the use of the method with a *set of differential equations*, a state-variable problem is solved.

The transform method is particularly well adapted to formulation in terms of *network functions* such as admittance, impedance, or transmittance, and a number of examples are given.

Network functions can be obtained by series and parallel *combination*, by the use of *loop* or *node* equations, or by various *theorems* and *techniques* that are developed in earlier chapters.

Inverse transformation to obtain the network output as a function of time is often the most lengthy part of a solution, and it can sometimes be avoided by recognizing that a good deal of information can be obtained by *inference* from the transform. Many of the rules of Chapter 14 are useful.

The *initial value theorem* and the *final value theorem* are introduced. *Proofs* of these theorems, and examples of their use, are given. A *restriction* of the final value theorem prevents its use if $sF(s)$ has poles *on the axis* of imaginaries or in the right half of the s plane.

Solutions based on network functions usually assume an initial condition of *rest* (no initial currents or charges). However, initial current and charge in series with an applied voltage can be included in an *excitation function*.

The transform method gives the response of a network to a disturbance of *any* given form, but some forms of applied voltage are so useful that they warrant special attention. The *step* of voltage, perhaps the most important, places an s in the denominator of the transform. The *time functions* that result from some of the most commonly used transforms have been computed and are available in *published curves*; one such family of curves is here reproduced.

The *ramp* function of voltage is another form of importance, though perhaps less useful than the step.

A discontinuous sine, cosine, or other *sinusoidal* function of voltage results when a switch is closed to connect an ac source. An example shows the natural component of current adding to the forced component to equal, at zero time, the *given initial condition* (in this example, zero initial current). As the step function is characteristic of switching a dc source, so the discontinuous sinusoid is characteristic of a switched ac source.

The next chapter will discuss the application of an *impulse* function, which is a mathematical model of a momentary jolt.

PROBLEMS

15-1. Two state equations are formulated but not solved in Problem 7-4 (page 179). Find their solution by Laplace transformation, unless this has already been done in Problem 7-5. The answer should agree with either Section 12 or 16 of Chapter 6, depending on the transform pair used.   §2

15-2. Two state equations are formulated but not solved in Problem 7-6 (page 179). Find their solution for i_1 by Laplace transformation, unless this has already been done in Problem 7-7. The answer is equation 10-9 of Chapter 6.   §2

15-3. Two state equations are formulated but not solved in Problem 7-9 (page 180). Find their solution for entering current by Laplace transformation, unless this has already been done in Problem 7-10. The answer is given for Problem 7-10.
§2

15-4. Two state equations are formulated but not solved in Problem 7-12 (page 181). Find their solution for entering current by Laplace transformation, unless this has already been done in Problem 7-13. (The answer is given for Problem 6-15.)
§2

15-5. Two state equations are formulated but not solved in Problem 7-14 (page 181). Find their solution for entering current by Laplace transformation, unless this has already been done in Problem 7-15. The answer is given for Problem 7-15.
§2

15-6. In the network of Figure 3a, find the transform of current I_2 by applying Thévenin's theorem. Find the equivalent Thévenin impedance and transform of voltage (both functions of s) for everything except L_2 and R_2. Then write I_2. Compare the result with equation 2-6.
§5

15-7. In the network of Figure 3a, find the transform of current I_2 by first changing the Y consisting of L_1, L_2, and R_{12} into an equivalent delta, as suggested in Chapter 10, and then writing the transform of current. Compare the result with equation 2-6 (Chapter 15), noting that there are several better ways to work the problem.
§5

15-8. State-variable equations are formulated in certain problems of Chapter 7. Now let us return to the same networks and formulate equations for the desired network currents by writing $Y(s)$, the admittance, to be multiplied by the Laplace transform of applied voltage. Inverse transformation then gives the current. Find four currents, as follows (but note that these same four currents are required in foregoing problems, to be computed by different methods, and you need not repeat work that you have already done):
§5

	Network is given with	Page No.:	Find:	Answer is given:
(a)	Problem 7-6	179	i_1	As Eq. 10-9 of Chapter 6
(b)	Problem 7-9	180	Entering current	To Problem 7-10
(c)	Problem 6-15	158	Entering current	To Problem 6-15
(d)	Problem 7-14	181	Entering current	To Problem 7-15

15-9. In the circuit (p. 412), $L = 10^{-2}$ henry, $R = 199$ ohms, $C = 10^{-6}$ farad. Initially, $i(0) = 0$ and $v_C(0) = 0$. Voltage $V = 10$ volts is applied by closing the switch at $t = 0$. Find $i(t)$ thereafter; compute $i(t)$, and sketch $i(t)$ with enough computed values to give reasonable accuracy.
§5*

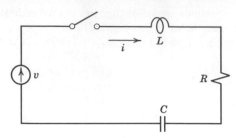

Problem 15-9.

15-10. Voltage $v = 15e^{-100t}$ is suddenly applied, by closing a switch at time $t = 0$, to a circuit consisting of $R = 3.0$ ohms and $L = 0.03$ henry in series. (*a*) Use Laplace transform methods to find the current. (*b*) Sketch voltage and current as functions of time. Find maximum current and the time at which it occurs.

§5

15-11. A capacitor of 40 nanofarads $(40 \cdot 10^{-9}$ farad) is initially charged to 25 volts. It is permitted to discharge through a circuit with 10 millihenrys of inductance and 250 ohms of resistance. Find the resulting current. Find Q of the circuit. §5*

15-12. An independent source of constant voltage $V = 10$ volts is connected at $t = 0$ to a circuit with $R = 2$ ohms, $L = 2$ henrys, and $C = 1.0$ farad in series. Initial current $i(0) = 0$ and initial capacitor voltage $v_C(0) = 0$. Find the current $i(t)$. Sketch $i(t)$. §5

15-13. Repeat Problem 15-12 but with the following initial conditions: initial current $i(0) = 1.0$ ampere and initial capacitor voltage $v_C(0) = 5$ volts in such a direction that the initial capacitor voltage opposes the source voltage. Compute and sketch the current. §5*

15-14. (*a*) For the lattice network shown, find the transfer function $V_2(s)/V_1(s)$; values are $R_1 = 10^6$, $R_2 = 10^6$, $C_1 = 10^{-6}$, $C_2 = 4 \times 10^{-6}$. Plot poles and zeros of the transfer function in the s plane. (*b*) Compute and plot the response $v_2(t)$ when the applied voltage $v_1(t)$ is a unit step. There is no initial charge, and there is no output current $(i_2 = 0)$. §6*

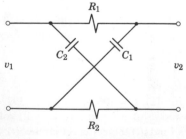

Problem 15-14.

15-15. It is found that $I(s)$, the transform for current entering a network, is as follows:

$$I(s) = \frac{(s + 3.5)}{(s + 0.5)(s^2 + 7s + 11)}$$

Without going through the process of inverse transformation to obtain $i(t)$, plot poles and zeros in the s plane and answer the following questions: (a) How many components, and of what form, add to give the total current $i(t)$? (b) Which of these components die away quickly, and which quite slowly? If not oscillatory, which are positive and which negative? (c) Which components may be considered as possibly negligible at relatively short times? at relatively long times? at all times? (d) What is the initial value of $i(t)$? (e) What is the initial value of its derivative? (f) What is the final value of $i(t)$? (g) Sketch, quite roughly, the current $i(t)$ according to the foregoing inferences. §8*

15-16. Repeat Problem 15-15 for this different $I(s)$:

$$I(s) = \frac{8(s + 2)}{s(s^2 + 5s + 9)} \qquad §8$$

15-17. Repeat Problem 15-15 for the following $I(s)$:

$$I(s) = \frac{(s + 1)(s + 3)^2}{s(s + 27)(s^2 + 9)} \qquad §8*$$

15-18. (a) Use the initial value theorem to find the three initial currents in the network of Figure 5c of Chapter 14. (b) Use the final value theorem to find the three steady-state currents in the same network. §8*

15-19. Given $F(s) = (2s + 1)/(s^3 + 2s^2)$, find the values of $f(t)$ at (a) $t = 0$, (b) $t = \infty$. Also, (c) check (a) and (b) by the initial and final value theorems. §8

15-20. A circuit with an admittance function of the form

$$Y(s) = H \frac{s - s_a}{s - s_1}$$

is to be designed. (a) Indicate the pole and zero in the s plane. (b) Devise a network that will provide such a pole and zero. (c) Write the admittance function for the network and solve for s_1 and s_a in numbers; give numerical values to whatever R's, L's, and C's you have decided to use.

Note: Like most design problems, this has more than one possible answer. §8*

15-21. A 1-volt battery is suddenly applied to the circuit that you have designed in Problem 15-20. (a) Find the initial and final values of battery current. (b) If you had designed a different network with the same pole and zero, would the resulting battery current be different? §8

15-22. Find the initial value of the second derivative of $f(t)$ given $F(s) = 1/s(s + \alpha)$. Use (a) the initial value theorem, and (b) inverse transformation followed by differentiation. §8*

15-23. Find the initial value, the initial slope, and the final value of $f(t)$ of pair 11, given the $F(s)$ of pair 11. Note that the initial values are more readily found from the initial value theorem even when the transform pair is known.   §8

15-24. Discuss the possibility of using the final value theorem to find the final derivative of $f(t)$, given $F(s)$.   §8*

15-25. A network devised as an electric wave filter is discussed in Section 25 of Chapter 12. The transfer function V_{out}/V_{in} is given in equation 25-3. Apply the final value theorem to find the final value of output voltage, there being no output current, when the input voltage is a unit step, $V_{in} = 1$ after $t = 0$. Explain the similarity of the result to equation 25-10.   §8*

15-26. The switch in the circuit of Figure 10a has been open for a long time. It is suddenly closed. Find the current through R_1 thereafter, using Laplace transform methods.   §10*

15-27. In the circuit shown, the switch is closed for a long time until $t = 0$. At that moment the switch is opened. Using Laplace transform methods, find current through the battery, $i(t)$. Sketch the battery current for a short time before and after the switch is opened. Find and sketch $v_2(t)$ also. Discuss the physical possibility.   §10

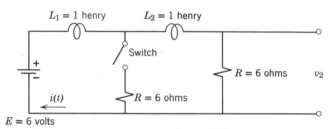

Problem 15-27.

15-28. A circuit consists of $L = 0.5$ henry, $C = 10^{-6}$ farad, and $R = 1414$ ohms in series. There is no source in the circuit, and there is no initial charge on the capacitor. Initial current in the inductance is 3.0 amperes. Find current thereafter. Sketch the current. (The transform pair of Problem 14-17 may be useful.)   §10*

15-29. Equation 9-5 gives a loop excitation function to be used in finding loop current from equation 9-6, which says $I(s) = $ (Excitation function)$/Z(s)$. Devise a node excitation function to be used in finding node voltage from an equation of the form $V(s) = $ (Node excitation function)$/Y(s)$.   §10*

15-30. A circuit is shown with Problem 6-8, page 157, in which a constant-current source supplies a conductance and a capacitance in parallel. There is initial charge on the capacitance. Using the node excitation function of Problem 15-29, find the source voltage $v(t)$ in this circuit.   §10*

15-31. A unit step voltage is to be applied to a network with the following admittance:

$$Y(s) = \frac{675}{s^2 + 18s + 225}$$

(a) Which, if any, of the precalculated curves of Figure 11a shows the resulting current? (b) Give the time scale. (c) What is the final steady current?   §12*

15-32. A ramp function of voltage is applied to a circuit in Section 13, with the result given in equation 13-6. (a) Draw the circuit. Map poles and zeros of $Y(s)$, and also of $I(s)$. Sketch $i(t)$, and also $v_C(t)$, the voltage across the capacitor. (b) If it is valid to do so, find from the initial and final value theorems: $i(0)$, $i'(0)$, $i(\infty)$, $i'(\infty)$, $v_C(0)$, $v_C'(0)$, $v_C(\infty)$, and $v_C'(\infty)$.   §13

15-33. A voltage is increasing continually, so that $v = 8000t$, beginning at time $t = 0$. This voltage is applied to a capacitor with a capacitance of 1.25 microfarads. Use the Laplace transform to find the current.   §13*

15-34. A circuit of R, L, and C in series is critically damped. $L = 1$ henry, $C = 1$ microfarad. The circuit is initially at rest: $i(0) = 0$ and $v_C(0) = 0$. Voltage applied at $t = 0$ is $v(t) = 2000 \sin 10^3 t + V$, where V is the undetermined amplitude of a direct component of voltage. (a) Find a value of V for which there is no natural component of current. (b) Find $v_C(t)$, capacitor voltage as a function of time.   §14*

15-35. Apply $v = V_m \cos \omega t$ to an LRC series circuit at $t = 0$. Initially there is zero stored energy (in L and C). Find i.   §14*

$$\omega_0 = \frac{1}{\sqrt{LC}} \qquad R = 2\sqrt{L/C}$$

$$Q = \frac{\omega_0 L}{R}$$

$$i(t) = A e^{s_1 t} + B e^{s_2 t} + C \cos(\omega t - \phi)$$

CHAPTER
16

TRANSLATION, CONVOLUTION, AND THE IMPULSE FUNCTION

1. THE IMPULSE FUNCTION

A sudden shock to a system can excite its natural response without causing any forced response. Such a sudden disturbance is used both physically and mathematically as a means of studying the natural response alone. Physically, like a wine glass ringing from the snap of a finger nail, a system responds in its natural mode to a momentary pulse that leaves no forced response.

The blow given to a golf ball is such a pulse. For the exceedingly short time that a club is in contact with a ball, the ball receives momentum; thereafter the motion of the ball is its natural response only, as the energy of the initial impulse is dissipated and the ball rolls more or less exponentially to rest.

Mathematically, the disturbing function (whether electrical or mechanical) is a function that gives one unit of impulse to a system though it acts through a negligibly short time. The letter δ is used to indicate this *Dirac* impulse function.

In mechanical terms, impulse is force times time. In electrical terms, the electrokinetic impulse is the product of voltage and time. Suppose, as in

Figure 1a, that 10 volts is applied for $\frac{1}{10}$ second; the electrokinetic impulse is 1 volt-second. If 20 volts is applied for $\frac{1}{20}$ second, the impulse is still 1 volt-second. If the time of application becomes short without limit, but with the impulse remaining 1 volt-second, the resulting function is called the δ function.

The δ function can be visualized as a force or voltage that is large without limit, applied for a vanishingly small time. It can only be indicated in Figure 1a by a vertical line with an arrowhead to suggest that it is infinitely high.

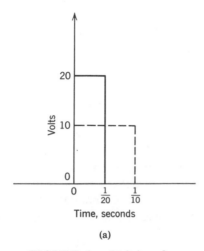

(a)

FIGURE 1a   Unit impulse.

Since δ is a function of time, it can be written $\delta(t)$. We say that $\delta(t)$ is 0 at all times except $t = 0$. When $t = 0$ the function cannot be specified, but it has unit impulse; that is, the δ function is *defined* by writing

$$\int_a^b \delta(t)\, dt = 1 \qquad \text{if } a \le 0 < b \tag{1-1}$$

The lower limit a can be zero or anything less than zero, and the upper limit b can be anything greater than zero. But if the limits are not on opposite sides of the impulse, the spike of force or voltage at $t = 0$, the value of the integral is zero.

It is often useful to have an impulse function for which the spike does not occur at $t = 0$ but at some other time which may be called τ. To define an impulse with its spike at τ the function is written $\delta(t - \tau)$, and the definition is

$$\int_a^b \delta(t - \tau)\, dt = 1 \qquad \text{if } a \le \tau < b \tag{1-2}$$

and if τ is not between a and b the integral is zero.

For our present purpose it will be convenient to use an impulsive spike that occurs at time $\tau = 0$, as in equation 1-1. It will also be convenient to let the limits of integration be zero and infinity, giving

$$\int_0^\infty \delta(t)\, dt = 1 \tag{1-3}$$

2. THE TRANSFORM OF δ

Clearly, $\delta(t)$ is a function of time, although a peculiar one. Perhaps we can find its Laplace transform. Since the Laplace transform in general is

$$\mathscr{L}f(t) = \int_0^\infty f(t)e^{-st}\, dt \tag{2-1}$$

we substitute the δ function for $f(t)$ to obtain its transform:

$$\mathscr{L}\, \delta(t) = \int_0^\infty \delta(t)e^{-st}\, dt \tag{2-2}$$

To evaluate this integral, notice that $\delta(t) = 0$ everywhere except when $t = 0$. Hence, at any time except $t = 0$, the product $\delta(t)e^{-st} = 0$. At the particular time when $t = 0$, the exponential $e^{-st} = 1$, and at this instant, therefore, $\delta(t)e^{-st} = \delta(t)$. Hence equation 2-2 may be written

$$\mathscr{L}\, \delta(t) = \int_0^\infty \delta(t)\, dt \tag{2-3}$$

But equation 1-3 tells us that this integral, by definition of the δ function, is 1. Therefore

$$\mathscr{L}\, \delta(t) = 1 \tag{2-4}$$

That is, the transform of the δ function is 1.

Here we have a new transform pair that can be added to our Table 14-2, as pair 13. It has a number of interesting implications.

In the first place, remembering that the transform is the spectrum of the time function, we see that the impulse function contains all frequencies equally. The value of the transform for every s, for every frequency, is 1. It is not easy to visualize the infinite spike of the δ function as the sum of an infinity of exponential components, of equal though infinitesimal amplitude and endless duration. Yet this is what the analysis tells us. By applying an impulse function to a network we are applying to it all frequencies at once, and in equal amplitude. The resulting current characterizes the network.

Second, the δ function has no value for any positive time. Hence, if voltage is δ there is no forced component of current, but only natural components.

The current that flows when a unit impulse of voltage is applied to a network is found as is current for any other applied voltage. When voltage is applied to a network at rest the transform of current is $I(s) = V(s)Y(s)$. For the unit impulse voltage, $V(s) = 1$ (equation 2-4), so the transform of current is merely $Y(s)$. Current is therefore the inverse transform of $Y(s)$, which we choose to call $y(t)$; that is, we let

$$\mathscr{L} y(t) = Y(s) \qquad (2\text{-}5)$$

Then $y(t)$ is the entering current when a unit impulse voltage is applied. (Note that an impulse of voltage gives a jolt *followed by a short circuit*.)

In a similar manner, if a transfer function (or, in general, a network function) relating voltage or current at any point in a network to voltage at a pair of terminals is $F(s)$, then the response at that point to a unit impulse of voltage at those terminals is $f(t)$ where

$$\mathscr{L} f(t) = F(s) \qquad (2\text{-}6)$$

The response to impulse voltage depends only on the network; it contains no forced component.

3. OPERATIONAL ANALYSIS

An interesting view of Table 14-2 can follow from this relation. In such a view the entry in the column headed $F(s)$ is a network function, and the natural response of that network is the time function in the column headed $f(t)$. The table is considered to be a list of natural responses.

Starting with a circuit for which the admittance is $F(s)$, one finds the response to a unit impulse by any means convenient. The classic solution of the differential equation is a likely method. Convolution, to be discussed in Section 5 of this chapter, may be used. The resulting time function is $f(t)$. The table can be constructed in this fashion, without reference to the Laplace integral, merely by listing the corresponding admittances and natural components (e.g., L. A. Manning).

This method is similar to that developed by Heaviside in the nineteenth century and elegantly presented by Bush in his *Operational Circuit Analysis*. Bush's chapter on the Fourier integral, following his presentation of the "direct operational method," was the predecessor of modern Laplace transformation.

4. TRANSLATION IN TIME

Moving the impulse function sideways so as to shift its spike from one value of t to another (for instance, from $t = 0$ to $t = \tau$) is an example of *translation* of the function. Any function can be so translated; if the variable is t, the means is to substitute $(t - \tau)$ for t, as in equation 1-2. Translation will be familiar from analytic geometry.

The question in which we are now interested is the following. If some function $f(t)$ transforms to $F(s)$, to what does the translated function $f(t - a)$ transform? We shall find that $f(t - a)$ transforms to $e^{-sa}F(s)$, provided that a is a positive real quantity. The development follows, but first there must be a few definitions.

The so-called step function or unit function is shown in Figure 4a with a step at $t = 0$; it is written $u(t)$, where

$$u(t) = \begin{cases} 0 & \text{if } t < 0 \\ 1 & \text{if } t > 0 \end{cases} \tag{4-1}$$

More generally the step in such a function can take place at any time; if it takes place at $t = a$, the symbol for the function is $u(t - a)$. Such a function is shown in Figure 4b, and

$$u(t - a) = \begin{cases} 0 & \text{if } t - a < 0, \text{ i.e., if } t < a \\ 1 & \text{if } t - a > 0, \text{ i.e., if } t > a \end{cases} \tag{4-2}$$

Any function multiplied by $u(t)$ becomes zero through negative time. Thus, if $f_1(t)$ is any function, as indicated in Figure 4c, the product $f_1(t)u(t)$ is zero through negative time, while through all positive time its value is the same as $f_1(t)$, as in Figure 4d.

If the function $f_1(t)$ and the step $u(t)$ are both translated to the right by changing t to $(t - a)$, the result is as shown in Figure 4e; everything has been moved to the right by a distance a.

Let us now return to transformation. By definition, the Laplace transform of any transformable function $f(\tau)$ is $F(s)$, where

$$\int_{\tau=0}^{\infty} f(\tau)u(\tau)e^{-s\tau}\, d\tau = F(s) \tag{4-3}$$

In this integral, $u(\tau)$ is unnecessary and is usually omitted, for its value is merely 1 through the entire region of integration, but here it is included, for it will be wanted later.

Let us now define $\tau = t - a$ and write

$$\int_{t-a=0}^{\infty} f(t - a)u(t - a)e^{-s(t-a)}\, d(t - a) = F(s) \tag{4-4}$$

Here t is a time variable, as is τ, and a is a constant, so $d(t - a) = dt$. Let us make this change in equation 4-4. Let us also multiply each side by e^{-sa};

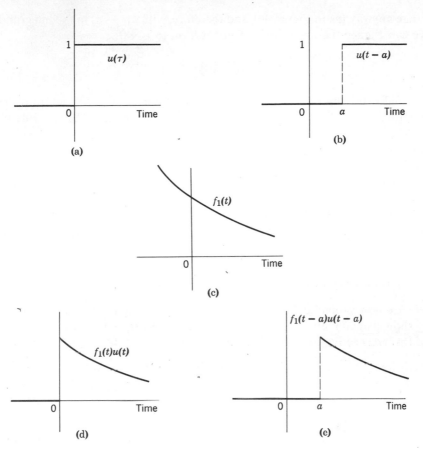

FIGURE 4 (a) A step function. (b) A translated step function. (c) Any function. (d) The function discontinuous at $t = 0$. (e) The function translated and discontinuous at $t = \alpha$.

since integration is with respect to t, not s, this exponential does not take part in the integration and can be put under the integral sign, giving

$$\int_{t=a}^{\infty} f(t - a)u(t - a)e^{-st}\, dt = e^{-sa}F(s) \qquad (4\text{-}5)$$

The lower limit of integration is $t - a = 0$, but this can equally well be written $t = a$, as is done in equation 4-5. The upper limit is $t - a = \infty$, and since a is a constant the upper limit of t is $t = \infty$.

However, if a is positive, it really makes no difference whether the lower limit of integration is $t = a$ or $t = 0$, for the integrand is zero through this range anyway (as in Figure 4e), and therefore, with $u(t - a)$ in the integrand, we can change the lower limit of integration to $t = 0$:

$$\int_{t=0}^{\infty} [f(t - a)u(t - a)]e^{-st}\, dt = e^{-sa}F(s) \tag{4-6}$$

Finally, this integral is recognized as the Laplace transformation of the function $f(t - a) \cdot u(t - a)$, for which the transform is $e^{-sa}F(s)$. Thus we have shown that translation of the time function through a corresponds to multiplication of the complex function by e^{-sa}, and this is pair IX of Table 14-7.

It is permissible to change the lower limit of equation 4-5 from $t = a$ to $t = 0$ only if a is a positive real quantity, as pointed out above, or if $f(t)$ is such a function that $f(t - a)u(t - a) = 0$ for all negative t. The latter is unlikely, so this real translation theorem is commonly restricted to *positive* values of a, a being real.

5. CONVOLUTION

We are now prepared to consider the meaning of the term *convolution*, and to show that convolution of two real functions corresponds to multiplication of their respective transforms. That is, we shall show that if

$$\mathscr{L}f_1(t) = F_1(s) \tag{5-1}$$

and

$$\mathscr{L}f_2(t) = F_2(s) \tag{5-2}$$

and if the process that is called convolution is indicated by the symbol $*$ and is defined by the following integral:

$$f_1(t) * f_2(t) = \int_0^t f_1(t - \tau)f_2(\tau)\, d\tau \tag{5-3}$$

then

$$\mathscr{L}[f_1(t) * f_2(t)] = F_1(s) \cdot F_2(s) \tag{5-4}$$

The word *convolve* means to roll together, and it will be seen that the two functions f_1 and f_2 are multiplied together in such a manner that one is continually moving (with time τ) relative to the other. A graphical interpretation of convolution can be devised (the integral of the product of functions as one function moves past the other), but an electric-circuit meaning will soon appear in which we shall interpret f_1 as admittance and f_2 as voltage.

Equation 5-4 is a remarkably interesting result because we are so often called upon to find the inverse transform of a quantity that is the product of two functions $F_1(s)$ and $F_2(s)$, for each of which, taken individually and separately, we know the inverse transforms $f_1(t)$ and $f_2(t)$. If convolution were an easy thing to do, this would be a splendid way to find the inverse transform of a product; however, regrettably, convolution, which is defined by equation 5-3, is not easy.

6. COMPLEX MULTIPLICATION

We wish to prove that convolution in the time domain transforms to multiplication in the complex domain, as stated in equation 5-4. We can first give a formal proof that is short and not hard to follow. Then an explanation will be considered, to give meaning to the symbols.

With the Laplace transform defined by the integral of equation 2-1, let it be given that

$$\mathscr{L}f_1(t) = F_1(s) \quad \text{and} \quad \mathscr{L}f_2(t) = F_2(s) \tag{6-1}$$

and

$$\mathscr{L}[f_1(t) * f_2(t)] = F(s) \tag{6-2}$$

We wish to find $F(s)$. Indeed, in order to prove that equation 5-4 is correct, we hope to show that $F(s) = F_1(s)F_2(s)$.

Equation 6-2 is written as the Laplace-transform integral (the integral of equation 2-1), and with the definition of equation 5-3 expressing convolution in the brackets the result is

$$F(s) = \int_{t=0}^{\infty} [f_1(t) * f_2(t)]e^{-st} \, dt$$

$$= \int_{t=0}^{\infty} \left[\int_{\tau=0}^{t} f_1(t-\tau)f_2(\tau) \, d\tau \right] e^{-st} \, dt \tag{6-3}$$

The unit function $u(t-\tau)$ is, by equation 4-2, equal to zero if $t < \tau$, which is to say if $\tau > t$. If we multiply the integrand of equation 6-3 by $u(t-\tau)$, it will be zero for all values of τ greater than t. If the integrand is zero when τ is greater than t, the upper limit of integration (of the inner integral) can be changed from t to ∞ without changing the value of the integral. Also, since e^{-st} is not a function of τ, it can be put within the inner integral. Then

$$F(s) = \int_{t=0}^{\infty} \left[\int_{\tau=0}^{\infty} f_1(t-\tau)f_2(\tau)u(t-\tau)e^{-st} \, d\tau \right] dt \tag{6-4}$$

The order of integration can be changed:†

$$F(s) = \int_{\tau=0}^{\infty} \left[\int_{t=0}^{\infty} f_1(t - \tau) f_2(\tau) u(t - \tau) e^{-st} \, dt \right] d\tau \qquad (6\text{-}5)$$

But $f_2(\tau)$ is not a function of t and need not be within the inner integral, so

$$F(s) = \int_{\tau=0}^{\infty} f_2(\tau) \left[\int_{t=0}^{\infty} f_1(t - \tau) u(t - \tau) e^{-st} \, dt \right] d\tau \qquad (6\text{-}6)$$

Comparison of equation 6-6 with equation 4-6 shows that the inner integral is the transform of a translated time function which (if $\tau > 0$) can be written $e^{-s\tau} F_1(s)$, and

$$F(s) = \int_{\tau=0}^{\infty} f_2(\tau) e^{-s\tau} F_1(s) \, d\tau \qquad (6\text{-}7)$$

$F_1(s)$, not being a function of τ, can be removed from the remaining integration, giving

$$F(s) = F_1(s) \int_{0}^{\infty} f_2(\tau) e^{-s\tau} \, d\tau \qquad (6\text{-}8)$$

Finally, by equation 6-1, the integral of this equation is $F_2(s)$, leaving

$$F(s) = F_1(s) \cdot F_2(s) \qquad (6\text{-}9)$$

Thus we have found $F(s)$ to be the product of $F_1(s)$ and $F_2(s)$, and equation 6-2 becomes

$$\mathscr{L}[f_1(t) * f_2(t)] = F(s) = F_1(s) \cdot F_2(s) \qquad (6\text{-}10)$$

This result is equation 5-4 which we wished to prove, and it is also pair X in Table 14-7. If $f(t)$ is the time function for which $\mathscr{L}f(t) = F(s)$, then

$$f(t) = f_1(t) * f_2(t) \qquad (6\text{-}11)$$

7. INVERSE TRANSFORMATION BY CONVOLUTION

By means of equation 6-10, convolution can be used to find the inverse transform of a function that is the product of two simpler and recognizable functions. As an example, suppose that the problem is to find $f(t)$ when

$$F(s) = \frac{1}{(s + \alpha)(s + \beta)} \qquad (7\text{-}1)$$

---

† This is permissible because the functions are Laplace transformable. For the meaning of "Laplace transformable," see page 122 of Gardner and Barnes. Ordinary functions are Laplace transformable if the Laplace transform is finite.

Although we suppose that we do not know $f(t)$, the transform of $F(s)$, we do recognize that

$$F(s) = F_1(s)F_2(s) \tag{7-2}$$

where

$$F_1(s) = \frac{1}{s + \alpha} \quad \text{so that} \quad f_1(t) = e^{-\alpha t} \tag{7-3}$$

and

$$F_2(s) = \frac{1}{s + \beta} \quad \text{so that} \quad f_2(t) = e^{-\beta t} \tag{7-4}$$

Then by equation 6-10, $f(t) = f_1(t) * f_2(t)$, and equation 5-3 gives this convolution as the integral:

$$f(t) = \int_0^t f_1(t - \tau)f_2(\tau) \, d\tau = \int_0^t e^{-\alpha(t-\tau)}e^{-\beta\tau} \, d\tau$$

$$= e^{-\alpha t} \int_0^t e^{(\alpha - \beta)\tau} \, d\tau = \frac{e^{-\alpha t}}{\alpha - \beta}[e^{(\alpha - \beta)t} - 1] \tag{7-5}$$

$$= \frac{1}{\alpha - \beta}(e^{-\beta t} - e^{-\alpha t})$$

We can thus obtain a new inverse transform, although it will be recognized that the result of this particular example is already known to us as pair 4.

8. CIRCUIT ANALYSIS BY CONVOLUTION

Convolution can be used to find current in a network. If voltage applied to the network is of *any* form $v(t)$, and if it is known that the current produced in the network by *unit impulse* voltage is $y(t)$, then current produced by $v(t)$ is $v(t) * y(t)$.

For example, the admittance of the circuit of Figure 10b is $Y(s) = 1/(Ls + R)$, so

$$y(t) = \frac{1}{L} e^{-(R/L)t} \tag{8-1}$$

To this circuit we apply the voltage (similar to Figure 4d):

$$v(t) = V_m e^{-t} \tag{8-2}$$

Current produced in the circuit by this voltage is given by convolution:

$$i(t) = y(t) * v(t) \tag{8-3}$$

To find the required $i(t)$, the convolution is written out as an integral similar to equation 5-3, giving:

$$i(t) = \int_0^t y(t - \tau)v(\tau)\, d\tau = \frac{V_m}{L} \int_0^t e^{-(R/L)(t-\tau)} e^{-\tau}\, d\tau \qquad (8\text{-}4)$$

Integration is now required. However, this particular integration has already been done in equation 7-5, so it need not be done again; the result is

$$i(t) = \frac{V_m}{R - L} (e^{-t} - e^{-(R/L)t}) \qquad (8\text{-}5)$$

Thus a circuit problem has been solved, using convolution to find $i(t)$ from $v(t)$ and $y(t)$.

Our proof of equation 8-3 as given in Section 6 uses transformation. However, Section 9 will now show how this proof can be accomplished entirely in the time domain.

9. CONVOLUTION AND SUPERPOSITION

Historically, convolution gave a means of finding the response of a linear system to any applied disturbing force by the integration of real time functions. Convolution was used by Heaviside in the nineteenth century, before Laplace transformation was applied to circuit analysis (see Bush), to find current resulting from a voltage other than a constant battery voltage.

The concept used by Heaviside and others after him was so interesting, and so helpful in understanding convolution, that it will now be explained. This explanation will not be a rigorous proof; it could be made so with sufficient care, but we have already proved the convolution theorem in equation 6-10 (using transformation), and another proof is not needed so much as a demonstration of plausibility.

Let us, therefore, think of a circuit to which a voltage is applied. The response of the circuit to an impulse, to a voltage of the form $v = \delta(t)$, is known to be $y(t)$; this response to an impulse is also the natural component of current. For this example, $y(t)$ is shown in Figure 9a. The voltage can be of any desired form; we shall assume that shown in Figure 9b and call it $v(\tau)$.

Superposition is to be used, and the voltage is approximated by a number of impulses. We first think of impulses of finite height and duration as in Figure 9c. We shall find the current produced by each impulse independently, and then add the individual responses to get the total current.

Each impulse of the approximation has a duration of $\Delta\tau$ and a height equal to the voltage at the time of the impulse. Thus the impulse at time τ_1 has a height of $v(\tau_1)$. The amount of impulse produced by the rectangular voltage at time τ_1 is the product $v(\tau_1)\,\Delta\tau$. The series of impulses can

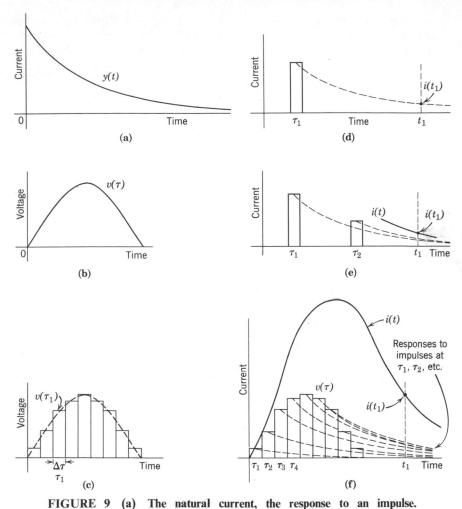

FIGURE 9 (a) **The natural current, the response to an impulse.** (b) **A voltage.** (c) **The voltage approximated by a series of impulses.** (d) **The response to one impulse.** (e) **The response to two impulses.** (f) **The response to the actual voltage as approximated by a series of impulses.**

approximate the actual voltage as closely as desired by letting the duration $\Delta\tau$ become small without limit. (Note that the instant at which a force is applied is called τ. The instant at which the response is determined will be called t, and both are instants of time.)

We know that the system response to a unit impulse is $y(t)$ as in Figure 9a. The response to a unit impulse applied at some later instant that can be called τ_1 is correspondingly delayed and can be written $y(t - \tau_1)$. If the impulse at τ_1 is not a *unit* impulse, the response is in proportion to the impulse. This is shown in Figure 9d. The response does not begin until time is greater than τ_1; at any later time t_1 the response is

$$i(t_1) = y(t_1 - \tau_1) \cdot (\text{Impulse at } \tau_1) \qquad (t_1 > \tau_1) \qquad (9\text{-}1)$$

If there are several impulses the response is the sum, as in Figure 9e, for the system is linear and superposition can be used. Thus the total response at t_1 is

$$i(t_1) = y(t_1 - \tau_1) \cdot (\text{Impulse at } \tau_1)$$
$$+ y(t_1 - \tau_2) \cdot (\text{Impulse at } \tau_2) + \cdots \qquad (9\text{-}2)$$

where the impulses are applied at $\tau_1, \tau_2, \ldots, \tau_n$ and $t_1 > \tau_n$. Any impulses applied at times greater than t_1 are disregarded, for they contribute nothing.

Equation 9-2 gives current at a specific time t_1. Generalizing, to give current at any instant t, we write

$$i(t) = y(t - \tau_1) \cdot (\text{Impulse at } \tau_1)$$
$$+ y(t - \tau_2) \cdot (\text{Impulse at } \tau_2) + \cdots \qquad (9\text{-}3)$$

where $t > \tau_n$, and this gives the response i as a function of time t.

This equation is more compact if written with a summation sign:

$$i(t) = \sum_{\tau_n = 0}^{\tau_n = t} y(t - \tau_n) \cdot (\text{Impulse at } \tau_n) \qquad (9\text{-}4)$$

The summation is interpreted to mean that the total response at time t is the sum of the responses to all the impulses that are applied between the starting time and the time t.

In Figure 9c there are many impulses. The amount of each impulse is $v(\tau_n) \, \Delta\tau$, where $v(\tau_n)$ is the height of an impulse that occurs at time τ_n, and $\Delta\tau$ is its breadth. We can write this amount of impulse into equation 9-4 to obtain the total current:

$$i(t) = \sum_{\tau_n = 0}^{\tau_n = t} y(t - \tau_n) v(\tau_n) \, \Delta\tau \qquad (9\text{-}5)$$

This is the current at any time t owing to the impulses of voltage applied at τ_1, τ_2, τ_3, and so on, between the starting time and the time t. More precisely, it is an approximation of that current. See Figure 9f.

If, now, the width of each rectangle of Figure 9c (or Figure 9f) is made less and less, letting $\Delta\tau$ approach $d\tau$, the sum of impulses approaches more closely the actual applied voltage $v(\tau)$. At the same time the summation of equation 9-5 approaches the integral:

$$i(t) = \int_{\tau = 0}^{\tau = t} y(t - \tau) v(\tau) \, d\tau \qquad (9\text{-}6)$$

Since, in the integral, τ has successively all values of time from 0 to t, the subscript n is not needed.

Thus we have obtained an integral that gives current in terms of the voltage applied to a circuit and the response of the circuit to an impulse. To obtain this integral we have used the concept of superposition.

If we now want to shorten the amount of writing, we can use the convention introduced in equation 5-3, which says that it means the same thing to write $y(t) * v(t)$ as to write the integral of equation 9-6. Then

$$i(t) = y(t) * v(t) \tag{9-7}$$

and when we speak of convolution, equation 9-6 is what we mean.

Thus we have introduced a physical interpretation of convolution. At the same time we have seen that a proof of equation 8-3 can be entirely in the time domain.

10. PIECEWISE FUNCTIONS

As a final study in translation, let us consider the transformation of a *composite* function, using the theorem from Section 4 for translation in time.

Voltages or other variables are often best described for a while by one algebraic function, and then by another. For instance $f(t)$, the solid-line function of Figure 10a, is zero until $t = 0$; it is then proportional to time until $t = t_1$, and is constant thereafter. The voltage applied to a circuit might readily approximate this function; let us see how it can be described for purposes of transformation.

At zero time the ramp function begins. If the slope is k, the ramp function is kt. To state explicitly that this ramp does not begin until zero time it can be written

$$ktu(t) \tag{10-1}$$

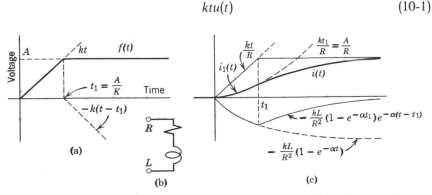

FIGURE 10 (a) A piecewise composite voltage. (b) The circuit. (c) The resulting current.

Then, at $t = t_1$, the ramp is stopped. This can be done by subtracting from the ramp function an equal function that is suppressed until t_1; that is, $ktu(t - t_1)$. This leaves the desired ramp from 0 to t_1 and nothing beyond t_1. For times greater than t_1 the constant kt_1 must be added; in the diagram the constant kt_1 is marked A. Then the voltage is:

$$f(t) = ktu(t) - ktu(t - t_1) + kt_1 u(t - t_1) \qquad (10\text{-}2)$$

$$= ktu(t) - k(t - t_1)u(t - t_1) \qquad (10\text{-}3)$$

If this function is a voltage applied to a network, its transform is needed. Using the real translation theorem of equation 4-6 (pair IX of Table 14-7), together with pair 12 from Table 14-2, equation 10-3 transforms to

$$F(s) = \frac{k}{s^2} - \frac{k}{s^2} e^{-st_1} \qquad (10\text{-}4)$$

Further steps of network analysis proceed as usual. As an example, let us suppose that the function of Figure 10a and equation 10-3 describes a voltage, and that this voltage is applied to a circuit of inductance L and resistance R in series. Equation 10-4 gives the voltage transform. Impedance of the circuit is $sL + R$. The voltage transform of equation 10-4 is divided by the impedance to get the current transform:

$$I(s) = \frac{k}{s^2(sL + R)} - \frac{ke^{-st_1}}{s^2(sL + R)} \qquad (10\text{-}5)$$

The desired time function of current requires the inverse transform of equation 10-5.

The first fraction of equation 10-5 is not unlike pair 3. Indeed, multiplying by $1/s$ and integrating the time function gives

$$\frac{1}{s^2(s + \alpha)} = \mathscr{L} \int_0^t \frac{1}{\alpha}(1 - e^{-\alpha t})\, dt = \mathscr{L}\left[\frac{t}{\alpha} + \frac{1}{\alpha^2}(e^{-\alpha t} - 1)\right] \qquad (10\text{-}6)$$

We have thus derived a new pair (it can be added to Table 14-2), and it is used to write, for the first fraction of equation 10-5,

$$i_1(t) = \frac{k}{L}\left[\frac{t}{\alpha} + \frac{1}{\alpha^2}(e^{-\alpha t} - 1)\right]u(t) \quad \text{where} \quad \alpha = \frac{R}{L} \qquad (10\text{-}7)$$

The inverse transform of the second fraction of equation 10-5 is the same time function, but, by the translation theorem of pair IX, it is delayed by t_1:

$$i_2(t) = -\frac{k}{L}\left[\frac{t - t_1}{\alpha} + \frac{1}{\alpha^2}(e^{-\alpha(t - t_1)} - 1)\right]u(t - t_1) \qquad (10\text{-}8)$$

The total current $i(t)$ is now the sum of equations 10-7 and 8:

$$i(t) = i_1(t) + i_2(t) \tag{10-9}$$

Because of the step functions $u(t)$ and $u(t - t_1)$ it is apparent that $i(t)$ can be considered in three intervals. Before $t = 0$, $i(t) = 0$. From $t = 0$ until $t = t_1$, $i(t)$ is given by $i_1(t)$ alone and hence by equation 10-7; Figure 10c shows the two components of equation 10-7 and also their sum, the heavy line, which is $i(t)$.

For times later than t_1, equations 10-7 and 8 are added. During this interval, $u(t) = 1$ and also $u(t - t_1) = 1$, and the sum of equations 10-7 and 8 simplifies to

$$i(t) = \frac{k}{R} t_1 - \frac{kL}{R^2} (1 - e^{-\alpha t_1}) e^{-\alpha(t - t_1)} \qquad \text{(for } t > t_1 \text{)} \tag{10-10}$$

Figure 10c shows this interval also. The first term of $i(t)$ is constant, and note that $kt_1/R = A/R$. Since $\alpha = R/L$, all of equation 10-10 except the final exponential is the value (from equation 10-7) of $i(t)$ at the instant $t = t_1$. Hence, in the interval beginning at t_1, current starts at the value it had at the end of the previous interval; one component of it then dies away exponentially with time after t_1, leaving finally the constant steady value of A/R at infinite time—which is of course what would be expected from the physical problem.

It should perhaps be mentioned that there is an alternative to expressing a piecewise composite function by the translation-in-time theorem. The alternative is to solve the problem for the first interval (as, for example, before t_1) and to find the current at t_1. Time after t_1 is then treated as a second problem, and the final current in the first interval is now used as the initial current in the second interval. Transformation in this second interval requires a new reference time; therefore time that was t_1 in the first interval becomes $t = 0$ in the second interval, but that is not troublesome. The two methods are really less different when actually performed than they sound, and the difference perhaps lies more in the neatness of notation than in the actual difficulty. It is hard to say which is simpler.

11. SUMMARY

The Dirac *impulse function* is defined.

The *transform* of the impulse function is found.

The response of a system to an impulse is seen to be the *natural* response of that system.

The unit step function $u(t - a)$ is used to provide a *discontinuity* where needed.

The theorem for *translation* in time states that if

$$\mathscr{L}f(t) = F(s), \quad \text{then} \quad \mathscr{L}f(t-a) = e^{-sa}F(s)$$

The *convolution* theorem states that the inverse transform of $F_1(s) \cdot F_2(s)$ is $f_1(t) * f_2(t)$ where

$$f_1(t) * f_2(t) = \int_0^t f_1(t-\tau)f_2(\tau)\, d\tau$$

The convolution theorem gives a means of finding the inverse transform of a *product* of functions of s.

If $f_1(t)$ is the response of a system to a unit impulse and $f_2(t)$ is the disturbing force, the total *response of the system* produced by the disturbing force is $f_1(t) * f_2(t)$.

The response of a network to an *impulsive* disturbance being known, its response to *any* disturbing function can be found. Any function can be considered the sum of an infinity of impulses. The total response is, by *superposition*, the sum of the infinity of responses to the impulses.

Composite functions can be expressed by means of the discontinuous unit function, $u(t)$ and $u(t - t_1)$, using the theorem for translation in time. Transformation of the translated functions leads, in the usual manner, to the solution of a problem.

PROBLEMS

16-1. (a) For the circuit given find the response to a unit impulse; that is, find the source current when $v(t) = \delta(t)$. (b) Find the transfer function $T(s) = V_C(s)/V(s)$, and find $v_C(t)$ when the input $v(t)$ is the unit impulse.   §2*

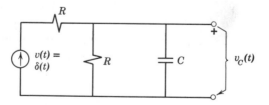

Problem 16-1.

16-2. The time function $f(t)$ is a unit step at time T (a positive time); before T it is zero, after T it is 1. Sketch $f(t)$ and find its Laplace transform $F(s)$ by integration in the Laplace-transform integral.   §4

16-3. A unit pulse $f(t)$ equals 1 from $t = 0$ to $t = T$; before $t = 0$ and after $t = T$, $f(t)$ is zero. Sketch $f(t)$, and find its Laplace transform $F(s)$ by integration in the Laplace-transform integral.   §4*

16-4. A time function $f(t)$ is an exponential discontinuous at the positive time T. Before $t = T$ the function is zero; after $t = T, f(t) = e^{-a(t-T)}$. Sketch $f(t)$, and find its Laplace transform $F(s)$. §4*

16-5. To a circuit of resistance and capacitance in series, a unit step of voltage (as in Problem 16-2) is applied at time T. Find the resulting current $i(t)$. (The answers to Problems 16-2 and 16-4 can be used if they have been computed.) §4*

16-6. Prove whether or not $f_1(t) * f_2(t) = f_2(t) * f_1(t)$. §6

16-7. Find, by integration, the convolution $e^{-\alpha t} * e^{-\alpha t}$. Which transform pair does this establish? For what physical problem could it be the solution? §6*

16-8. $\mathscr{L}f(t) = 1/(s + \alpha)(s + \beta)^2$. Find $f(t)$ by convolution. §7*

16-9. Use the convolution theorem to obtain $f(t)$ from

$$F(s) = (s + 1)/s(s^2 + 4) \qquad \S 7$$

16-10. A step function of voltage is applied to a circuit; that is, $v(t) = u(t)$. The circuit consists of resistance R and inductance L in series. Find current by three different methods: (a) by real integration, (b) by partial fractions, and (c) by convolution. The three *methods* must be clear and distinct, for the answer is of course well known. §8*

16-11. A voltage $v(t) = V_m(1 - e^{-t})$ is applied at $t = 0$ to a circuit consisting of resistance R and capacitance C in series. Use convolution to find the current. §8*

16-12. Use the convolution integral to find current in the circuit of Problem 16-1 when the voltage is the step function $v(t) = u(t)$. §8

16-13. The voltage in Section 10 rises from zero to a steady final value A at a rate k through time t_1. Let k increase; the limit as k increases is a step function of voltage. What limit is approached by current in equation 10-10 as k becomes infinite? §10*

16-14. A gate function of voltage, $v(t) = u(t) - u(t - t_0)$, is applied to a circuit that consists of resistance R and inductance L in series. Find the resulting current by whatever method is easiest; sketch $i(t)$ and its components, showing t_0. §10*

16-15. Voltage $v(t) = 1 - t/T$ for values of time between 0 and T (that is, $0 < t < T$) where T is a positive constant. Voltage is zero before $t = 0$ and after $t = T$. (a) Find the Laplace transform of this $v(t)$. (b) This voltage is applied to a circuit of R and C in series. Find the resulting current $i(t)$ by making use of the transform developed in part (a), and sketch the voltage and current. The capacitor is uncharged until $t = 0$. §10*

16-16. Solve part (b) of Problem 16-15 by this different method: Find current from $t = 0$ until $t = T$. Then formulate a different problem for time later than $t = T$, the final v_C of the first problem being the initial v_C of the second problem. Show that the results of Problems 16-15 and 16-16 are the same. §10

CHAPTER

17

THREE-PHASE SYSTEMS

1. POWER SYSTEMS†

When a customer turns on an electric light, or pushes the button to start an industrial motor, he expects electric power to be supplied at once and to continue steadily as long as he wants it. Moreover, he expects to get the power at low cost, and hence he demands generation and transmission at high efficiency.

These requirements of an unvarying supply at high efficiency characterize the power industry; a communication system, on the other hand, can convey information only by means of changes, and it obtains maximum power transfer by permitting low efficiency.

When power systems were first proposed, a number of questions had to be answered. Perhaps the first question was whether to use direct or alternating current. A sharp difference of opinion in the 1880's was decided in favor of alternating current because transformers could so easily be used to change voltage; this, giving greater efficiency of transmission and distribution, far outweighed the advantage of smooth-running, variable-speed motors that operated on direct current.

Then the frequency of alternation had to be decided. At the low frequency of 15 or 20 cycles per second (15 or 20 Hertz) the voltage drop in transmission-line inductance is less serious, but lights flicker and the vibra-

---

†Chapter 17 on three-phase systems can follow Chapter 11 on transformers whenever desired; there is no dependence on intervening chapters.

tion of machines is distressing. At 133 cycles per second (133 Hertz), transformers and some electrical machines can be smaller for the same power, but electric traction is difficult. A compromise of 60 cycles per second was finally adopted quite generally in North America, and 50 cycles in much of the rest of the world (with several exceptions). Higher frequency, however, is common for use within aircraft, where transmission lines are short and lightness of machines is vital, and frequencies of the order of 400 Hertz are not uncommon.†

    The pulsing of power flow inherent with alternating current is the cause of vibration and noise that contrast unpleasantly with the smooth operation of direct current. This can be overcome by using three-phase systems. Also, a three-phase generator allows more efficient use of the machine structure. Essentially, three generating windings can be placed on the armature of one machine, as suggested in Figure 1a. Voltages generated are shown in Figure 1b, and the phasors of these voltages in Figure 1c.

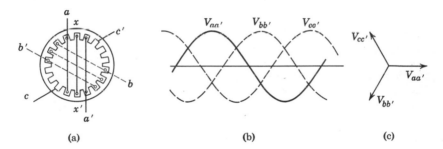

(a)   (b)   (c)

FIGURES 1a,b,c   Three-phase generation.

2. THREE-PHASE OPERATION

A three-phase generator is similar in many ways to the three separate single-phase generators shown in Figure 2a. There is a load on each of these generators, and phasors of the three load currents are shown in Figure 2d. The three voltages are equal in amplitude and equally spaced in time (120 degrees apart), while the three loads are equal in magnitude and power

---

† It is interesting that direct-current transmission became practical during the 1960's, with the advent of high-voltage tubes or valves for rectification from alternating current at the sending end followed by inversion to alternating current at the receiving end of the line. A number of intersystem links, long overhead lines or short under-sea cables, now avoid the problems of electrical reactance by transmitting at zero frequency. Such a line connecting the Columbia River region of Oregon with the Los Angeles area, 1,000 miles southward, presents an interesting study in practical operation.

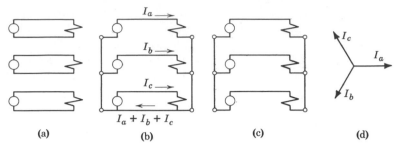

FIGURES 2a,b,c,d   Three-phase connections.

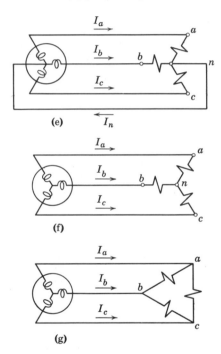

FIGURES 2e,f,g   Y and Δ loads.

factor; hence the three currents are equal in amplitude and are 120 degrees apart in phase angle as shown. This is a *balanced* system.

Figure 2a shows six wires from generators to loads: three outgoing and three return. As a measure of economy, the three return wires can be combined into the single return wire of Figure 2b. This single return wire must carry the sum of the three load currents. But the sum of the three load currents is zero at every instant as can be seen from the phasors of Figure 2d, or from the sum of three waves similar to those of Figure 1b, or by mathematical addition of the currents or their transforms. This return wire can surely be small. In fact it can be vanishingly small in an ideally balanced system and it can even be omitted entirely, as in Figure 2c.

In actual practice, some loads can be considered ideally balanced. A three-phase motor, for instance, usually has only three terminals to which its three windings are connected. The opposite ends of the three windings are connected together within the machine, like the arms of a Y, with no provision for any connection from this common neutral to any external circuit.

Often, though, it is not certain that loads will be balanced. For instance, the electric lights of a building can be connected to constitute a three-phase load, balanced if all the lights are turned on. Here the actual degree of balance depends on choice and chance, and each lamp is therefore connected from one of the three main lines to a fourth *neutral* wire which (as in Figure 2b) provides an actual physical connection to the generator—or perhaps to an intervening transformer bank.

Figure 2e shows a three-phase four-wire system. Figure 2f shows a three-phase three-wire system that can be used to supply balanced currents when the three loads are equal. If, in a three-wire system, the three loads are not equal, the line-to-neutral voltages at the loads must be unbalanced, this being a necessary result of forcing the line currents to add to zero. Hence, to avoid unbalanced voltages at the loads, the four-wire system of Figure 2e is used unless it is quite certain that the three loads will be equal.

A more advantageous three-wire system is shown in Figure 2g. Here loads are connected between lines in a Δ configuration. These Δ-*connected* loads need not be accurately balanced. As an extreme example, it would be possible to connect load between points *a* and *b* with the other phases unloaded, and power would be supplied by the generator to this single load.

3. TRANSFORMER CONNECTIONS

Most power systems use transformers first to raise and then to lower voltage on the lines. The generator voltage, usually about 10 to 30 kilovolts, is "stepped up" to perhaps 100 to 500 kilovolts for transmission. The transmission voltage is reduced for distribution; major distribution lines may operate at 13 to 69 kilovolts, with local distribution at 2.4 to 14 kilovolts. Delivery is finally made to the consumer at 120 or 240 volts.

A *bank* of three single-phase transformers (Figure 3a) is commonly used for three-phase transformation. It is significant for our present discussion that the coils are completely insulated from each other, and hence the three *primary* (or input) coils and the three *secondary* (or output) coils can be interconnected independently.

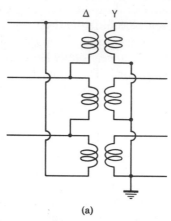

(a)

FIGURE 3a   A bank of three single-phase transformers.

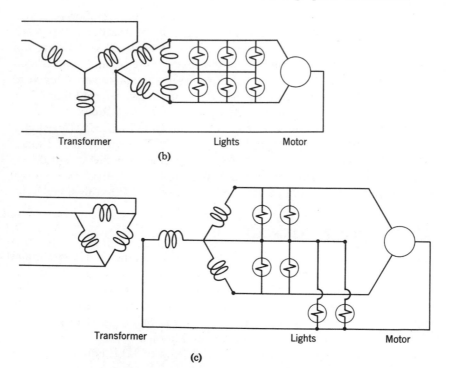

FIGURES 3b,c   Two systems of distribution.

It is usually advantageous to connect one set of transformer coils in a Δ and the other in a Y. Figure 3a shows such an arrangement. The junction of the Y is connected to a fourth *neutral* wire, which may also be connected to ground. This Δ-Y connection is quite typical of transformer installations at generating stations, the Δ-connected primary being supplied from a Y-connected generator, and the Y-connected secondary feeding the sending end of a three-phase transmission line.

An industrial customer probably needs three-phase power for motors. Two types of distribution systems are used. A transformer bank with Δ secondary (Figure 3b), 240 volts line to line, may supply power to motors, while 120-volt lamps and other single-phase loads are connected between the outer wires and a mid-tap of one or more of the Δ-connected transformers.

A common alternative system (Figure 3c) is to connect the transformer secondaries as a Y to supply 120 volts line to neutral and 208 volts line to line. Either the line-to-neutral voltage is a little high for single-phase lighting loads, or the line-to-line voltage is a little low for three-phase motor loads, but for metropolitan installations it is often a satisfactory compromise.

As a residential consumer ordinarily does not require three-phase power, transformers in residential districts may be used as three separate single-phase supplies. *Single-phase three-wire* systems are common; the middle of the three wires is from a mid-tap at the center of the transformer secondary winding. Voltage between outer wires is 240 volts, with 120 volts between either of the outer wires and the mid-tap or neutral wire, which is usually grounded.

4. A SINGLE-PHASE MODEL

When a three-phase system is balanced, the method of analysis always used (although not always clearly stated) is to determine the voltage, current, and power distribution in one phase only, knowing that the other phases will have corresponding quantities exactly the same except for a time difference of one-third cycle.

Figure 2c suggests that one phase of a balanced system in which all the apparatus is Y-connected can be represented as in Figure 4a. To this extent we merely duplicate one-third of the three-phase system. What, however, is to be done to complete the circuit between n, the load neutral, and n', the generator neutral?

The actual situation in the three-phase system is that current going out to the load in phase a divides at point n and returns through wires b and c. No current passes from n to n' through any neutral wire or in the ground; there is no neutral current; there is no voltage drop in impedance of a neutral wire,

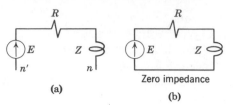

FIGURES 4a,b   Single-phase representation of a three-phase system.

nor any power loss in resistance. The single-phase equivalent circuit of Figures 4a and b must therefore close the gap from n to n' with a hypothetical conductor of *zero impedance.*[†]

If all generators, transformers, and loads were Y-connected, the analysis of a balanced three-phase system would thus become simple and familiar. In fact, however, there are Δ connections of loads and transformers that require consideration. Let us, therefore, consider the relations between line-to-line and line-to-neutral voltages, and between load currents and line currents in a Δ load. What impedance of a balanced Y load is equivalent to the given impedance of a balanced Δ load?

5. LINE-TO-LINE AND LINE-TO-NEUTRAL VOLTAGES

The voltage from line to line, being the sum of two of the voltages from line to neutral, is (in a balanced system) greater than a line-to-neutral voltage by $\sqrt{3}$ in amplitude, and different in phase angle by 30 degrees. The relations are worked out in Figure 5.

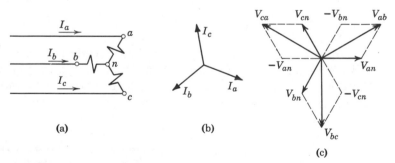

FIGURE 5   **(a)**   A balanced Y load.   **(b)**   The three-line (or phase) currents.   **(c)**   The voltages both line-to-neutral and line-to-line.

[†] A conductor of zero resistance can be visualized without trouble. A conductor of zero inductance, surrounded by no magnetic field, is more difficult. In fact, transmission-line reactance in a three-phase system is not clearly attributable to the magnetic field linking any one wire and must be treated more abstractly. See Skilling (3).

6. LINE CURRENTS AND Δ LOAD CURRENTS

The current in each line is the sum of two of the load currents in a Δ-connected load. Each line current is therefore greater in amplitude than each load current by $\sqrt{3}$ (the load being balanced), and differs in phase angle by 30 degrees. The relations are shown in Figure 6.

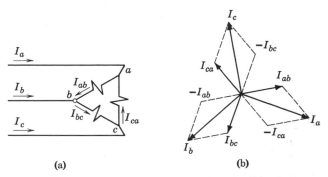

(a)   (b)

FIGURE 6   **(a)**   **A balanced Δ load.**   **(b)**   **The relation of load currents to line currents.**

7. EQUIVALENT Y AND Δ LOADS

A Y-connected load can be equivalent to a Δ-connected load in the sense that it receives the same line current, and consumes the same power. For a balanced system, the necessary relation is that the impedance of each of the three loads connected in Y must be one-third of the impedance of each of the three loads connected in Δ.

To show the equivalence, let the three equal Y-connected loads of Figure 5a be compared with the three equal Δ-connected loads of Figure 6a with the requirement that $Z_{ab} = 3Z_{an}$. Line-to-line voltage V_{ab} remains the same. In Figure 5, the line current is:

$$I_a = I_{an} = \frac{V_{an}}{Z_{an}} \tag{7-1}$$

Figure 5c shows that the line-to-line voltage is:

$$V_{ab} = \sqrt{3}V_{an}\underline{/30°} \tag{7-2}$$

For the Δ-connected load of Figure 6:

$$I_a = \sqrt{3}I_{ab}\underline{/-30°} = \sqrt{3}\frac{V_{ab}}{Z_{ab}}\underline{/-30°} \tag{7-3}$$

and with V_{ab} given by equation 7-2 and $Z_{ab} = 3Z_{an}$ this becomes

$$I_a = \frac{V_{an}}{Z_{an}} \qquad (7\text{-}4)$$

Thus the line currents in the two cases are identical. Three-phase power is discussed in the next section.

This equivalence is, indeed, just a special case of the Y-Δ transformation theorem in Section 9 of Chapter 10. It is special because here the load is balanced; the three load impedances are taken to be equal. If they were not equal, so the load was unbalanced, the theorem from Chapter 10 would still apply, but it would not be useful in the present study of a balanced network.

8. THREE-PHASE POWER

Three comments will now be made about power; in view of previous discussion these comments are nearly self-evident and proofs will not be given but will be kept for problems.

First, the power consumed by the three equal loads of a balanced three-phase system is three times the power consumed by each of the loads alone. This is true whether the loads are connected Y or Δ. For the Y-connected loads of Figure 5a:

$$P_{\text{total}} = 3\,|\,V_{an}\,|\,|\,I_a\,|\,(\text{load p.f.}) \qquad (8\text{-}1)$$

For the Δ-connected load of Figure 6a:

$$P_{\text{total}} = 3\,|\,V_{ab}\,|\,|\,I_{ab}\,|\,(\text{load p.f.}) \qquad (8\text{-}2)$$

In both equations, "load p.f." is the power factor of each load.

Second, it is often easier to measure or compute power in terms of line current and line-to-line voltage, and it can be seen from the respective phasor diagrams that either of the foregoing equations for balanced loads can be written:

$$P_{\text{total}} = \sqrt{3}\,|\,V_{ab}\,|\,|\,I_a\,|\,(\text{load p.f.}) \qquad (8\text{-}3)$$

Note that "load p.f." refers to each load, not to an angle between line current and line-to-line voltage.[†]

Third, the *instantaneous* power to each of the three loads of a three-phase system is pulsing as the alternating current passes through peaks and zeros,

---

[†] Note also that whereas equations 8-1 and 8-2 are valid for any wave form, power factor being defined as in equation 24-9 of Chapter 4, equation 8-3 is derived with the assumption that voltages (in the former) or currents (in the latter) are sinusoidal.

but the three pulses of power are equally spaced through the time of each cycle so that (in a balanced, sinusoidal three-phase system) they add to a constant value. Thus the total instantaneous power to a balanced three-phase load is constant, not pulsing as it is in each one of the three phases by itself.

This constant flow of power can be shown by expressing the power to each phase analytically and adding the three; the time-varying terms vanish from the final expression for power, leaving the instantaneous power p (as well as the time-average power P) expressed by either equation 8-1 or 8-2; V and I are rms values of sinusoidal voltage and current in a balanced three-phase system.

A similar smooth flow of total power, unchanging with time, is found in other balanced polyphase systems, but not in a single-phase system. That is why it is possible to interchange power between one balanced polyphase system and another by transformation (and without energy storage) but not to supply a polyphase system from a single-phase system.

9. EXAMPLE: BALANCED LOADS

Balanced loading is reasonably well approximated in practical power systems, and a few examples will be given to illustrate an almost unlimited range of balanced three-phase problems.

Figure 9a shows a three-phase feeder with two loads. One load is a bank of lamps connected line-to-neutral; each lamp is rated 500 watts, 120 volts. The other load has three coils connected in Δ, each coil having the impedance given in the diagram; the physical nature of this load need not concern us, but it might well be a bank of transformers with magnetizing current supplied to the Δ-connected primary windings (shown in the diagram) and with the secondaries (not shown) carrying no current and hence not entering into the problem.

Line-to-line voltage on the feeder is 220 volts. Find current in the feeder lines, and total power.

Solution: Let us compute current and power in one phase (phase a), representing the loads as in Figure 9b. The line-to-neutral voltage, with magnitude $220/\sqrt{3} = 127.0$ volts, may be taken as reference, its angle therefore being zero.

First the lamp impedance is computed. It is pure resistance, and from the power and voltage rating:

$$Z_s = R = \frac{(\text{Rated } V)^2}{\text{Rated } P} = \frac{120^2}{500} = 28.8 \text{ ohms}$$

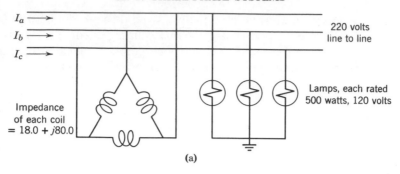

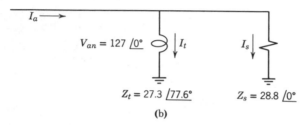

FIGURES 9a,b Loads on a three-phase feeder, and their single-phase representation.

Next the impedance to represent the coils is computed. The line-to-line impedance is given, and the equivalent line-to-neutral impedance is one-third as much:

$$Z_t = \frac{18.0 + j80.0}{3} = \frac{82.0\underline{/77.6°}}{3} = 27.3\underline{/77.6°}$$

Since lamp and coil impedances are in parallel, it is convenient to convert to admittances:

$$Y_s = \frac{1}{28.8} = 0.0347$$

$$Y_t = \frac{1}{27.3\underline{/77.6°}} = 0.0366\underline{/-77.6°} = 0.0080 - j0.0357$$

Adding to obtain total admittance:

$$Y = 0.0427 - j0.0357$$

The line current (line a) is now found:

$$I_a = YV_{an} = (0.0427 - j0.0357)(127.0)$$
$$= 5.42 - j4.54 = 7.06\underline{/-39.9°} \text{ amperes}$$

The power carried by phase a is

$$P_a = (127.0)(7.06)\cos 39.9° = 688 \text{ watts}$$

and total power to the three-phase feeder is three times 688 or

$$P = 2064 \text{ watts}$$

10. EXAMPLE: A Y-CONNECTED TRANSFORMER

Transformers present special problems. A transformer has impedance. It also, by virtue of having more turns in one of its windings than in the other, "transforms" voltages and currents. A transformer bank may be represented in the single-phase equivalent circuit by impedances in the line, with the transforming characteristics accounted for by "referring" all voltages, currents, and impedances to a single voltage level, as in Figure 4d, of Chapter 11. If a transformer has a turn ratio of n to 1, secondary voltages may be referred to the primary side by multiplying by n. Currents are converted by dividing by n, impedances by multiplying by n^2. The following example will show an application of the method.

A distribution substation (Figure 10a) supplies power through approximately a mile of line (No. 1 AWG copper) and a transformer bank to an induction motor load. Impedance of the line, one wire to neutral, is given in the figure; impedance of one of the transformers of the three-phase bank is

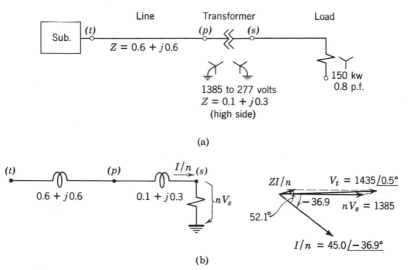

(a)

(b)

FIGURE 10 **(a)** One-line diagram of a distribution circuit. **(b)** Single-phase representation of the system of (a).

given *referred to the high-voltage side*. The three-phase load requires a maximum of 150 kilowatts, 0.80 power factor, at 480 volts, its rated terminal voltage. What voltage is necessary at the substation terminals, at (t) in the diagram? Is the difference between no-load voltage and full-load voltage at the consumer's terminals, at (s), excessive?

Figure 10a is a "one-line diagram" of the three-phase system. The line from substation to transformer to load indicates a three-phase circuit.

The pair of zigzag lines is the symbol for a transformer, and the symbols below show that it is a Y-Y bank with the neutrals grounded. The load is shown to be Y-connected with isolated neutral, which is usual for an induction motor.

Solution: The first step is to draw a diagram of a single phase of the three-phase system (following the general method of Section 4). Let us simplify the problem in this example by assuming the magnetizing current to the transformer to be negligible; this means that the shunt or central member of the equivalent T carries negligible current (has negligible admittance) and so may be omitted from the circuit. All that remains, then, to represent the transformer is its series impedance. This is a common representation.

Figure 10b shows the load impedance as yet unknown. But load current can be found. By equation 8-3, power delivered at the load terminals (at s) is

$$P = \sqrt{3}\,|V_s|\,|I|\,(\text{p.f.})$$

so

$$|I| = \frac{P}{\sqrt{3}\,|V_s|\,(\text{p.f.})} = \frac{150{,}000}{\sqrt{3}(480)(0.8)} = 225 \text{ amperes}$$

The value desired, of course, is not the actual motor current, but the current referred to the primary side, I/n:

$$\frac{|I|}{n} = \frac{225}{5} = 45 \text{ amperes}$$

The angle of this current is found from the power factor:

$$\text{Angle} = \cos^{-1} 0.8 = 36.9°$$

Arbitrarily taking the angle of the voltage at (s) to be zero, as in the diagram of Figure 10b, and knowing that current to an induction motor surely *lags* behind the applied voltage, the current (referred to the primary side) is

$$\frac{I}{n} = 45.0\underline{/-36.9°}$$

This also is shown in the diagram.

Now the voltage drop in the transmission system can be found: the voltage drop between the source at (t) and the load at (s). Total impedance is line impedance plus transformer impedance:

$$Z = (0.6 + j0.6) + (0.1 + j0.3) = (0.7 + j0.9) = 1.14\underline{/52.1°}$$

The voltage drop (see Figure 10b) is

$$Z\left(\frac{I}{n}\right) = (1.14\underline{/52.1°})(45.0\underline{/-36.9°}) = 51.3\underline{/15.2°} = 49.5 + j13.4$$

This voltage drop is to be added to the voltage at the motor terminals (referred to the primary side) to find V_t, the substation voltage:

$$V_t = 1385 + (49.5 + j13.4) = 1435\underline{/0.5°}$$

This, the line-to-neutral voltage at the generator, is the answer to the first part of the problem. Line-to-line voltage is probably more interesting; its magnitude is $1435\sqrt{3} = 2485$ volts.

To answer the second part, notice that voltage at the receiving end of the transmission system will rise from $2400/n$ volts to $2485/n$ volts when load is removed. The rise is $85/2400$, or 3.54%. This is not good; lights on the same feeder would flicker rather badly with this much voltage change. A 2% variation is considered good voltage regulation; 4% is not often tolerated.

11. EXAMPLE: A Δ-Y TRANSFORMER BANK

The preceding example relates to a line supplying power to an induction motor through a Y-Y transformer bank. In the following example a similar motor is supplied by a similar line, but through a transformer bank connected Δ-Y as in Figure 11a. This is actually a more probable transformer connection for several practical reasons, one of which will be mentioned in the next section.

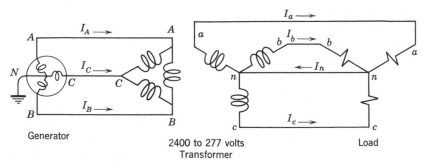

FIGURE 11a   A Δ-Y transformer bank.

The problem is to find voltage V_{AN} at the generator when the induction motor load is 150 kilowatts at 0.8 power factor, the motor voltage being the rated value of 480 volts line to line. Each transformer is rated 2400 to 277 volts and its impedance as measured on the high-voltage side (presumably by the familiar short-circuit test) is $Z = 0.3 + j0.9$ ohm. The line impedance from generator to transformer is (as before) $0.6 + j0.6$ ohm.

Solution: First, a single-line diagram of this three-phase system can be drawn. The single-line diagram will look very like Figure 10a for the previous example, but the diagram must this time indicate a Δ-Y transformer bank instead of a Y-Y bank.

Now what of the transformer impedance? In the one-line diagram, and in the equivalent single-phase diagram of Figure 10b, all impedances are given on a 1385-volt base, which is the rated line-to-neutral voltage on the high-voltage side of the transformer bank. We must use this same base in the present Δ-Y example. What we know about the transformer is its impedance as measured at the terminals of the 2400-volt winding, and what we want is the impedance that can be used in our equivalent single-phase model on a base of 1385 volts. We therefore multiply by the square of the voltage-base ratio:

$$Z \text{ (on 1385-volt base)} = \left(\frac{1385}{2400}\right)^2 (0.3 + j0.9) = 0.1 + j0.3 \text{ ohm}$$

and this is the value we must use in our model. Since it is, in fact, the same as the transformer impedance entered in Figure 10b, it is evident that the transformer bank in the present example is equivalent to the transformer bank in the previous example although it is composed of different transformers differently connected.

The rest of the solution of our problem is identical in the two examples, so the solution of the previous example can be taken as the solution of this example, too. That is, the required voltage V_{AN} has, under load, a magnitude of 1435 volts as compared to its value of 1385 volts when there is no load.

But what of the angle? The Δ-Y transformer bank introduces an angular shift of -30 degrees. Let us understand that in Figure 11a the three-phase transformer bank consists of three single-phase transformers, and that two coils drawn parallel to each other are the primary and secondary coils of one single-phase transformer. Thus phase *an* of the load is supplied from the coil *an* of one of the transformer secondaries, and the primary winding of this transformer is *BC*. Now V_{BC} is a line-to-line voltage that leads by 30 degrees the line-to-neutral voltage V_{BN} (this 30-degree relation is shown in Figure 5c, but with lower-case subscripts: V_{bc} leads V_{bn} by 30 degrees). Since V_{BC} is in phase (approximately) with V_{an} because these are voltages from the two coils of the same transformer, the voltage V_{BN} lags 30 degrees behind V_{an}. That is, because of the Δ-Y transformer connection one of the line-to-neutral voltages at the generator lags 30 degrees behind one of the line-to-neutral voltages at the load.† In addition to this 30-degree shift caused by the Δ-Y connection there is also the angular difference caused by impedance drop in the transformer and line. This was found to be 0.5 degree, with the generator voltage leading. Therefore, adding the two, the actual generator voltage must lag 29.5 degrees behind the corresponding motor voltage. (See Problem 17-4.)

Is it necessary to know the angular relations between voltages on the two sides of a transformer bank? If there is interconnection between two banks on both primary and secondary sides, and this is not unusual, the angle is essential. For example, a Y-Y bank of transformers cannot be connected in parallel with a Δ-Y bank even though the transformer ratios are such that the magnitudes of voltages are equal, for the angles cannot be right.

12. UNBALANCED LOADS

Can a three-phase system supply an unbalanced load? Can it, that is, be of practical use in delivering power to three single-phase loads that are not equal? For certain three-phase connections the answer is no, so these connections are usually avoided. For instance, loads arranged in Y with no neutral connection cannot operate independently of each other. But many three-phase connections are perfectly well able to supply unbalanced loads, and these are more commonly used.

An example of such a system can be seen in Figure 11a with a neutral conductor connected, as shown, between the load and the transformer bank. Let us consider the case of extreme unbalance in which the motor at the terminals *abc* is replaced by a single-phase load between terminals *a* and *n*, with no connections at all to terminals *b* and *c*. Current to this single-phase load is supplied from the transformer secondary *an*, with corresponding current in the transformer primary winding *BC*. The transformer primary winding is supplied over lines *B* and *C* from the generator windings *BN* and *CN*. Note that current must be zero in the primary coils *AB* and *CA* because it is zero in their secondary windings. Note, also, that there is no current in the grounded neutral of the generator; indeed, because of the transformer Δ, there is no circuit in which such neutral current might flow.

Computation of currents and voltages with unbalanced loads can often be carried out by the methods that have been presented in foregoing chapters.

---

† Two assumptions: generated voltages are in positive sequence *abc* (not *acb*), and transformer polarity is as stated (not the opposite).

However in complicated power networks, and particularly when the mutual inductances of rotating machines are to be considered, the elegant method of symmetrical components is tremendously useful.[†] This method converts an unbalanced problem into three simultaneous balanced problems. There is perhaps some analogy to Laplace analysis that converts a transients problem into simultaneous steady-state problems.

13. SUMMARY

A system to convey *power* (contrasted to a system to convey information) is characterized by *steady* voltage and *high efficiency*.

Alternating current is used because transformers permit high efficiency. *Three-phase* operation is more efficient than single-phase. DC transmission with three-phase ac generation and distribution is used in special situations.

The *three equal loads* of a balanced three-phase system can be connected either Y or Δ. Transformers are often connected Δ-Y or Y-Δ, although Y-Y or Δ-Δ banks are possible.

Analysis of one phase of a balanced three-phase system gives currents and voltages of all three phases. The single-phase model assumes a return conductor without impedance.

Balanced *line-to-line voltages* are greater than line-to-neutral voltages by $\sqrt{3}$ and differ in angle by 30 degrees. With a Δ-connected balanced load, *line currents* are greater than load currents by $\sqrt{3}$ and differ in angle by 30 degrees.

Total *power* to a three-phase load is the sum of the powers to the three loads. The total power P to a balanced load is three times the power of each phase. It is sometimes more useful that

$$P = \sqrt{3} V_{\text{line--to--line}} I_{\text{line}} (\text{load p.f.})$$

The *instantaneous* total power p is constant with time.

An *example* computes currents with Y and Δ loads.

A second example computes currents and voltages in a system containing a Y-Y *transformer bank*.

A third example computes currents and voltages in a system containing a Δ-Y *transformer bank*, including consideration of phase angle shifts.

It is probable that the loads on the three phases of a large power system will be approximately balanced. However, it is advisable to use load and transformer connections that permit good operation with *unbalanced* loads. Loads that are arranged in Δ, or in Y with connected neutral, and Δ-Y or Y-Δ transformer banks, are often advantageous. Fortunately most loads may be considered balanced.

[†] A short discussion is given in Chapter 21 of Skilling (7).

PROBLEMS

17-1. The figure shows a three-phase line with a Δ-connected load and a Y-connected load. If each of these loads is balanced, the *total* load on the line is balanced. However, this is not the case, for $R_4 = R_5 = 1$ ohm, but $R_6 = 2$ ohms. Find R_1, R_2, and R_3 that will now give a balanced *total* load on the line. (*a*) First, assume $R_1 = \infty$ (open circuit), and find R_2 and R_3 to balance the total load. (*b*) Then assume $R_1 = 5$, and find R_2 and R_3 to balance the total load.   §2*

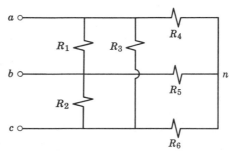

Problem 17-1.

17-2. Figure 3a shows a Δ-Y transformer bank. Redraw the diagram and letter the lines on each side. Specify polarity of the transformer windings by placing dots beside corresponding ends of the coils. One of the line currents on the Y side is 10 amperes, and the currents are balanced. Draw a phasor diagram of line currents on each side of the transformer, giving the magnitude and angle of each. (The transformer turn ratio is 1 to 1, and unity coefficient of coupling may be assumed.)   §6

17-3. The transformer bank of Problem 17-2 (Figure 3a) has 100 volts from line to line on the Δ side. When there is no load on the transformer, what is the line-to-line voltage on the Y side? Draw a phasor diagram showing voltages on each side and giving magnitude and angle of each.   §6*

17-4. The generator in Figure 11a produces balanced voltages, and (for use in this present problem) the load is disconnected. Draw a phasor diagram showing three line-to-neutral voltages at the load terminals, three line-to-line voltages at the load terminals, three line-to-line voltages at the generator and three line-to-neutral voltages at the generator. It will be understood that in the figure the windings that are parallel to each other, such as *an* and *BC*, constitute a single-phase transformer but it will be noted that these windings are not dotted to indicate polarity; draw diagrams assuming both possible polarities.   §6

17-5. Derive equation 8-3 for power from equation 8-1, and also from equation 8-2. Show whether the footnote following equation 8-3 is or is not correct.   §8

17-6. Find, as suggested in Section 8, an expression for instantaneous power p in a balanced three-phase system with sinusoidal voltages and currents; show that p does not vary with time.   §8

17-7. Referring to Figure 3c, the lighting load is balanced: from each line to neutral there are two lamps in parallel, each lamp rated 500 watts at 120 volts. (Assume constant resistance for the lamps.) The motor is to draw 5.00 kilowatts; its power factor may be estimated at 0.80, inductive. Output voltage of the transformer bank is maintained constant at 208 volts, balanced, line to line. The distribution line from the transformer to the load (where both lamps and motor are located) is 1500 feet long. Resistance of the distribution line wire is 0.403 ohm per thousand feet of wire (AWG No. 6 copper), and the inductive reactance of the line is 0.143 ohm per thousand feet of wire. Find the voltage at the load.   §9*

17-8. Each of the transformers of Figure 3a has 1.0 ohm of reactance, negligible resistance and exciting current, and unity turn ratio. A balanced unity-power-factor load on the Y side receives 20.0 amperes in each line. Line-to-line voltage on the Δ side is maintained at 125 volts. Find the terminal voltage on the Y side. Draw a phasor diagram showing all voltages and currents, giving magnitude and angle of each.   §11*

17-9. The equipment is the same as in Problem 17-7: the same lamps are used, and the motor draws 5.00 kilowatts at 0.80 power factor. The same distribution line is in use. This time, however, 120 volts is maintained at the load terminals, line to neutral. Reactance of each transformer of the Δ-Y bank is 0.50 ohm referred to the secondary side. The voltage rating of each transformer is 2400 to 120 volts. Find line currents and line-to-line voltages on the primary side of the transformer bank.   §11

17-10. Draw connection diagrams of a Y-Y and a Y-Δ bank of transformers. Connect the primary Y's of both to the same three-phase supply. Draw phasor diagrams including the secondary voltages of each bank; discuss the possibility of interconnection on the secondary side.   §11*

17-11. Draw connection diagrams of a Y-Δ and a Δ-Y bank of transformers. Connect the primaries (Y of one, Δ of the other) to the same three-phase supply, and, if this is possible, connect the secondaries together so that the transformers will operate in parallel. How must transformer turn ratios be related?   §11*

17-12. A Δ-Δ transformer bank is supplying a balanced three-phase load of pure resistance. One of the three transformers is removed, leaving what is called an open-Δ connection. How much is the resistance (I^2R) loss in the transformer bank increased? Draw a phasor diagram.   §12*

17-13. A three-phase load, Y connected, is supplied from a balanced source of voltage over three line wires; there is no neutral connection. The load from line a to the midpoint of the Y, called R_{an}, is purely resistive, and $R_{an} = 1$ ohm. Similarly, $R_{bn} = 2$ ohms and $R_{cn} = 2$ ohms. One of the balanced voltages applied from line to line is $V_{ab} = 120\underline{/0°}$. Find the three line currents. Show all currents and voltages in a phasor diagram.   §12*

APPENDIX

1

DETERMINANTS

This appendix explains the use of determinants in solving simultaneous linear algebraic equations. A more complete treatment will be found in almost any advanced algebra book.

A determinant is an array of numbers or letters written in a square, between vertical lines. The determinant has a value that is found by multiplying the elements of the determinant together in a specified pattern. If the elements of the determinant are numbers, the value of the determinant is a number. Evaluation is as follows:

Second-order determinant:

$$\begin{vmatrix} a_{11} & a_{12} \\ a_{21} & a_{22} \end{vmatrix} = a_{11}a_{22} - a_{12}a_{21} \tag{1}$$

Third-order determinant:

$$\begin{vmatrix} a_{11} & a_{12} & a_{13} \\ a_{21} & a_{22} & a_{23} \\ a_{31} & a_{32} & a_{33} \end{vmatrix} = a_{11}\begin{vmatrix} a_{22} & a_{23} \\ a_{32} & a_{33} \end{vmatrix} - a_{21}\begin{vmatrix} a_{12} & a_{13} \\ a_{32} & a_{33} \end{vmatrix} + a_{31}\begin{vmatrix} a_{12} & a_{13} \\ a_{22} & a_{23} \end{vmatrix}$$

$$= a_{11}a_{22}a_{33} + a_{12}a_{23}a_{31} + a_{13}a_{21}a_{32} - a_{11}a_{23}a_{32}$$
$$- a_{12}a_{21}a_{33} - a_{13}a_{22}a_{31} \tag{2}$$

Fourth-order determinant:

$$\begin{vmatrix} a_{11} & a_{12} & a_{13} & a_{14} \\ a_{21} & a_{22} & a_{23} & a_{24} \\ a_{31} & a_{32} & a_{33} & a_{34} \\ a_{41} & a_{42} & a_{43} & a_{44} \end{vmatrix} = a_{11}\begin{vmatrix} a_{22} & a_{23} & a_{24} \\ a_{32} & a_{33} & a_{34} \\ a_{42} & a_{43} & a_{44} \end{vmatrix} - a_{21}\begin{vmatrix} a_{12} & a_{13} & a_{14} \\ a_{32} & a_{33} & a_{34} \\ a_{42} & a_{43} & a_{44} \end{vmatrix}$$

$$+ a_{31}\begin{vmatrix} a_{12} & a_{13} & a_{14} \\ a_{22} & a_{23} & a_{24} \\ a_{42} & a_{43} & a_{44} \end{vmatrix} - a_{41}\begin{vmatrix} a_{12} & a_{13} & a_{14} \\ a_{22} & a_{23} & a_{24} \\ a_{32} & a_{33} & a_{34} \end{vmatrix} \tag{3}$$

The *minor* of a determinant is itself a determinant. The minor M_{pq} of a determinant is found by striking out the pth row and the qth column of the determinant. The above expansions of determinants are written in terms of elements multiplied by minors. Thus, if D signifies the determinant, equation 2 can be written:

$$D = a_{11} M_{11} - a_{21} M_{21} + a_{31} M_{31} \tag{4}$$

In general, for a determinant of any order,

$$D = a_{11} M_{11} - a_{21} M_{21} + a_{31} M_{31} - \cdots \tag{5}$$

Alternatively,

$$D = a_{11} M_{11} - a_{12} M_{12} + a_{13} M_{13} - \cdots \tag{6}$$

The minor together with its appropriate sign is called the *cofactor* of the element with the same subscript pair. Thus the cofactor of a_{12} is $-M_{12}$. The cofactor (except for sign) remains when both the row and the column containing the element are struck out. The sign is determined by the following rule: If the sum of the subscripts is even the sign is $+$; if odd it is $-$. (Hence the sign of the cofactor of a_{12} is $-$, for $1 + 2 = 3$, an odd number.) To state the rule symbolically, let the cofactor of a_{pq} be A_{pq}; then

$$A_{pq} = (-1)^{p+q} M_{pq}$$

The easiest way to apply this rule (if subscripts are not given in the form shown here) is to think of the determinant as the checkerboard of Figure 1. Start in the

+	−	+	−	+
−	+	−	+	−
+	−	+	−	+
−	+	−	+	−
+	−	+	−	+

FIGURE 1   Sign of the cofactor.

upper left-hand corner and say plus. Move either horizontally or vertically to the next square and say minus, to the next and say plus, and so on, until, reaching the element in question, you give the correct sign of its cofactor. The diagram shows the pattern that is counted out as you proceed.

A *working rule* for evaluating determinants of second or third order (but not higher orders) is to multiply together terms on diagonal lines, adding together those on lines that slope down to the right, and subtracting those on lines that slope down to the left. The diagonals of a second-order determinant are shown in Figure 2, and equation 1 shows the application of the rule. The diagonals of a third-order determinant are shown in Figure 3, and equation 2 shows the application of the rule. The last member of equation 2 is fully written out according to this rule; the principal diagonal, $a_{11} a_{22} a_{33}$, gives the first term, the next diagonal is $a_{12} a_{23}$ and this must be completed with a_{31} to give the second term, and so on.

FIGURE 2 **Pattern for expanding a second-order determinant.**

FIGURE 3 **Pattern for expanding a third-order determinant.**

PROPERTIES OF DETERMINANTS

The following theorems can readily be proved.

a. If a_{pq} of one determinant equals a_{qp} of another determinant for all values of p and q, the determinants are equal. This means that the rows of a determinant may be made columns, and the columns rows, without altering the value of the determinant. Thus

$$
\begin{vmatrix} 7 & 5 & 0 \\ 2 & -1 & 8 \\ 6 & 4 & 2 \end{vmatrix} = \begin{vmatrix} 7 & 2 & 6 \\ 5 & -1 & 4 \\ 0 & 8 & 2 \end{vmatrix} = -14 + 240 + 0 - [224 + 20 + 0] = -18
$$

b. Multiplying all the elements of any one row (or of any one column) by any quantity m multiplies the value of the determinant by m. (Rules c and d follow from this, letting m be -1 and 0, respectively.)

c. Changing the signs of every element of a row (or of a column) changes the sign of the value of the determinant.

d. If all elements of one column are zero, or if all elements of one row are zero, the value of the determinant is zero. Thus,

$$
\begin{vmatrix} 0 & 3 \\ 0 & 2 \end{vmatrix} = 0
$$

e. If all the elements of one row are m times the corresponding elements of another row, the value of the determinant is zero. (The same is true of columns.) Thus,

$$
\begin{vmatrix} 2 & 4 & 1 \\ 3 & 6 & 1 \\ 1 & 2 & 1 \end{vmatrix} = 12 + 4 + 6 - [4 + 12 + 6] = 0
$$

f. If two rows (or two columns) are identical, the value of the determinant is zero. (This follows from rule e, letting $m = 1$.)

g. Interchanging any two rows (or any two columns) of a determinant changes the sign of the value.

SOLUTION OF SIMULTANEOUS LINEAR EQUATIONS

This is known as Cramer's rule. As an example, consider three equations:

$$a_{11}x + a_{12}y + a_{13}z = A$$
$$a_{21}x + a_{22}y + a_{23}z = B \tag{7}$$
$$a_{31}x + a_{32}y + a_{33}z = C$$

Write the determinant which, for convenience, is called D:

$$\begin{vmatrix} a_{11} & a_{12} & a_{13} \\ a_{21} & a_{22} & a_{23} \\ a_{31} & a_{32} & a_{33} \end{vmatrix} = D \tag{8}$$

If $D = 0$ the equations have, as a rule, no solution. If D is not zero the equations are solved by writing

$$x = \frac{N_1}{D} \qquad y = \frac{N_2}{D} \qquad z = \frac{N_3}{D} \tag{9}$$

where N_1, N_2, and N_3 (the numerators) are themselves determinants formed from D by substituting A, B, and C for the coefficients of the desired unknown:

$$N_1 = \begin{vmatrix} A & a_{12} & a_{13} \\ B & a_{22} & a_{23} \\ C & a_{32} & a_{33} \end{vmatrix} \quad N_2 = \begin{vmatrix} a_{11} & A & a_{13} \\ a_{21} & B & a_{23} \\ a_{31} & C & a_{33} \end{vmatrix} \quad N_3 = \begin{vmatrix} a_{11} & a_{12} & A \\ a_{21} & a_{22} & B \\ a_{31} & a_{32} & C \end{vmatrix} \tag{10}$$

The same method is applied if there are more (or less) than three non-homogeneous linear equations.

For homogeneous equations, such as

$$a_{11}x + a_{12}y + a_{13}z = 0$$
$$a_{21}x + a_{22}y + a_{23}z = 0 \tag{11}$$
$$a_{31}x + a_{32}y + a_{33}z = 0$$

the value of each N is zero. This gives $x = y = z = 0$. There is no other solution unless $D = 0$. If $D = 0$, write all but one of the equations in the form

$$a_{11}x + a_{12}y = -a_{13}z$$
$$a_{21}x + a_{22}y = -a_{23}z \tag{12}$$

which have the solution

$$x = \frac{\begin{vmatrix} -a_{13} & a_{12} \\ -a_{23} & a_{22} \end{vmatrix}}{\begin{vmatrix} a_{11} & a_{12} \\ a_{21} & a_{22} \end{vmatrix}} z \qquad y = \frac{\begin{vmatrix} a_{11} & -a_{13} \\ a_{21} & -a_{22} \end{vmatrix}}{\begin{vmatrix} a_{11} & a_{12} \\ a_{21} & a_{22} \end{vmatrix}} z \tag{13}$$

provided the denominator of equations 13 is not zero. Thus all but one of the unknowns can be evaluated in terms of the remaining unknown. Any value may be assigned to the remaining unknown (to z in equations 13), and the other unknowns will be proportional to it.

APPENDIX
2

BIBLIOGRAPHY

Atabekov, G. I., *Linear Electrical Networks*, "Energy" Publishers, Moscow, 1964.

Blum, J. J., *Introduction to Analog Computation*, Harcourt, Brace and World, Inc., New York, 1969.

Bode, H. W., *Network Analysis and Feedback Amplifier Design*, D. Van Nostrand Co., Princeton, N.J., 1945.

Bush, V., *Operational Circuit Analysis*, John Wiley and Sons, New York, 1929.

Campbell, G. A., and R. M. Foster, *Fourier Integrals for Practical Applications*, D. Van Nostrand Co., Princeton, N.J., 1948.

Carslaw, H. S., *Introduction to the Theory of Fourier's Series and Integrals*, Dover Publications, New York, 1930.

Cheng, D. K., *Analysis of Linear Systems*, Addison-Wesley Publishing Co., Reading, Mass., 1959.

Churchill, R. V., *Operational Mathematics*, 2nd ed., McGraw-Hill Book Co., New York, 1958.

Clark, R. N., *Introduction to Automatic Control Systems*, John Wiley and Sons, New York, 1962.

Cruz, J. B., Jr., and M. E. Van Valkenburg, *Introductory Signals and Circuits*, Blaisdell Publishing Co., Waltham, Mass., 1967.

DeRusso, P. M., R. J. Roy, and C. M. Close, *State Variables for Engineers*, John Wiley and Sons, New York, 1965.

Dorf, R. C., *Modern Control Systems*, Addison-Wesley Publishing Co., Reading, Mass., 1967.

Eshbach, O. W. (ed.), *Handbook of Engineering Fundamentals*, 2nd ed., John Wiley and Sons, New York, 1952.

Frazier, R. H., *Elementary Electric-Circuit Theory*, McGraw-Hill Book Co., New York, 1945.

Gardner, M. F., and J. L. Barnes, *Transients in Linear Systems*, John Wiley and Sons, New York, 1942.

Gibbons, J. F., *Semiconductor Electronics*, McGraw-Hill Book Co., New York, 1966.

Goldman, S., *Transformation Calculus and Electrical Transients*, Prentice-Hall, Englewood Cliffs, N.J., 1949.

Guillemin, E. A. (1), *Introductory Circuit Theory*, John Wiley and Sons, New York, 1953.

Guillemin, E. A. (2), *The Mathematics of Circuit Analysis*, John Wiley and Sons, New York, 1949.

Guillemin, E. A. (3), *Communication Networks*, John Wiley and Sons, New York, Vol. I, 1931, Vol. II, 1935.

Guillemin, E. A. (4), *Synthesis of Passive Networks*, John Wiley and Sons, New York, 1957.

Guillemin, E. A. (5), *Theory of Linear Physical Systems*, John Wiley and Sons, New York, 1963.

Gupta, S. C., *Transform and State Variable Methods in Linear Systems*, John Wiley and Sons, New York, 1966.

Harman, W. W., and D. W. Lytle, *Electrical and Mechanical Networks*, McGraw-Hill Book Co., New York, 1962.

Huelsman, L. P., *Digital Computations in Basic Circuit Theory*, McGraw-Hill Book Co., New York, 1968.

Institute of Electrical and Electronics Engineers, *IEEE Standard Dictionary of Electrical and Electronics Terms*, John Wiley and Sons, New York, 1972.

Jaeger, J. C., *An Introduction to the Laplace Transformation*, Methuen Monographs on Physical Subjects, John Wiley and Sons, New York, 1949.

Jahnke, E., and F. Emde, *Tables of Functions with Formulae and Curves*, Dover Publications, New York, 1943.

Kennelly, A. E., *Chart Atlas of Complex Hyperbolic and Circular Functions*, 3rd ed., Harvard University Press, Cambridge, Mass., 1924.

Kron, G., *Tensor Analysis of Networks*, John Wiley and Sons, New York, 1942.

Lawden, D. F., *Mathematics of Engineering Systems*, John Wiley and Sons, New York, 1955.

Le Corbeiller, P., *Matrix Analysis of Electric Networks*, Harvard University Press, Cambridge, Mass., and John Wiley and Sons, New York, 1950.

LePage, W. R., and S. Seely, *General Network Analysis*, McGraw-Hill Book Co., New York, 1952.

Linvill, J. G., *Models of Transistors and Diodes*, McGraw-Hill Book Co., New York, 1963.

Linvill, J. G., and J. F. Gibbons, *Transistors and Active Circuits*, McGraw-Hill Book Co., New York, 1961.

Manning, L. A., *Electrical Circuits*, McGraw-Hill Book Co., New York, 1966.

Maxwell, J. C., *Electricity and Magnetism*, Clarendon Press, Oxford, 1873 (3rd ed., 1904).

Middendorf, W. H., *Analysis of Electric Circuits*, John Wiley and Sons, New York, 1956.

M.I.T. Electrical Engineering Staff, *Electric Circuits*, John Wiley and Sons, New York, 1940.

Oliver, B. M., *Square Wave and Impulse Testing of Linear Systems*, Hewlett-Packard Co., Palo Alto, California, 1966.

Pettit, J. M., and M. M. McWhorter, *Electronic Amplifier Circuits*, McGraw-Hill Book Co., New York, 1961.

Pipes, L. A., and S. A. Hovanessian, *Matrix-Computer Methods in Engineering*, John Wiley and Sons, New York, 1969.

Ramo, S., J. R. Whinnery, and T. Van Duzer, *Fields and Waves in Communication Electronics*, John Wiley and Sons, New York, 1965.

Rohrer, R. A., *Circuit Theory: An Introduction to the State Variable Approach*, McGraw-Hill Book Co., 1970.

Rybner, J., *Nomograms of Complex Hyperbolic Functions*, Jul. Gjellerups Forlag, Copenhagen, 1947 (Scandinavian Book Service, New York).

Sah, A. P.-T., *Dyadic Circuit Analysis*, International Textbook Co., Scranton, Pa., 1939.

Skilling, H. H. (1), *Fundamentals of Electric Waves*, 2nd ed., John Wiley and Sons, New York, 1948.

Skilling, H. H. (2), *Exploring Electricity*, The Ronald Press Co., New York, 1948.

Skilling, H. H. (3), *Electric Transmission Lines*, McGraw-Hill Book Co., New York, 1951.

Skilling, H. H. (4), *Transient Electric Currents*, 2nd ed., McGraw-Hill Book Co., New York, 1952.

Skilling, H. H. (5), *Electromechanics*, John Wiley and Sons, New York, 1962.

Skilling, H. H. (6), "An Operational View," *American Scientist* (December 1964), Vol. 52, No. 4, pp. 388A–396A.

Skilling, H. H. (7), *Electrical Engineering Circuits, 2nd ed.*, John Wilcy and Sons, New York, 1965.

Skilling, H. H. (8), *Do You Teach?*, Holt, Rinehart and Winston, New York, 1969.

Smith, R. J., *Circuits, Devices, and Systems: A First Course in Electrical Engineering*, 2nd ed., John Wiley and Sons, New York, 1971.

Steinmetz, C. P., "Complex Quantities and Their Use in Electrical Engineering," *Proceedings of the International Electrical Congress*, page 33 (Chicago, 1893), published by the American Institute of Electrical Engineers, 1894.

Terman, F. E. (1), *Radio Engineers' Handbook*, McGraw-Hill Book Co., New York, 1943.

Terman, F. E. (2), *Electronic and Radio Engineering*, 4th ed., McGraw-Hill Book Co., New York, 1955.

Thompson, S. P., *Calculus Made Easy*, 2nd ed., The Macmillan Co., New York, 1914.

Tuttle, D. F., *Network Synthesis*, John Wiley and Sons, New York, 1958.

Van Valkenburg, M. E. (1), *Introduction to Modern Network Syntheses*, John Wiley and Sons, New York, 1960.

Van Valkenburg, M. E. (2), *Network Analysis*, 2nd ed., Prentice-Hall, Englewood Cliffs, N.J., 1964.

Weber, E., *Linear Transient Analysis*, John Wiley and Sons, New York, Vol. I, 1954.

Wilson, E. B., *Advanced Calculus*, Ginn and Co., Boston, 1912.

APPENDIX
3

TRIGONOMETRIC FUNCTIONS

0°–6°

Deg.	Sin	Tan	Cot	Cos	Deg.
0.0	.00000	.00000	∞	1.0000	**90.0**
.1	.00175	.00175	573.0	1.0000	89.9
.2	.00349	.00349	286.5	1.0000	.8
.3	.00524	.00524	191.0	1.0000	.7
.4	.00698	.00698	143.24	1.0000	.6
.5	.00873	.00873	114.59	1.0000	.5
.6	.01047	.01047	95.49	0.9999	.4
.7	.01222	.01222	81.85	.9999	.3
.8	.01396	.01396	71.62	.9999	.2
.9	.01571	.01571	63.66	.9999	89.1
1.0	.01745	.01746	57.29	0.9998	**89.0**
.1	.01920	.01920	52.08	.9998	88.9
.2	.02094	.02005	47.74	.9998	.8
.3	.02269	.02269	44.07	.9997	.7
.4	.02443	.02444	40.92	.9997	.6
.5	.02618	.02619	38.19	.9997	.5
.6	.02792	.02793	35.80	.9996	.4
.7	.02967	.02968	33.69	.9996	.3
.8	.03141	.03143	31.82	.9995	.2
.9	.03316	.03317	30.14	.9995	88.1
2.0	.03490	.03492	28.64	0.9994	**88.0**
.1	.03664	.03667	27.27	.9993	87.9
.2	.03839	.03842	26.03	.9993	.8
.3	.04013	.04016	24.90	.9992	.7
.4	.04188	.04191	23.86	.9991	.6
.5	.04362	.04366	22.90	.9990	.5
.6	.04536	.04541	22.02	.9990	.4
.7	.04711	.04716	1.20	.9989	.3
.8	.04885	.04891	20.45	.9988	.2
.9	.05059	.05066	19.74	.9987	87 1
3.0	.05234	.05241	19.081	0.9986	**87.0**
.1	.05408	.05416	18.464	.9985	86.9
.2	.05582	.05591	17.886	.9984	.8
.3	.05756	.05766	17.343	.9983	.7
.4	.05931	.05941	16.832	.9982	.6
.5	.06105	.06116	16.350	.9981	.5
.6	.06279	.06291	15.895	.9980	.4
.7	.06453	.06467	15.464	.9979	.3
.8	.06627	.06642	15.056	.9978	.2
.9	.06802	.06817	14.669	.9977	86.1
4.0	.06976	.06993	14.301	0.9976	**86.0**
.1	.07150	.07163	13.951	.9975	85.9
.2	.07324	.07344	13.617	.9973	.8
.3	.07498	.07519	13.300	.9972	.7
.4	.07672	.07695	12.996	.9971	.6
.5	.07846	.07870	12.706	.9969	.5
.6	.08020	.08046	12.429	.9968	.4
.7	.08194	.08221	12.163	.9966	.3
.8	.08368	.08397	11.909	.9965	.2
.9	.08542	.08573	11.664	.9963	85.1
5.0	.08716	.08749	11.430	0.9962	**85.0**
.1	.08889	.08925	11.205	.9960	84.9
.2	.09063	.09101	10.988	.9959	.8
.3	.09237	.09277	10.780	.9957	.7
.4	.09411	.09453	10.579	.9956	.6
.5	.09585	.09629	10.385	.9954	.5
.6	.09758	.09805	10.199	.9952	.4
.7	.09932	.09981	10.019	.9951	.3
.8	.10106	.10158	9.845	.9949	.2
.9	.10279	.10334	9.677	.9947	84.1
6.0	.10453	.10510	9.514	0.9945	**84.0**
Deg.	Cos	Cot	Tan	Sin	Deg.

84°–90°

6°–12°

Deg.	Sin	Tan	Cot	Cos	Deg.
6.0	.10453	.10510	9.514	0.9945	**84.0**
.1	.10626	.10687	9.357	.9943	83.9
.2	.1080	.10863	9.205	.9942	.8
.3	.10973	.11040	9.058	.9940	.7
.4	.11147	.11217	8.915	.9938	.6
.5	.11320	.11394	8.777	.9936	.5
.6	.11494	.11570	8.643	.9934	.4
.7	.11667	.11747	8.513	.9932	.3
.8	.11840	.11924	8.386	.9930	.2
.9	.12014	.12101	8.264	.9928	83.1
7.0	.12187	.12278	8.144	0.9925	**83.0**
.1	.12360	.12456	8.028	.9923	82.9
.2	.12533	.12633	7.916	.9921	.8
.3	.12706	.12810	7.806	.9919	.7
.4	.12880	.12988	7.700	.9917	.6
.5	.13053	.13165	7.596	.9914	.5
.6	.13226	.13343	7.495	.9912	.4
.7	.13399	.13521	7.396	.9910	.3
.8	.13572	.13698	7.300	.9907	.2
.9	.13744	.13876	7.207	.9905	82.1
8.0	.13917	.14054	7.115	0.9903	**82.0**
.1	.14090	.14232	7.026	.9900	81.9
.2	.14263	.14410	6.940	.9898	.8
.3	.14436	.14588	6.855	.9895	.7
.4	.14608	.14767	6.772	.9893	.6
.5	.14781	.14945	6.691	.9890	.5
.6	.14954	.15124	6.612	.9888	.4
.7	.15126	.15302	6.535	.9885	.3
.8	.15299	.15481	6.460	.9882	.2
.9	.15471	.15660	6.386	.9880	81.1
9.0	.15643	.15838	6.314	0.9877	**81.0**
.1	.15816	.16017	6.243	.9874	80.9
.2	.15988	.16196	6.174	.9871	.8
.3	.16160	.16376	6.107	.9869	.7
.4	.16333	.16555	6.041	.9866	.6
.5	.16505	.16734	5.976	.9863	.5
.6	.16677	.16914	5.912	.9860	.4
.7	.16849	.17093	5.850	.9857	.3
.8	.17021	.17273	5.789	.9854	.2
.9	.17193	.17453	5.730	.9851	80.1
10.0	.1736	.1763	5.671	0.9848	**80.0**
.1	.1754	.1781	5.614	.9845	79.9
.2	.1771	.1799	5.558	.9842	.8
.3	.1788	.1817	5.503	.9839	.7
.4	.1805	.1835	5.449	.9836	.6
.5	.1822	.1853	5.396	.9833	.5
.6	.1840	.1871	5.343	.9829	.4
.7	.1857	.1890	5.292	.9826	.3
.8	.1874	.1908	5.242	.9823	.2
.9	.1891	.1926	5.193	.9820	79.1
11.0	.1908	.1944	5.145	0.9816	**79.0**
.1	.1925	.1962	5.097	.9813	78.9
.2	.1942	.1980	5.050	.9810	.8
.3	.1959	.1998	5.005	.9806	.7
.4	.1977	.2016	4.959	.9803	.6
.5	.1994	.2035	4.915	.9799	.5
.6	.2011	.2053	4.872	.9796	.4
.7	.2028	.2071	4.829	.9792	.3
.8	.2045	.2089	4.787	.9789	.2
.9	.2062	.2107	4.745	.9785	78.1
12.0	.2079	.2126	4.705	0.9781	**78.0**
Deg.	Cos	Cot	Tan	Sin	Deg.

78°–84°

Reprinted from *Handbook of Chemistry and Physics*, C. D. Hodgman, R. C. Weast, and S. M. Selby, Chemical Rubber Publishing Company, Cleveland, Ohio, 38th ed., 1956, pp. 128–131.

Appendix 3

12°–18°

Deg.	Sin	Tan	Cot	Cos	Deg.
12.0	0.2079	0.2126	4.705	0.9781	**78.0**
.1	.2096	.2144	4.665	.9778	77.9
.2	.2113	.2162	4.625	.9774	.8
.3	.2130	.2180	4.586	.9770	.7
.4	.2147	.2199	4.548	.9767	.6
.5	.2164	.2217	4.511	.9763	.5
.6	.2181	.2235	4.474	.9759	.4
.7	.2198	.2254	4.437	.9755	.3
.8	.2215	.2272	4.402	.9751	.2
.9	.2233	.2290	4.366	.9748	77.1
13.0	0.2250	0.2309	4.331	0.9744	**77.0**
.1	.2267	.2327	4.297	.9740	76.9
.2	.2284	.2345	4.264	.9736	.8
.3	.2300	.2364	4.230	.9732	.7
.4	.2317	.2382	4.198	.9728	.6
.5	.2334	.2401	4.165	.9724	.5
.6	.2351	.2419	4.134	.9720	.4
.7	.2368	.2438	4.102	.9715	.3
.8	.2385	.2456	4.071	.9711	.2
.9	.2402	.2475	4.041	.9707	76.1
14.0	0.2419	0.2493	4.011	0.9703	**76.0**
.1	.2436	.2512	3.981	.9699	75.9
.2	.2453	.2530	3.952	.9694	.8
.3	.2470	.2549	3.923	.9690	.7
.4	.2487	.2568	3.895	.9686	.6
.5	.2504	.2586	3.867	.9681	.5
.6	.2521	.2605	3.839	.9677	.4
.7	.2538	.2623	3.812	.9673	.3
.8	.2554	.2642	3.785	.9668	.2
.9	.2571	.2661	3.758	.9664	75.1
15.0	0.2588	0.2679	3.732	0.9659	**75.0**
.1	.2605	.2698	3.706	.9655	74.9
.2	.2622	.2717	3.681	.9650	.8
.3	.2639	.2736	3.655	.9646	.7
.4	.2656	.2754	3.630	.9641	.6
.5	.2672	.2773	3.606	.9636	.5
.6	.2689	.2792	3.582	.9632	.4
.7	.2706	.2811	3.558	.9627	.3
.8	.2723	.2830	3.534	.9622	.2
.9	.2740	.2849	3.511	.9617	74.1
16.0	0.2756	0.2867	3.487	0.9613	**74.0**
.1	.2773	.2886	3.465	.9608	73.9
.2	.2790	.2905	3.442	.9603	.8
.3	.2807	.2924	3.420	.9598	.7
.4	.2823	.2943	3.398	.9593	.6
.5	.2840	.2962	3.376	9588	.5
.6	.2857	.2981	3.354	.9583	.4
.7	.2874	.3000	3.333	.9578	.3
.8	.2890	.3019	3.312	.9573	.2
.9	.2907	.3038	3.291	.9568	73.1
17.0	0.2924	0.3057	3.271	0.9563	**73.0**
.1	.2940	.3076	3.251	.9558	72.9
.2	.2957	.3096	3.230	.9553	.8
.3	.2974	.3115	3.211	.9548	.7
.4	.2990	.3134	3.191	.9542	.6
.5	.3007	.3153	3.172	.9537	.5
.6	.3024	.3172	3.152	.9532	.4
.7	.3040	.3191	3.133	.9527	.3
.8	.3057	.3211	3.115	.9521	.2
.9	.3074	.3230	3.096	.9516	72.1
18.0	0.3090	0.3249	3.078	0.9511	**72.0**
Deg.	Cos	Cot	Tan	Sin	Deg.

72°–78°

18°–24°

Deg.	Sin	Tan	Cot	Cos	Deg.
18.0	0.3090	0.3249	3.078	0.9511	**72.0**
.1	.3107	.3269	3.060	.9505	71.9
.2	.3123	.3288	3.042	.9500	.8
.3	.3140	.3307	3.024	.9494	.7
.4	.3156	.3327	3.006	.9489	.6
.5	.3173	.3346	2.989	.9483	.5
.6	.3190	.3365	2.971	.9478	.4
.7	.3206	.3385	2.954	.9472	.3
.8	.3223	.3404	2.937	.9466	.2
.9	.3239	.3424	2.921	.9461	71.1
19.0	0.3256	0.3443	2.904	0.9455	**71.0**
.1	.3272	.3463	2.888	.9449	70.9
.2	.3289	.3482	2.872	.9444	.8
.3	.3305	.3502	2.856	.9438	.7
.4	.3322	.3522	2.840	.9432	.6
.5	.3338	.3541	2.824	.9426	.5
.6	.3355	.3561	2.808	.9421	.4
.7	.3371	.3581	2.793	.9415	.3
.8	.3387	.3600	2.778	.9409	.2
.9	.3404	.3620	2.762	.9403	70.1
20.0	0.3420	0.3640	2.747	0.9397	**70.0**
.1	.3437	.3659	2.733	.9391	69.9
.2	.3453	.3679	2.718	.9385	.8
.3	.3469	.3699	2.703	.9379	.7
.4	.3486	.3719	2.689	.9373	.6
.5	.3502	.3739	2.675	.9367	.5
.6	.3518	.3759	2.660	.9361	.4
.7	.3535	.3779	2.646	.9354	.3
.8	.3551	.3799	2.633	.9348	.2
.9	.3567	.3819	2.619	.9342	69.1
21.0	0.3584	0.3839	2.605	0.9336	**69.0**
.1	.3600	.3859	2.592	.9330	68.9
.2	.3616	.3879	2.578	.9323	.8
.3	.3633	.3899	2.565	.9317	.7
.4	.3649	.3919	2.552	.9311	.6
.5	.3665	.3939	2.539	.9304	.5
.6	.3681	.3959	2.526	.9298	.4
.7	.3697	.3979	2.513	.9291	.3
.8	.3714	.4000	2.500	.9285	.2
.9	.3730	.4020	2.488	.9278	68.1
22.0	0.3746	0.4040	2.475	0.9272	**68.0**
.1	.3762	.4061	2.463	.9265	67.9
.2	.3778	.4081	2.450	.9259	.8
.3	.3795	.4101	2.438	.9252	.7
.4	.3811	.4122	2.426	.9245	.6
.5	.3827	.4142	2.414	.9239	.5
.6	.3843	.4163	2.402	.9232	.4
.7	.3859	.4183	2.391	.9225	.3
.8	.3875	.4204	2.379	.9219	.2
.9	.3891	.4224	2.367	.9212	67.1
23.0	0.3907	0.4245	2.356	0.9205	**67.0**
.1	.3923	.4265	2.344	.9198	66.9
.2	.3939	.4286	2.333	.9191	.8
.3	.3955	.4307	2.322	.9184	.7
.4	.3971	.4327	2.311	.9178	.6
.5	.3987	.4348	2.300	.9171	.5
.6	.4003	.4369	2.289	.9164	.4
.7	.4019	.4390	2.278	.9157	.3
.8	.4035	.4411	2.267	.9150	.2
.9	.4051	.4431	2.257	.9143	66.1
24.0	0.4067	0.4452	2.246	0.9135	**66.0**
Deg.	Cos	Cot	Tan	Sin	Deg.

66°–72°

24°–30°

Deg.	Sin	Tan	Cot	Cos	Deg.
24.0	0.4067	0.4452	2.246	0.9135	**66.0**
.1	.4083	.4473	2.236	.9128	65.9
.2	.4099	.4494	2.225	.9121	.8
.3	.4115	.4515	2.215	.9114	.7
.4	.4131	.4536	2.204	.9107	.6
.5	.4147	.4557	2.194	.9100	.5
.6	.4163	.4578	2.184	.9092	.4
.7	.4179	.4599	2.174	.9085	.3
.8	.4195	.4621	2.164	.9078	.2
.9	.4210	.4642	2.154	.9070	65.1
25.0	0.4226	0.4663	2.145	0.9063	**65.0**
.1	.4242	.4684	2.135	.9056	64.9
.2	.4258	.4706	2.125	.9048	.8
.3	.4274	.4727	2.116	.9041	.7
.4	.4289	.4748	2.106	.9033	.6
.5	.4305	.4770	2.097	.9026	.5
.6	.4321	.4791	2.087	.9018	.4
.7	.4337	.4813	2.078	.9011	.3
.8	.4352	.4834	2.069	.9003	.2
.9	.4368	.4856	2.059	.8996	64.1
26.0	0.4384	0.4877	2.050	0.8988	**64.0**
.1	.4399	.4899	2.041	.8980	63.9
.2	.4415	.4921	2.032	.8973	.8
.3	.4431	.4942	2.023	.8965	.7
.4	.4446	.4964	2.014	.8957	.6
.5	.4462	.4986	2.006	.8949	.5
.6	.4478	.5008	1.997	.8942	.4
.7	.4493	.5029	1.988	.8934	.3
.8	.4509	.5051	1.980	.8926	.2
.9	.4524	.5073	1.971	.8918	63.1
27.0	0.4540	0.5095	1.963	0.8910	**63.0**
.1	.4555	.5117	1.954	.8902	62.9
.2	.4571	.5139	1.946	.8894	.8
.3	.4586	.5161	1.937	.8886	.7
.4	.4602	.5184	1.929	.8878	.6
.5	.4617	.5206	1.921	.8870	.5
.6	.4633	.5228	1.913	.8862	.4
.7	.4648	.5250	1.905	.8854	.3
.8	.4664	.5272	1.897	.8846	.2
.9	.4679	.5295	1.889	.8838	62.1
28.0	0.4695	0.5317	1.881	0.8829	**62.0**
.1	.4710	.5340	1.873	.8821	61.9
.2	.4726	.5362	1.865	.8813	.8
.3	.4741	.5384	1.857	.8805	.7
.4	.4756	.5407	1.849	.8796	.6
.5	.4772	.5430	1.842	.8788	.5
.6	.4787	.5452	1.834	.8780	.4
.7	.4802	.5475	1.827	.8771	.3
.8	.4818	.5498	1.819	.8763	.2
.9	.4833	.5520	1.811	.8755	61.1
29.0	0.4848	0.5543	1.804	0.8746	**61.0**
.1	.4863	.5566	1.797	.8738	60.9
.2	.4879	.5589	1.789	.8729	.8
.3	.4894	.5612	1.782	.8721	.7
.4	.4909	.5635	1.775	.8712	.6
.5	.4924	.5658	1.767	.8704	.5
.6	.4939	.5681	1.760	.8695	.4
.7	.4955	.5704	1.753	.8686	.3
.8	.4970	.5727	1.746	.8678	.2
.9	.4985	.5750	1.739	.8669	60.1
30.0	0.5000	0.5774	1.732	0.8660	**60.0**
Deg.	Cos	Cot	Tan	Sin	Deg.

60°–66°

30°–36°

Deg.	Sin	Tan	Cot	Cos	Deg.
30.0	0.5000	0.5774	1.7321	0.8660	**60.0**
.1	.5015	.5797	1.7251	.8652	59.9
.2	.5030	.5820	1.7182	.8643	.8
.3	.5045	.5844	1.7113	.8634	.7
.4	.5060	.5867	1.7045	.8625	.6
.5	.5075	.5890	1.6977	.8616	.5
.6	.5090	.5914	1.6909	.8607	.4
.7	.5105	.5938	1.6842	.8599	.3
.8	.5120	.5961	1.6775	.8590	.2
.9	.5135	.5985	1.6709	.8581	59.1
31.0	0.5150	0.6009	1.6643	0.8572	**59.0**
.1	.5165	.6032	1.6577	.8563	58.9
.2	.5180	.6056	1.6512	.8554	.8
.3	.5195	.6080	1.6447	.8545	.7
.4	.5210	.6104	1.6383	.8536	.6
.5	.5225	.6128	1.6319	.8526	.5
.6	.5240	.6152	1.6255	.8517	.4
.7	.5255	.6176	1.6191	.8508	.3
.8	.5270	.6200	1.6128	.8499	.2
.9	.5284	.6224	1.6066	.8490	58.1
32.0	0.5299	0.6249	1.6003	0.8480	**58.0**
.1	.5314	.6273	1.5941	.8471	57.9
.2	.5329	.6297	1.5880	.8462	.8
.3	.5344	.6322	1.5818	.8453	.7
.4	.5358	.6346	1.5757	.8443	.6
.5	.5373	.6371	1.5697	.8434	.5
.6	.5388	.6395	1.5637	.8425	.4
.7	.5402	.6420	1.5577	.8415	.3
.8	.5417	.6445	1.5517	.8406	.2
.9	.5432	.6469	1.5458	.8396	57.1
33.0	0.5446	0.6494	1.5399	0.8387	**57.0**
.1	.5461	.6519	1.5340	.8377	56.9
.2	.5476	.6544	1.5282	.8368	.8
.3	.5490	.6569	1.5224	.8358	.7
.4	.5505	.6594	1.5166	.8348	.6
.5	.5519	.6619	1.5108	.8339	.5
.6	.5534	.6644	1.5051	.8329	.4
.7	.5548	.6669	1.4994	.8320	.3
.8	.5563	.6694	1.4938	.8310	.2
.9	.5577	.6720	1.4882	.8300	56.1
34.0	0.5592	0.6745	1.4826	0.8290	**56.0**
.1	.5606	.6771	1.4770	.8281	55.9
.2	.5621	.6796	1.4715	.8271	.8
.3	.5635	.6822	1.4659	.8261	.7
.4	.5650	.6847	1.4605	.8251	.6
.5	.5664	.6873	1.4550	.8241	.5
.6	.5678	.6899	1.4496	.8231	.4
.7	.5693	.6924	1.4442	.8221	.3
.8	.5707	.6950	1.4388	.8211	.2
.9	.5721	.6976	1.4335	.8202	55.1
35.0	0.5736	0.7002	1.4281	0.8192	**55.0**
.1	.5750	.7028	1.4229	.8181	54.9
.2	.5764	.7054	1.4176	.8171	.8
.3	.5779	.7080	1.4124	.8161	.7
.4	.5793	.7107	1.4071	.8151	.6
.5	.5807	.7133	1.4019	.8141	.5
.6	.5821	.7159	1.3968	.8131	.4
.7	.5835	.7186	1.3916	.8121	.3
.8	.5850	.7212	1.3865	.8111	.2
.9	.5864	.7239	1.3814	.8100	54.1
36.0	0.5878	0.7265	1.3764	0.8090	**54.0**
Deg.	Cos	Cot	Tan	Sin	Deg.

54°–60°

36°–40.5°   **40.5°–45°**

Deg.	Sin	Tan	Cot	Cos	Deg.		Deg.	Sin	Tan	Cot	Cos	Deg.
36.0	0.5878	0.7265	1.3764	0.8090	**54.0**		**40.5**	0.6494	0.8541	1.1708	0.7604	**49.5**
.1	.5892	.7292	1.3713	.8080	53.9		.6	.6508	.8571	1 1667	.7593	.4
.2	.5906	.7319	1.3663	.8070	.8		.7	.6521	.8601	1.1626	.7581	.3
.3	.5920	.7346	1.3613	.8059	.7		.8	.6534	.8632	1.1585	.7570	.2
.4	.5934	.7373	1.3564	.8049	.6		.9	.6547	.8662	1.1544	.7559	49.1
.5	.5948	.7400	1.3514	.8039	.5		**41.0**	0.6561	0.8693	1.1504	0.7547	**49.0**
.6	.5962	.7427	1.3465	.8028	.4		.1	.6574	.8724	1.1463	.7536	48.9
.7	.5976	.7454	1.3416	.8018	.3		.2	.6587	.8754	1.1423	.7524	8
.8	.5990	.7481	1.3367	.8007	.2		.3	.6600	.8785	1.1383	.7513	.7
.9	.6004	.7508	1.3319	.7997	53.1		.4	.6613	.8816	1.1343	.7501	.6
37.0	0.6018	0.7536	1.3270	0.7986	**53.0**		.5	.6626	.8847	1.1303	.7490	.5
.1	.6032	.7563	1.3222	.7976	52.9		.6	.6639	.8878	1.1263	.7478	.4
.2	.6046	.7590	1.3175	.7965	.8		.7	.6652	.8910	1.1224	.7466	.3
.3	.6060	.7618	1.3127	.7955	.7		.8	.6665	.8941	1.1184	.7455	.2
.4	.6074	.7646	1.3079	.7944	.6		.9	.6678	.8972	1.1145	.7443	48.1
.5	.6088	.7673	1.3032	.7934	.5		**42.0**	0.6691	0.9004	1 1106	0.7431	**48.0**
.6	.6101	.7701	1.2985	.7923	.4		.1	.6704	.9036	1.1067	.7420	47 9
.7	.6115	.7729	1.2938	.7912	.3		.2	.6717	.9067	1.1028	.7408	8
.8	.6129	.7757	1.2892	.7902	.2		.3	.6730	.9099	1.0990	.7396	.7
.9	.6143	.7785	1.2846	.7891	52.1		.4	.6743	.9151	1.0951	.7385	.6
38.0	0.6157	0.7813	1.2799	0.7880	**52.0**		.5	.6756	.9163	1.0913	.7373	.5
.1	.6170	.7841	1.2753	.7869	51.9		.6	.6769	.9195	1.0875	.7361	.4
.2	.6184	.7869	1.2708	.7859	.8		.7	.6782	.9228	1.0837	.7349	.3
.3	.6198	.7898	1.2662	.7848	.7		.8	.6794	.9260	1.0799	.7337	.2
.4	.6211	.7926	1.2617	.7837	.6		.9	.6807	.9293	1.0761	.7325	47.1
.5	.6225	.7954	1.2572	.7826	.5		**43.0**	0.6820	0.9325	1.0724	0.7314	**47.0**
.6	.6239	.7983	1.2527	.7815	.4		.1	.6833	.9358	1 0686	.7302	46.9
.7	.6252	.8012	1.2482	.7804	.3		.2	.6845	.9391	1.0649	.7290	8
.8	.6266	.8040	1.2437	.7793	.2		.3	.6858	.9424	1.0612	.7278	.7
.9	.6280	.8069	1.2393	.7782	51.1		.4	.6871	.9457	1.0575	.7266	.6
39.0	0.6293	0.8098	1.2349	0.7771	**51.0**		.5	.6884	.9490	1.0538	.7254	.5
.1	.6307	.8127	1.2305	.7760	50.9		.6	.6896	.9523	1.0501	.7242	.4
.2	.6320	.8156	1.2261	.7749	.8		.7	.6909	.9556	1.0464	.7230	.3
.3	.6334	.8185	1.2218	.7738	.7		.8	.6921	.9590	1.0428	.7218	.2
.4	.6347	.8214	1.2174	.7727	.6		.9	.6934	.9623	1.0392	.7206	46.1
.5	.6361	.8243	1.2131	.7716	5		**44.0**	0.6947	0.9657	1.0355	0.7193	**46.0**
.6	.6374	.8273	1.2088	.7705	.4		.1	.6959	.9691	1.0319	.7181	45.9
.7	.6388	.8302	1.2045	.7694	.3		.2	.6972	.9725	1.0283	.7169	.8
.8	.6401	.8332	1.2002	.7683	.2		.3	.6984	.9759	1.0247	.7157	.7
.9	.6414	.8361	1.1960	.7672	50.1		.4	.6997	.9793	1.0212	.7145	.6
40.0	0.6428	0.8391	1.1918	0.7660	**50.0**		.5	.7009	.9827	1.0176	.7133	.5
.1	.6441	.8421	1.1875	.7649	49.9		.6	.7022	.9861	1.0141	.7120	.4
.2	.6455	.8451	1.1833	.7638	.8		.7	.7034	.9896	1.0105	.7108	.3
.3	.6468	.8481	1.1792	.7627	.7		.8	.7046	.9930	1.0070	.7096	.2
.4	.6481	.8511	1.1750	.7615	.6		.9	.7059	.9965	1.0035	.7083	45.1
40.5	0.6494	0.8541	1.1708	0.7604	**49.5**		**45.0**	0.7071	1.0000	1.0000	0.7071	**45.0**
Deg.	Cos	Cot	Tan	Sin	Deg.		Deg.	Cos	Cot	Tan	Sin	Deg.

49.5°–54°   **45°–49.5°**

CHANGING FORM OF COMPLEX QUANTITIES

Example. Change $10\underline{/60°}$ from polar form to rectangular form:

$$10\underline{/60°} = 10 \cos 60° + j10 \sin 60°$$
$$= 10(0.5000) + j10(0.8660) = 5.00 + j8.66$$

Example. Change $4 + j3$ from rectangular to polar form. First divide the larger component by the smaller: $4/3 = 1.333$.

Then find 1.333 under "cot" in table and read: Angle = 36.9°.

Under "cos" read, in same line, 0.7997. Then divide the larger component: $4/0.7997 = 5.00$. Therefore $4 + j3 = 5.00\underline{/36.9°}$.

APPENDIX
4

ANSWERS TO SELECTED PROBLEMS

Problems for which answers (or partial answers) are given below are marked in the text with asterisks.

CHAPTER 1

1-1. Compare Section 2 of Chapter 15.

1-2. Compare Section 2 of Chapter 15.

1-7. A voltage wave of triangles plus rectangles.

1-8. A voltage wave of trapezoids (shape of Wave C) plus rectangular impulses; as current Wave C approaches Wave A, rectangular impulses of voltage become higher and narrower but keep same area (see Chapter 16, Section 1); the limit is impossible as requiring infinite voltage.

1-9. Similar to Problem 1-7 and part of 1-8.

1-10. (a) Sloping straight lines, 500 to 1000, 0 to -500, repeated. (b) Parabola 0 to 750, straight to 1250, parabola to 250, straight to -250, parabola to 0, repeated.

1-11. (a) 10 volts, 0.05 joules; (b) 100 amp, 5 joules.

1-13. 2 min, 13 sec.

CHAPTER 2

2-3. 0.745 amp.

2-4. Wave A: rms = 1 amp. Wave B: rms = $1/\sqrt{3} = 0.577$ amp.

2-5. RMS of stepped wave = 1.672 amp; rms of trapezoidal wave = 1.711 amp. The latter rms value is larger (about $2\frac{1}{2}$%), although *average* values are equal.

2-7.

Wave:	(a)	(b)	(c)
RMS value	1	$1/\sqrt{3}$	$K/\sqrt{5}$
Half-cycle average	1	$1/3$	$K/3$
Crest factor	1	$\sqrt{3}$	$\sqrt{5}$

2-9. (a) 0.5 amp, (b) 1.5 amp, (c) 1.58 amp.

2-10. $1/\sqrt{2} = 0.707$.

2-12. (a) No, for equation 11-10 applies to a load with constant resistance, and (b) this load does not have constant resistance; if it did, sinusoidal voltage would produce sinusoidal current. This load might be a rectifier, a filter, and a resistive load cascaded.

CHAPTER 3

3-3. From Euler's theorem, equation 5-4, or perhaps more convincingly by computing 8 or 9 terms of the series of equation 5-1.

3-7. (a) 4. (b) Magnitude 1.78, angle 15° or 105° or −75° or −165°. (c) Magnitude 1.82, angle −9° or 63° or 135° or −81° or −153°.

3-10. (a) −250 + j250. (b) Magnitude 2, angle 18° or 90° or 162° or 234° or 306°. None conjugate.

3-11. $\ln A + j\alpha$.

3-12. 1.49 at angles of 23.6°, 143.6°, and 263.6°.

3-14. 10 cos 25t and 10 sin 25t, respectively.

CHAPTER 4

4-3. $Z = 35.6\underline{/-32.4°}$, $V = 356\underline{/-0.4°}$ rms volts, $I = 10\underline{/32°}$ rms amps, $i = 14.14 \cos(377t + 32\pi/180)$.

4-5. $V = 250/\sqrt{2}\underline{/0°}$; $Z = 25.0\underline{/73.7°}$; $I = 10/\sqrt{2}\underline{/-73.7°}$; $i = 10 \cos(\omega t - 73.7°)$.

4-6. $V = 177\underline{/0°}$, $Z = 26.0\underline{/67.4°}$ ohms, $I = 6.8\underline{/-67.4°}$, $i = 9.61 \cos(377t - 67.4\,\pi/180)$ amp.

4-9. $2090\underline{/33.6°}$, $2380\underline{/18.2°}$, $2440\underline{/12.5°}$ ohms at 60, 120, 180 Hz, respectively.

4-10. 0.188 μf.

4-11. 45.1 mh.

4-13.

$f =$	800	1000	1200 Hz
$\mathscr{R}e\,Z =$	454	400	349 ohms
$\mathscr{I}m\,Z =$	−30.5	0	45.5 ohms

4-14. No, but two circuits that can be used are:

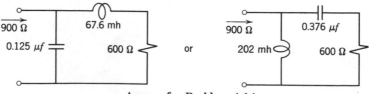

Answer for Problem 4-14.

4-18. $f = 500$ 1000   1500   Hz

       $G = 28.8 \cdot 10^{-4}$ $9.21 \cdot 10^{-4}$ $4.30 \cdot 10^{-4}$ mho

       $B = -45.2 \cdot 10^{-4}$ $-28.9 \cdot 10^{-4}$ $-20.3 \cdot 10^{-4}$ mho

       $|B/G| = Q$ is a straight line through the origin.

4-19. $\frac{5}{6}$ ohm.

4-21. 0.743 amp.

4-22. 0.850 amp.

4-23. 712.5 Hz, 8.0 ohms, 0.125 mho.

4-24. (a) $8 + j43$. (b) 8. (c) $0.023/\!-79.5°$. (d) 0.0042.

4-26. 725 and 699 Hz; $8 \pm j8$; $0.0885/\!\pm 45°$.

4-29. $f = 600$ 650   700   712   750   800 Hz

       $|Z| = 645$ 1192   4580   6250   2020   947 ohms

4-30. 1.59 mh, 15.9 picofarads, 50 ohms.

4-31. 1.13 mh, 35.4 ohms.

4-32. $R_1 = R_2 = \sqrt{L/C}$.

4-35. 35.8 μf.

4-36. (a) 14.7 amp, p.f. 0.0678. (b) 13.00, 0.520. (c) 15.00, 0.20.

4-37. 10 kilowatts; No. 2 is motoring because in-phase current is against electromotive force.

4-38. $I_1 - 5.15/\!-63.4°$ amp, $P_1 = 265$ watts,

       $Q_1 = 530$ vars, $S_1 = 595/63.4°$ volt-amp.

       $I_2 = 10.3/63.4°$ amp, $P_2 = 530$ watts,

       $Q_2 = -1060$ vars, $S_2 = 1180/\!-63.4°$ volt-amp.

       $I = 8.31/33.7°$ amp, $P = 795$ watts, $Q = -530$ vars,

       $S = 935/\!-33.7°$ volt-amp.

CHAPTER 5

5-1. For $s = -200, -140, -100, -80, -50, 0, 100$

       $Y(s) = -0.1, -0.167, \infty, 0.5, 0.2, 0.1, 0.04$

5-2. $F = 1/(RCs + 1)$

5-4. Poles at $-1/\tau_1$ and at $-1/3\tau_1$, zeros at 0 and ∞,

       a minimum at

       $s = -1/\sqrt{3}\,\tau_1$ where $|Y| = 5.56\tau_1/L$

       and a maximum at

       $s = 1/\sqrt{3}\,\tau_1$ where $|Y| = 0.405\tau_1/L$.

5-8. Poles; (all values times 10^4) (a) $-1.866, -0.134$, (b) $-1.707, -0.293$. (c) $-1.0 \pm j1.0$. (d) Straight lines crossing at -1.0. There is a zero at the origin.

5-11. Can be 6 ohms in series with 1 henry, both in parallel with n ohms (n any positive number); or 2 ohms in series with $\frac{1}{2}$ henry, in parallel with 2 ohms, both in series with 2 ohms; etc.

5-14. Poles at 0, 1000, ∞; zeros at 699, 4590.

5-15. Yes. Write its impedance function; this shows 2 pairs of imaginary zeros, 1 pair of imaginary poles, and poles at $s = 0$ and $s = \infty$.

CHAPTER 6

6-1. (a) 1.33 amp. (b) -0.49 amp. (c) 0.84 amp.

6-3. 2.79 amp.

6-5. (a) Maximum transient if switch is closed when the steady component is maximum; i.e., if $\theta = \varphi + \pi/2$ where $\tan \varphi = \omega L/R$. (b) Zero transient if switch is closed when the steady component is zero; i.e., if $\theta = \varphi$.

6-6. (a) Yes. (b) $t = 2.47 \cdot 10^{-3}$ sec.

6-7. $v = 20(1 - e^{-(t/2)10^{6}})$.

6-9. $v(t) = I/G + (v_0 - I/G)e^{-t/\tau}$.

6-10. $v = t - 1 + 2e^{-t}$. Could be $G = 1$ and $C = 1$ in parallel, with $i = t$.

6-12. (a) 1.92 amp; (b) 9.24 joules; (c) 250 joules; (d) 261 joules.

6-15. (a) $Y_{in}(s) = 3(s + 1)/2(s + 4.64)(s + 0.862)$.
    (e) $i(t) = 3V/8 - 0.311Ve^{-4.64t} - 0.063Ve^{-0.862t}$.

6-17. (a) $Y(s) = \dfrac{s}{L(s - s_1)(s - s_2)}$

     with two poles, at $s = s_1 = -2.76 \cdot 10^4$ and at $s = s_2 = -7.24 \cdot 10^4$, and zeros at $s = 0$ and $s = \infty$.
    (b) $i(t) = (3.85 \cdot 10^{-2})e^{s_1 t} - (2.85 \cdot 10^{-2})e^{s_2 t}$ amp.
    (c) $|Z(\omega)| = \sqrt{R^2 + (\omega L - 1/\omega C)^2}$.

6-19. $i(t) = 1.5e^{-133t} - 1.5e^{-67t}$.

6-20. (a) $i = k_1 e^{-t/\tau} + k_2 te^{-t/\tau}$. (b) At $t = 0$, $i = 0.10$ amp, $di/dt = 0$. (c) A second-order pole at -10^4; a zero at 0.

6-21. $i = 0.1e^{-t/\tau} + 1000te^{-t/\tau}$.

6-22. (a) $i = 0.131e^{-3750t} \sin 3810t$;
    (b) $R = 25$, $s = -1250 \pm j5340$;
           $R = 75$, $s = -3750 \pm j3810$;
           $R = 100$, $s = -5000 \pm j1900$;
           $R = 150$, $s = -2260$ and -12740.
     Three pairs of poles plot on a semicircle, the fourth on the horizontal axis.

6-24. (a) Change in the current curve is hardly appreciable; oscillations on the tail of the curve cannot be seen because current is so small by the time it reverses. (b) The two poles are very slightly above and below the second-order pole of the critical case. (c) Exact computation becomes harder, requiring the small difference of two large quantities; for that reason it may be best to assume exactly critical damping as a good approximation.

CHAPTER 7

7-2. (a) $c = -R$, $d = 1$. (b) $v_L = Ve^{-t/\tau}$ where $\tau = L/R$.

7-4. $a_{11} = -R/L$, $a_{12} = -1/L$, $b_1 = 1/L$
     $a_{21} = 1/C$, $a_{22} = 0$, $b_2 = 0$.

7-6. $a_{11} = -8$, $a_{12} = 4$, $b_1 = 0$, $a_{21} = 4$, $a_{22} = -8$, $b_2 = 2$.

7-10. $0.667 - 0.0665e^{-264t} - 0.600e^{-1136t}$.

7-11. $a = -1/RC$ where $R = 1$ ohm, $b = 1/R_{12}C$; $i_{12} = 3 + 3e^{-10^6t}$.

7-13. The answer is given for Problem 6-15.

7-15. $i = 10e^{-10^4t}\sin(1.58 \cdot 10^5 t)$.

CHAPTER 8

8-1. $I_1 = 0.0396$, $I_2 = -0.1915$, $I_3 = 0.1519$ amp.

8-2. $I_a = 2.13\underline{/13.9°}$, $I_b = 1.40\underline{/126.5°}$, $I_c = 2.97\underline{/-12.1°}$.

8-3. $Er/(4.025 + 4.25r)$; approximately, for small unbalance.

8-5. (a)

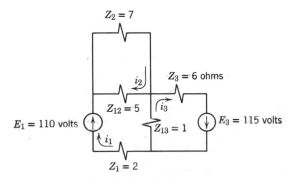

Answer for Problem 8-5.

(b) 19.55 amp.

8-7. $-j11.1$ ohms.

8-8. 33.3 ohms.

8-9. (a) $\quad j100I_1 - j100I_{ab} \qquad\qquad = V_1$

$\quad -j100I_1 + j180I_{ab} - j100I_2 \qquad = 0$

$\qquad\qquad - j100I_{ab} + (33.3 + j100)I_2 = 0$

(b) 33.3 ohms.

8-10. $V_{AB} = 3.3$ volts.

8-11. Same as Problem 8-10.

8-13.

$$V_b = \frac{\begin{vmatrix} 1 & 2 & 0 \\ -1 & 0 & j\frac{1}{2} \\ 0 & 0 & j\frac{1}{2} \end{vmatrix}}{\begin{vmatrix} 1 & -1 & 0 \\ -1 & 1-j\frac{1}{2} & j\frac{1}{2} \\ 0 & j\frac{1}{2} & j\frac{1}{2} \end{vmatrix}} = j2$$

8-15. (a) 12 elements, 8 nodes, 8 junctions, 12 branches. (b) 7 independent nodes, 5 independent loops.

8-16. (a) A bridge with shunted source of independent current. (b) Yes. (c) Yes.

CHAPTER 9

9-1. (a) 10 ohms; (b) 0.225 mho; (c) 0.0830 mho.

9-2.

$$[I] = \begin{bmatrix} 10.58 \\ 6.5 \\ 5.7 \end{bmatrix}$$

9-4. (b) $z_{BA} = \dfrac{10s}{(2.05)10^{-3}s + 1}$;   (c) $V_B = 97.1\underline{/4.45°}$.

9-6.

$$\begin{bmatrix} 7 & 7 & 6 \\ 9 & 6 & 1 \\ 7 & 4 & 5 \end{bmatrix}.$$

9-7.

$$\begin{bmatrix} 7 & 2 & 3 \\ 4 & 4 & 4 \\ 17 & 12 & 7 \end{bmatrix}.$$

9-8. 52.

9-10.

$$[F]^{-1} = \frac{1}{10}\begin{bmatrix} 5 & 3 & -1 \\ -5 & 1 & 3 \\ 5 & -3 & 1 \end{bmatrix}.$$

9-11. $[U]$.

9-13. (a) Loop 1 alone: source, 3 ohms of R. Mutual: $-j4$ ohms of C. Loop 2 alone: 8 ohms of R and $j8$ ohms of L. Arrows show I_1 and I_2 in opposite directions through mutual C. Alternative answer: Loop 1 alone: source, 3 ohms of R and $-j8$ ohms of C. Mutual: $j4$ ohms of L. Loop 2 alone: 3 ohms of R. Arrows show I_1 and I_2 in the same direction through the mutual L.

   (b) $[y] = \dfrac{1}{41}\begin{bmatrix} 3 + j4 & -j4 \\ -j4 & 3 - j4 \end{bmatrix}$ mho

   (c) $I_2 = -j9.75$ amp.

9-14. $I_1 = 18.5$, $I_2 = 9.4$, $I_3 = 10.5$.

9-17. See Problem 9-18; these are special cases.

9-18. $[T] = \begin{bmatrix} 1 & 0 \\ 0 & n \end{bmatrix}.$

CHAPTER 10

10-1. $I_{AB} = 2$, $I_{BC} = 3$, $I_{BO} = -1$ amp.

10-4. 4.5 amps.

10-5. (a) $V' = 60.0\underline{/0°}$, $Z' = 43.3\underline{/0°}$. (b) $V' = 100$, $Z' = 30$.

10-8. $V_{AB} = 87.1$ volts.

10-10. $V_\theta = 141\underline{/45°}$; $Z_\theta = 1 + j1$.

10-12. (b) $V_\theta = 50 - j0.32$ millivolts; $Z_\theta = 4.98 - j0.165$ ohms.

10-14. Current is $j1$, upward, for any R. The circuit changes constant alternating voltage to constant alternating current. The loop method is among possible solutions.

10-16. $Z_1 = j1$ ohm (use same L), $Z_2 = -j1$ ohm (use same C), $Z_3 = 0.5$ ohm (need new R).

10-17. $X = -1/3$ ohm (purely capacitive); at one frequency only.

10-18. (a) 4.37 ohms. (b) 2.91 amp.

10-19. 0.121 amp, right to left.

10-20. $R_1 = 3.5$ ohms, $L_1 = 3$ hen, $R_2 = 7$ ohms, $L_2 = 6$ hen, $R_3 = 1.5$ ohms, $C_4 = 4$ farads, $G_4 = 4$ mhos $= 1/R_4$

10-22.

	I_{ca}	I_{ao}	I_{cb}	I_{bo}	I_{oc}	I_{ab}
Source in o-c	52.5	48.0	48.0	52.5	100.5	4.5
Source in a-b	45.5	−45.7	−41.0	50.2	4.5	91.2

(all in milliamperes)

These equal values illustrate reciprocity.

10-25.

$R_1/R_s = 0$	$\frac{1}{2}$	1	2	∞
$P(R_s/E^2) = 0$	1/4.5	1/4	1/4.5	0

CHAPTER 11

11-2. $-750 \sin \omega t \pm 1200 \cos \omega t$; the relative direction of winding of the coils is needed to resolve the $\pm$ uncertainty.

11-3. $j\omega(L_{11} L_{22} - M^2)/(L_{11} + L_{22} + 2M)$.

11-4. 0.90, $L_{in} = 5.0$ mh.

11-7. (a) 23.5 mh. (b) 45.0 mh. (c) 0.523.

11-8. (a) $[R_1 + j\omega(L_1 + L_2 + 2L_{12})]I_1 - j\omega(L_{13} + L_{23})I_2 = E$
$-j\omega(L_{13} + L_{23})I_1 + (R_2 + j\omega L_3)I_2 = 0$

(b) Inductance in Circuit 1 only:

$$L_1 + L_2 + 2L_{12} - L_{13} - L_{23}$$

Inductance in Circuit 2 only: $L_3 - L_{13} - L_{23}$
Inductance in both (mutual): $L_{13} + L_{23}$

11-9. $I_1 = 0.442\underline{/-90°}$.

11-10. The following symbols refer to Figure 4b:

	L_1	aM	a^2L_2
If $a = 1$	−2.7	3	27 hen.
If $a = 10$	−29.7	30	2970 hen.

11-11. (a) 0.0530 amp, 196 volts. (b) 0.624 amp, 0.291 amp.

11-12. $X_1 + a^2X_2 = 0.053$ ohms, $M = 3.05$ henrys.

CHAPTER 12

12-3. (a) $y_{11} = 1/5$, $y_{22} = 8/15$, $y_{12} = -1/5$. (b) $z_{11} = 8$, $z_{22} = 3$, $z_{12} = 3$.

12-5. (a) $I_1 = 0$, $I_2 = 10/3$ amp. (b) $I_1 = 1$, $I_2 = 2/3$. (c) $V_1 = 55$ volts, $V_2 = 30$.
(d) $V_1 = 25$, $V_2 = 0$.

12-7. $i_1(t) = 1 = u(t)$, $i_2(t) = -1 + (5/3)t$

12-8. (a) $y_{11} = 0.05$ mho,   (b) $z_{11} = 55.6$ ohms,
$y_{21} = -0.04$. $z_{12} = 4.44$.

12-10. $y_{12} = -162(s + 2000)/(s + 3050)(s + 2120)$. There is a zero at $s = -2000$
and poles at $s = -2120$ and $s = -3050$.

12-11. $A = 1 + Y_r Z$, $B = Z$, $C = Y_s + Y_r + Y_s Y_r Z$, $D = 1 + Y_s Z$.

12-12. $\begin{bmatrix} 7 & 48 \\ 1 & 7 \end{bmatrix}$. Symmetrical, for $A = D$.

12-13. (a) 707 ohms, 708 ohms, $+j101$ ohms. (b) 707 ohms.

12-14. $z_{01} = z_{02} = 14.14$ ohms.

12-17. (a) $z_{01} = 399$,   (b) $z_{01} = 469$,
$z_{02} = 599$. $z_{02} = 1618$.

12-18. (a) 1. (b) $-j1$. (c) $-j\sqrt{2}$. (d) $-(3 + 2\sqrt{2})$. Alternatively, (c) $+j\sqrt{2}$,
(d) $-3 + 2\sqrt{2}$.

12-21.

$\omega =$	10^6	10^5	10^3 rad/sec
$\alpha =$	0	0	∞ nepers
$\beta =$	-0.01414	-0.148	0 radians
db =	0	0	∞

CHAPTER 13

13-2. $\frac{1}{4} + \sqrt{2}/\pi \cos \omega t + 1/\pi \cos 2\omega t + \sqrt{2}/3\pi \cos 3\omega t$
$- \sqrt{2}/5\pi \cos 5\omega t + \cdots$, $\omega = 500\pi$.

13-5. $(8/\pi)(1 - \sqrt{2}/\pi) = 1.4004$

13-6. Even part $= \frac{1}{2}[f(t) + f(-t)]$;
odd part $= \frac{1}{2}[f(t) - f(-t)]$

13-9. $a_0 = 15$, $a_1 = 10\sqrt{2}/\pi$, $a_2 = -10/\pi$, $a_3 = 10\sqrt{2}/3\pi$, $a_4 = 0$, $a_5 = -2\sqrt{2}/\pi$

13-10. $\frac{2}{\pi}\left(\cos \omega t - \frac{1}{3}\cos 3\omega t + \cdots + \sin \omega t + \frac{1}{3}\sin 3\omega t + \cdots\right)$.

13-11. $a_0 = 2/3$, $a_1 = -4/\pi^2$, $a_2 = 1/\pi^2$, $a_3 = -4/9\pi^2$, $b_1 = b_2 = b_3 = 0$.

13-13. $103 \sin x + 3.82 \sin 3x + 0.82 \sin 5x + 0.30 \sin 7x + \cdots$.

13-16. $i = \dfrac{8}{\omega_1 L\pi^2}\left(-\cos \omega_1 t + \dfrac{1}{3^3}\cos 3\omega_1 t - \dfrac{1}{5^3}\cos 5\omega_1 t\right.$
$\left. + \dfrac{1}{7^3}\cos 7\omega_1 t - \cdots\right)$

13-17. 60 Hz, 25.3%; 120 Hz, 2.4%.

13-18.

Amplitude of	60 Hz	120 Hz	(relative to dc component)
Original filter	25.3%	2.4%	
Doubled elements	5.72%	0.59%	
Tandem sections	5.13%	0.092%	

13-19. 1.3 mv rms.

13-20. The following are exact values; values computed from series will be less because of omitted higher harmonics. Error of 2 or 3% for square and saw-tooth waves is expected; others will be closer. Square, 1.00; saw-tooth, $1/\sqrt{3} = 0.577$; triangular, $1/\sqrt{3} = 0.577$; rectified half-wave, $\frac{1}{2} = 0.5$; rectified full wave, $1/\sqrt{2} = 0.707$.

13-21. (a) $21.2 \cos(\omega t - 45°) - 3.16 \cos 3(\omega t - 23.8°)$. (c) 230 watts.

13-24. (a) $Z_n = 1 + jn$. (b) $I_n = 2V_{max}/jn\pi(1 + jn)$.

$$(c) \quad I = -\frac{2V_{max}}{\pi}\left(\cdots + \frac{1}{9 + j3}e^{-j3\omega_1 t} + \frac{1}{1 + j1}e^{-j\omega_1 t} + \frac{1}{1 - j1}e^{j\omega_1 t}\right.$$

$$\left. + \frac{1}{9 - j3}e^{j3\omega_1 t} + \cdots\right).$$

13-25. $f(t) = \cdots + 0.101e^{-j3\pi t/10} + 0.1514e^{-j2\pi t/10}$
$\qquad + 0.187e^{-j\pi t/10} + 0.20 + 0.187e^{j\pi t/10}$
$\qquad + 0.1514e^{j2\pi t/10} + 0.101e^{j3\pi t/10} + \cdots$

13-27. $g(\omega) = \dfrac{\tau}{j\pi}\dfrac{[\sin(\omega\tau/2)]^2}{\omega\tau/2}$. See sketch.

13-28. $\dfrac{\tau}{2\pi}\dfrac{\sin^2(\omega\tau/2)}{(\omega\tau/2)^2}$. See sketch.

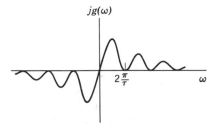

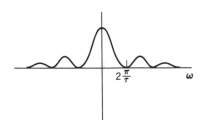

Answer for Problem 13-27.   **Answer for Problem 13-28.**

13-30. $\dfrac{\alpha}{\pi(\alpha^2 + \omega^2)}$

13-31. $g(\omega) = \dfrac{1}{2\pi(1 + j\omega)^2} \qquad |g| = \dfrac{1}{2\pi(1 + \omega^2)}$

13-34. $g(\omega) = \dfrac{j}{2\pi\omega}[(\cos 2\omega\tau - 1) - j\sin 2\omega\tau]; \; |g| = \dfrac{\tau}{\pi}\dfrac{\sin \omega\tau}{\omega\tau}$.

13-35. $i(t) = [V_m/(R - \alpha L)](e^{-\alpha t} - e^{-Rt/L}) \qquad$ for $t > 0$

CHAPTER 14

14-1. $0.3e^{-4 \cdot 10^6 t}$ amp.

14-2. $i(t) = 7.1 \cdot 10^{-3} e^{-625t} \sin 7050t$.

14-3. $F(s) = \dfrac{s + \alpha}{(s - s_1)(s - s_2)}$; $s_1, s_2 = -\alpha \pm j\omega$.

14-4. $I(s) = \dfrac{3}{s + 6} + \dfrac{12}{s(s + 6)}$; $i(t) = 2 + e^{-6t}$.

14-7. $x(t) = \cos mt$.

14-9. (a) $x(t) = 1$. (b) $x(t) = 2\sqrt{2} \sin(t/\sqrt{2})$, $y(t) = 2[1 - \cos(t/\sqrt{2})]$.

14-14. (Courtesy R. N. Clark)

(a) $\dfrac{1}{s + 1} + \dfrac{2}{s + 2} + \dfrac{3}{s + 3}$

(b) $\dfrac{2}{s} + \dfrac{2}{s + 2} - \dfrac{2}{s + 1 + j\sqrt{3}} - \dfrac{2}{s + 1 - j\sqrt{3}}$

(c) $\dfrac{1 + j2}{s + 5 + j5} + \dfrac{1 - j2}{s + 5 - j5}$.

14-15. $0.0224e^{-100t} \sin 2230t$, $Q = 11.15$.

14-17. $e^{-\alpha t}(1 - \alpha t)$.

14-18. $i(t) = (1/2L)t^2 e^{-\alpha t}$.

14-20. $f(t) = \dfrac{1}{s_1 s_2} + \dfrac{1}{s_1(s_1 - s_2)} e^{s_1 t} + \dfrac{1}{s_2(s_2 - s_1)} e^{s_2 t}$.

14-22. $0.0530 \sin 377t$.

CHAPTER 15

15-9. $i = 10^3 t e^{-10^4 t}$

15-11. $i(t) = 0.0517 e^{-1.25 \cdot 10^4 t} \sin 4.84 \cdot 10^4 t$; $Q_0 = 2$.

15-13. $i(t) = e^{-t/2}(\cos t/2 + 4 \sin t/2)$.

15-14. (a) Poles at -0.25 and -1.00. Zeros at -0.50 and at $+0.50$. (Note the zero in the right half-plane, and its effect.) (b) Output voltage is zero until $t = 0$, then it drops (discontinuously) to -1.0, then rises as the sum of exponentials (time constants 1 and 4 seconds) to a final value of $+1.0$.

15-15. (a) 3 exponential components, (b) 1 decreases more slowly than the others, (c) that from the pole nearest the zero is perhaps negligible at all times. (d) $i(0) = 0$. (e) $i'(0) = 1$. (f) $i(\infty) = 0$.

15-17. (a) 1 constant, 1 exponential, 1 undamped oscillation, (b) exponential dies quickly, constant and exponential are positive, (c) the exponential becomes negligible at long times. (d) $i(0) = 1$. (e) $i'(0) = -20$. (f) There is none.

15-18. As in Figure 5d of Chapter 14.

15-20. (a) A pole at s_1 and a zero at s_a, both negative real. (b) A possible solution has resistance R' in series with R and L in parallel, for which (c) $s_a = -R/L$ and $s_1 = -RR'/(R + R')L$.

15-22. $f''(0) = -\alpha$.

15-24. Necessarily zero.

15-25. $L_b/(L_b + \frac{1}{2}L_a)$.

15-26. $12 - 7e^{-10t}$.

15-28. $3e^{-1414t}(1 - 1414t)$.

15-29. Node excitation function $= I(s) + Cv(0_+) - i_L(0)/s$.

15-30. $200 - 181.8e^{-104t}$.

15-31. (a) $\zeta = 0.6$ (b) Divide the numerals by 15 to get seconds. (c) 3 amp.

15-33. 10 milliamps after $t = 0$.

15-34. (a) $V = 1000$ volts. (b) $v_c(t) = 1000 - 1000 \cos 1000t$.

15-35. $i = \dfrac{V_m}{|Z|} \cos(\omega t - \varphi) + Ae^{s_1 t} + Be^{s_2 t}$, where

$$|Z| = \sqrt{R^2 + (\omega L - 1/\omega C)^2}, \qquad \varphi = \tan^{-1}\left[\left(\omega L - \frac{1}{\omega C}\right)/R\right]$$

$$s_1 = -\frac{R}{2L} - \sqrt{\frac{R^2}{4L^2} - \frac{1}{LC}}, \qquad s_2 = -\frac{R}{2L} + \sqrt{\frac{R^2}{4L^2} - \frac{1}{LC}}$$

$$A = \frac{V_m}{s_1 - s_2}\left(\frac{1}{L} + \frac{s_2}{|Z|}\cos\varphi - \frac{\omega}{|Z|}\sin\varphi\right)$$

$$B = \frac{V_m}{s_2 - s_1}\left(\frac{1}{L} + \frac{s_1}{|Z|}\cos\varphi - \frac{\omega}{|Z|}\sin\varphi\right)$$

CHAPTER 16

16-1. (a) $\delta/R - (e^{-2t/RC})/CR^2$. (b) $T(s) = \dfrac{1}{CR(s + 2/RC)}$.

16-3. $F(s) = (1/s)(1 - e^{-sT})$.

16-4. $F(s) = e^{-sT}/(s + a)$.

16-5. $i(t) = (1/R)(e^{-(t-T)/RC}$ for $t > T$, and 0 for $t < T$.

16-7. Pair 5. It could result from an exponential voltage applied to R and C in series, the time constants being equal.

16-8. $f(t) = \dfrac{e^{-\alpha t} + [(\alpha - \beta)t - 1]e^{-\beta t}}{(\alpha - \beta)^2}$.

16-10. $i(t) = (1 - e^{-Rt/L})/R$.

16-11. $\dfrac{V_m C}{RC - 1}(e^{-t/RC} - e^{-t})$.

16-13. $(A/R)(1 - e^{-\alpha t})$.

16-14. $i(t) = (1/R)\{u(t) - u(t - t_0) - [e^{-Rt/L}u(t) - e^{-R(t-t_0)/L}u(t - t_0)]\}$.

16-15. (a) $\mathscr{L}v(t) = \dfrac{1}{s} + \dfrac{1}{s^2T}(e^{-sT} - 1)$.

(b) $i(t) = \dfrac{C}{T}[(T\alpha + 1)e^{-\alpha t} - 1 + u(t - T) - e^{-\alpha(t-T)}u(t - T)]$.

CHAPTER 17

17-1. (a) $R_2 = R_3 = 5$ ohms. (b) $R_2 = R_3 = 5/2$ ohms.

17-3. $V_{ab} = 173\underline{/30°}$, $V_{bc} = 173\underline{/-90°}$, $V_{ca} = 173\underline{/150°}$ is a possible answer.

17-7. $V_{an} = 103.6\underline{/-1.1°}$; $|V_{ab}| = 179.5$ volts.

17-8. Line-to-neutral voltage, Y side, $= 123\underline{/0°}$ volts (taking reference angle from current). Line-to-line voltage, Y side, $= 214\underline{/30°}$ volts. Line current, delta side, $= 34.6\underline{/-30°}$ amp.

17-10. Interconnection on secondary side is impossible because of phase differences.

17-11. Possible if polarity of connection (or of winding) is reversed in one bank. Turn ratio in Y-Δ, $n:1$, in Δ-Y, $3n:1$.

17-12. Loss is doubled.

17-13. $I_a = 30(1.5 - j0.866)$, $I_b = 15(-2.5 - j0.866)$, $I_c = 15(-0.5 + 3j0.866)$.

INDEX

$$\int \sin^2 ax \, dx = \frac{x}{2} - \frac{\sin 2ax}{4a}$$

$$\int \cos^2 ax \, dx = \frac{x}{2} + \frac{\sin 2ax}{4a}$$

$$\int \sin ax \cos ax \, dx = \frac{1}{2a} \sin^2 ax$$

$$\int x \sin ax \, dx = \frac{1}{a^2}(\sin ax - ax \cos ax)$$

$$\int x \cos ax \, dx = \frac{1}{a^2}(\cos ax + ax \sin ax)$$

$$\int x^2 \sin ax \, dx = \frac{1}{a^3}(2ax \sin x + 2\cos ax - a^2 x^2 \cos$$

$$\int x^2 \cos ax \, dx = \frac{1}{a^3}(2ax \cos ax - 2\sin ax + a^2 x^2 \sin$$

$$\int \sin ax \sin bx \, dx = \frac{\sin(a-b)x}{2(a-b)} - \frac{\sin(a+b)x}{2(a+b)} \quad a^2 \neq$$

$$\int \cos ax \cos bx \, dx = \frac{\sin(a-b)x}{2(a-b)} + \frac{\sin(a+b)x}{2(a+b)} \quad a^2 \neq b$$

$$\int \sin ax \cos bx \, dx = -\frac{\cos(a-b)x}{2(a-b)} - \frac{\cos(a+b)x}{2(a+b)}$$

$$\int x e^{ax} \, dx = \frac{1}{a^2} e^{ax}(ax - 1)$$

$$\int x^2 e^{ax} \, dx = \frac{1}{a^3} e^{ax}(a^2 x^2 - 2ax + 2)$$

$$\int e^{ax} \sin bx \, dx = \frac{1}{a^2+b^2} e^{ax}(a \sin bx - b \cos bx)$$

$$\int e^{ax} \cos bx \, dx = \frac{1}{a^2+b^2} e^{ax}(a \cos bx + b \sin bx)$$